Two Manuals Combined into One Book

Military Working Dogs

And

Military Role in Civil Disturbance Operations

Department of Defense

Pentagon
Publishing

Field Manual
No. 3-19.17

Headquarters
Department of the Army
Washington, DC, 6 July 2005

Military Working Dogs

Contents

Figures

Tables

Preface

The last Army dog field manual (FM) was published in 1977. It reflected military working dog (MWD) doctrine developed during the Vietnam era. Although useful at that time, much of the information has since become obsolete. Today, MWD teams are employed in dynamic ways never before imagined.

Today's MWD team is a highly deployable capability that commanders have used around the world from Afghanistan to Africa and from the Balkans to Iraq. These specialized teams aid commanders in stability and support operations as well as in warfighting. Being modular and mobile makes these teams very agile. As situations dictate, MWD teams are quick to arrive and able to conduct various operations. Their versatility allows for effective transformation at all echelons among readiness for deployment and operations on the ground, through redeployment and back to readiness.

The highly aggressive dog tactics of the 1960s and 1970s are long gone. Today's MWD program effectively employs expertly trained and motivated handlers coupled with highly intelligent breeds of dogs. These teams are continuously rotating between their assigned duties and deployments worldwide to perform joint operations, multiechelon tasks, and interagency missions.

This FM addresses the current capabilities of the Military Police Working Dog Program as well as the potential for future applications. As technology and world situations change, the MWD team will continue the transformation process and give commanders the full-spectrum capabilities needed to be combat multipliers on the battlefield as well as persuasive force protection and antiterrorism assets.

Appendix A complies with current Army directives, which state that the metric system will be incorporated into all new publications.

This publication applies to the Active Army, the Army National Guard (ARNG)/the Army National Guard of the United States (ARNGUS), and the United States Army Reserve (USAR).

The proponent of this publication is United States Army Training and Doctrine Command (TRADOC). Send comments and recommendations on *Department of the Army (DA) Form 2028 (Recommended Changes to Publications and Blank Forms)* directly to Commandant, United States Army Military Police School, ATTN: ATSJ-DD, 401 MANSCEN Loop, Suite 2060, Fort Leonard Wood, Missouri 65473-8926.

Unless otherwise stated, masculine pronouns do not refer exclusively to men.

Note: For the purposes of this manual, a dog team refers to one handler and one MWD unless specified otherwise.

Chapter 1

The Military Working Dog Program

Dogs have been used for the protection of life and property since ancient times. From these beginnings, dog training and employment has been continuously refined to produce a highly sophisticated and versatile extension of the soldier's own senses. Even the most complex machines remain unable to duplicate the operational effectiveness of a properly trained MWD team.

THE ROLE OF THE MILITARY WORKING DOG

1-1. MWDs provide a valuable asset to military police, infantry, special forces, the Department of Defense (DOD), and other government agencies. The MWD's senses of sight, smell, and hearing enhance his detection capabilities and provide commanders with a physical and psychological deterrent to criminal activity. Properly trained MWDs can prevent an intruder or suspect from escaping. When necessary, the MWD provides an added dimension of physical force as an alternative to the use of deadly force. Public knowledge of MWD team capabilities provides military police and various security forces with a formidable deterrent wherever the MWD team is employed.

1-2. MWDs are key resources for use in the military police combat support role. MWDs are trained for scouting, patrolling, and performing building and area searches. Commanders can use *Table 1-1* to determine what MWD type best suits their needs. Some MWDs have also been trained to track although this is not a required area of expertise. All of these skills contribute to the successful completion of the five military police functions across the full spectrum of military operations.

Table 1-1. MWD Mission-to-Type Matrix

Mission	*MWD Type*		
	PEDD	**PNDD**	**PD**
A. Area Security Operations			
(1) Area and reconnaissance	X	X	X
(2) Screening and surveillance	X	X	X
(3) Base/air-base defense	X	X	X
(4) Cordon and search	X		
(5) Checkpoints	X	X	X
(6) Roadblocks	X	X	X
(7) Response force	X	X	X
(8) Critical site, asset, HRP security	X[1]	X	X
B. Maneuver and Mobility Support Operations			
(1) Maneuver support	X	X	X
(2) Mobility support	X	X	X
C. I/R Operations			
(1) EPW and CI	X[2]	X[2]	X[2]
(2) Evacuation	X	X	X
D. Law and Order Operations			
(1) Force protection	X	X	X
(2) Military police investigations		X	

Table 1-1. MWD Mission-to-Type Matrix (Continued)

Mission	*MWD Type*		
	PEDD	**PNDD**	**PD**
(3) Customs support	X	X	
(4) Redeployment operations	X	X	
(5) Suspicious/unattended packages	X[3]		
(6) Health and welfare inspections	X	X	
(7) Crowd control	X[4]	X[4]	X[4]
(8) Alarm responses	X	X	X
(9) Bomb threats	X		
(10) Public MWD demonstrations	X	X	X
E. PIO	X	X	X

[1] PEDDs will not be used to provide security in ASPs, as the ammunition and explosives contained at the ASP will detract from the MWD's detection ability and may distract the MWD from his patrol function.
[2] MWDs will not be used in a correction/detention facility to ensure custody of prisoners.
[3] PEDDs can be used to search the area around a suspicious/unattended package for secondary devices. At no time will a PEDD or handler be used to search the package itself.
[4] Direct confrontation with demonstrators is not recommended, but is authorized with the commander's approval when lesser means of force have been unsuccessful.

1-3. MWDs are a unique item; they are the only living item in the Army supply system. Like other highly specialized equipment, MWDs complement and enhance the capabilities of the military police. MWD teams enable the military police to perform its mission more effectively and with significant savings of manpower, time, and money.

HISTORICAL OVERVIEW OF THE MILITARY WORKING DOG

1-4. Dogs have served in active service at the sides of their handlers for decades. They have been heroes, showing bravery under fire, saving lives (often losing their own), and bringing comfort to the injured and infirmed. The first recorded American use of military dogs was during the Seminole War of 1835 and again in 1842. In Florida and Louisiana, the Army used Cuban bred bloodhounds for tracking. During the US Civil War, dogs were used as messengers, guards, and unit mascots.

1-5. The Army Quartermaster Corps began the US Armed Forces' first war dog training during World War II. By 1945, they had trained almost 10,000 war dogs for the Army, Navy, Marine Corps, and Coast Guard. Fifteen war dog platoons served overseas in World War II. Seven platoons saw service in Europe and eight in the Pacific.

1-6. MWDs were trained at Fort Carson, Colorado, organized into scout dog platoons, and used in the Korean conflict for sentry duty and support of combat patrols. In 1957, MWD training moved to Lackland Air Force base (LAFB), Texas, with the Air Force managing the program.

1-7. Throughout the Vietnam Conflict, the Military Police Corps used dogs with considerable success. Most of these were sentry dogs used to safeguard critical installations such as ports and airfields. A new dimension in canine utilization was realized when marijuana detector dog teams were trained and deployed to assist military police in suppressing illicit drug traffic. Sentry and marijuana detector dog teams were then deployed worldwide in support of military police. An important outgrowth of the conflict was the development of canine research and development efforts. These ongoing efforts were able to initiate the first steps toward developing a more intelligent and stronger military dog, training dogs to detect specific drugs and explosives, developing multiple-purpose dogs, and employing tactical dogs by electronic remote control.

1-8. In the 1990s and early 2000s, MWDs were deployed around the globe in military operations such as Just Cause, Desert Shield and Desert Storm, Uphold Democracy, and Enduring Freedom and Iraqi Freedom. These teams were effectively utilized to enhance the security of critical facilities and areas, as well as bolster force protection and antiterrorism missions, allowing commanders to use military police soldiers and other assets more effectively elsewhere. MWD operations in Haiti showed the first large-scale utilization of MWDs since Vietnam. In the Balkans, military police deployed with MWDs to provide force protection and antiterrorism support and area security throughout the US base camps. MWDs trained as explosive detector dogs conducted vehicle sweeps and building searches to ensure the safety of US soldiers. US Army MWD teams operated beside Iraqi police officials to help in law and order missions and restore the war torn country into a peaceful region.

1-9. MWDs are force multipliers. Installation commanders should include MWDs when planning for force protection and antiterrorism countermeasures. They expand the individual soldier's effectiveness in the face of those that would come against him. They provide an immense physical and psychological effect when used as a show of force in day-to-day operations. The various uses of MWDs have been effectively employed in many aspects of military police missions. MWDs are utilized effectively at gates, camps and bases, and checkpoints and for random searches for narcotics and explosive devices. MWDs are also utilized for other missions in support of combat, combat support, and combat service support units.

1-10. MWD teams are assigned to perform high-risk personnel (HRP) security missions. MWD teams perform narcotic detection missions on the Mexico and US border to assist US Customs. They have been deployed on the United States and Canada border to perform law and order operations. The canine can easily detect what his handler cannot see, hear, or smell. The MWD's controllable aggressive behavior, coupled with the physical and psychological effects he creates, makes him and his handler an impressive and unmatchable team.

ORGANIZATION AND STRUCTURE OF MILITARY WORKING DOG UNITS

1-11. The US Army MWD program is an organization with structure and responsibilities that reach from the DOD level down to the handler level. For information concerning organization and responsibilities above the local level, see *Army Regulation (AR) 190-12*.

KENNEL MASTER TEAM

1-12. The kennel master (KM) team is comprised of two noncommissioned officers (NCOs) (one sergeant first class [SFC], who is the KM and one staff sergeant [SSG], who is the plans NCO). The KM team is modular and provides technical supervision of MWD teams assigned to the military police brigade. This team may be assigned to the corps military police brigade, echelon above corps (EAC) military police brigade, or the internment and resettlement (I/R) brigade. In addition to technical supervision, this team plans and coordinates MWD operations and support requirements. The KM team is also responsible for MWD team proficiency training. This team supervises the storage and accountability of the narcotic and explosive training aids as well as all assigned equipment.

PATROL EXPLOSIVE DETECTOR DOG TEAM

1-13. The patrol explosive detector dog (PEDD) team is comprised of three NCO MWD handlers (one SSG, and two sergeants [SGTs]) and three PEDDs. This team detects explosives and contraband in support of combat, combat support, and combat service support units and the five military police functions. It provides MWD patrol support (including I/R operations) when not employed with explosive detection. This team is normally assigned to the corps military police brigade, EAC military police brigade, military police I/R brigade, or the I/R command and control (C2) detachment. This entire modular team can be deployed together as a unit with all assigned equipment or, based on the specific mission, a handler and an assigned MWD may be deployed apart from the team.

PATROL NARCOTIC DETECTOR DOG TEAM

1-14. The patrol narcotic detector dog (PNDD) team is comprised of three NCO MWD handlers (one SGT and two specialists [SPC]) and three PNDDs. This team detects narcotics and other drugs in support of combat, combat support, and combat service support units and the five military police functions. It provides MWD patrol support (including I/R operations) when not employed with narcotics detection. This team is normally assigned to the corps military police brigade, EAC military police brigade, military police I/R brigade, or the I/R C2 detachment. This entire modular team can be deployed together as a unit with all assigned equipment or, based on the specific mission, a handler and an assigned MWD can be deployed apart from the team.

INDIVIDUAL DUTIES AND RESPONSIBILITIES

1-15. There are five levels of responsibility in the MWD program. The terminal level is the program manager and the beginning level, the handler. This is not to be confused with the operational chain of command or standards of grade. These individuals have specific duties in managing and performing the MWD program.

PROGRAM MANAGER

1-16. The program manager executes major Army command (MACOM) level responsibilities for the MWD program. He is the equivalent of an operations NCO and is an appointed certification authority whose duties include—

- Conducting MWD team patrol and detector certifications.
- Requisitioning personnel and MWDs.
- Providing policy and guidelines for the MACOM.
- Forecasting annual operational budgets.
- Coordinating MWD support for the office of the Secretary of Defense, US Secret Service (USSS), US Customs Service (USCS), and Department of State missions.
- Serving as the MACOM provost marshal's (PM's) advisor for the employment of MWDs in contingency operations, force protection and antiterrorism support, and homeland defense.

KENNEL MASTER

1-17. The KM is the senior NCO in charge of the installation's MWD program, consisting of 6 to 30 handlers and respective MWDs. KM duties are the equivalent of a platoon sergeant's duties when 9 or more handlers and MWDs are assigned. When 8 or fewer handlers and MWDs are assigned, the duties are equal to those of a squad leader. In addition to traditional leadership responsibilities, KM duties include—

- Monitoring all training to ensure that MWD teams are ready for annual certifications and worldwide deployments in support of various operational missions.
- Advising the commander on the employment of MWD teams.
- Ensuring that team proficiency is maintained.
- Preparing for MACOM and veterinary inspections.
- Forecasting annual operational budgets.
- Assisting the MACOM program manager in requisitioning personnel, MWDs, and equipment.
- Managing the daily maintenance and upkeep of kennel facilities.
- Monitoring MWDs' health and welfare.
- Managing procurements and accounts for explosive and narcotic training aids.
- Advising installation and unit force protection and antiterrorism officers on the use of MWDs.
- Coordinating MWD support for the USSS, the USCS, DOD protective services units, combat support operations, the Army Recruiting Command, and other civilian agencies.

MILITARY WORKING DOG PLANS NONCOMMISSIONED OFFICER

1-18. The MWD plans NCO is part of the KM team and performs the duties of the maneuver mobility support operations (MMSO) and training NCOs. The plans NCO is selected for his ability and overall knowledge of training, mentoring, and deploying MWD assets. This NCO is responsible for—

- Developing training plans.
- Coordinating and planning the unit mission-essential task list (METL).
- Supervising the execution and tracking of the deployment plan.
- Tracking the movement of MWD teams while they are deployed.
- Ensuring that deployed soldier's issues are resolved.
- Assisting in the supervision of the installation MWD program.
- Acting as the KM when the KM is absent.
- Preparing short- and long-term training outlines and risk assessments.
- Validating and supervising proficiency training.
- Overwatching preparation for the deployment of MWD teams.

SENIOR MILITARY WORKING DOG HANDLER

1-19. The senior MWD handler is the senior NCO of the PEDD or PNDD teams. His duties are the equivalent of the squad leader of the PEDD team or the team leader of the PNDD team. In addition to traditional leadership responsibilities, he is responsible for ensuring that—

- Annual certification training is conducted.
- MWD teams are ready for deployments.
- Team proficiency is maintained.
- Handlers are proficient with their assigned MWDs.
- Coordination for MWD team support for the USSS, the USCS, combat support operations, the Army Recruiting Command, and other civilian agencies is completed.

MILITARY WORKING DOG HANDLER

1-20. The handler provides the daily care and grooming for his assigned MWD. He ensures that proficiency skills are maintained in his assigned MWD. The handler and MWD perform either PNDD operations or PEDD operations. The handler and MWD are all trained for patrol but the variant for narcotics detector or explosive detector depends on the MWD team's training. The handler also—

- Fills out training records daily.
- Grooms the assigned MWD daily.
- Trains the assigned MWD daily.
- Maintains MWD kennels daily.
- Performs additional kennels duties.

Note: For more information about additional skill identifier (ASI) Z6 standards of grade, refer to *DA Pamphlet (Pam) 611-21*.

FUNDAMENTALS FOR THE USE OF MILITARY WORKING DOGS

1-21. An MWD can be trained to respond consistently to certain sensory stimuli (sights, sounds, odors, or scents) and alert the handler. Under most circumstances, a properly trained MWD can smell, hear, and visually detect movement better than a person.

1-22. Commanders and supervisors should consider the capabilities of the MWD team when identifying and evaluating possible MWD employment. Not all MWD teams are able or equipped to perform all missions equally (see *Table 1-1,* page 1-1).

USE OF PATROL DOGS

1-23. All MWD teams are trained to perform patrol missions. One benefit of employment is the deterrence of criminal activity. The obvious presence and well-published activities of MWD teams can successfully deter trespassers, vandals, violent persons, and other would-be criminals. When the desired task is to deter unauthorized or suspicious individuals, an MWD team should be assigned to a location during day or night.

1-24. The patrol dog's (PD's) contribution to the law and order effort is most effective when the team is utilized as a walking patrol. The law enforcement duties that a PD team can perform as a walking patrol include checking or clearing buildings and patrolling parking lots, family housing areas, and troop billet areas. Mobility significantly increases the potential area of coverage. The best and most effective method is the ride awhile, walk awhile method. This concept helps cover large patrol areas, and the periodic exercise helps keep the team alert. All PD teams can be utilized for—

- Antiterrorism operations.
- Force protection and antiterrorism efforts.
- Special-reaction team (SRT) operations.
- Assistance in military police investigations.
- Walking and mobile patrols.
- Alarm responses.
- Building checks and searches.
- The identification and apprehension of individuals.
- The protection of HRPs.
- The protection of funds.
- Open-area searches.
- Civil disturbances.
- Perimeter security.
- External intrusion detection.
- Response forces.
- Access control points (ACPs) support.
- High-risk targets (HRT) support.
- Mission-essential vulnerable areas (MEVA) support.
- I/R operations.

1-25. Many of the tasks above can be performed in combat support or force protection and antiterrorism operations. In addition, PD teams can and should be utilized as a tool for area defense, providing warning and response. PD teams are effective on combat patrols and as listening posts and reconnaissance team members. The employment of PD teams should not be overlooked during I/R or enemy prisoner of war (EPW) operations. PDs are trained to apprehend suspects at or near a crime scene, stop those who may attempt to escape, and protect their handlers from harm. Some PDs may be able to locate suspects who have left the scene of a crime.

1-26. The PD team should be utilized to provide a psychological deterrent and a show of additional force protection and antiterrorism measures. PD teams should be placed on roving patrols at ACPs, vulnerable areas, and the location of HRP. These missions should increase as force protection and antiterrorism levels increase.

USE OF PATROL NARCOTIC DETECTOR DOGS

1-27. Military police commanders, planners, and supervisors have a very effective means of detecting the possession and/or use of controlled substances with the employment of a PNDD team. PNDDs are trained to recognize the scent of certain illegal substances through a program of practice and reward.

1-28. In addition to PD missions, PNDD dogs can be utilized for the following:

- Criminal Investigation Division (CID) and military police investigator drug investigations.
- USCS inspections.
- Health and welfare inspections.
- Barracks searches.
- Workplace searches.
- Random gate inspections.
- Area searches.
- Vehicle searches.
- Postal inspections.
- I/R operations.
- Aircraft and luggage searches.
- Predeployment and redeployment searches.
- Other US government agency support.

1-29. When the PNDD responds on any of the substances he is trained to detect, his handler takes the appropriate action. PNDD teams perform a valuable service by helping to rid the military community of illegal drugs and the problems associated with drug and controlled substance abuse.

USE OF PATROL EXPLOSIVE DETECTOR DOGS

1-30. PEDD teams are very useful in many searches or investigations involving explosives. The search for explosives is critically important to ensure that the community is safe and protected from would-be terrorists. PEDD teams are particularly useful if there is a need to locate one or more of these explosive items, which may have been hidden in an area.

1-31. The deterrent value and the detection capabilities of PEDD teams are a very effective countermeasure to terrorism. Public knowledge that PEDD teams are in use is a deterrent to persons who may try to use explosives illegally. The knowledge that explosives can be detected by PEDD teams at entry points or in places where explosives have been hidden can prevent a person from attempting to commit terrorist acts.

1-32. In addition to PD missions, PEDD dogs can be utilized for—

- Bomb threat incidents.
- Suspicious/unattended package incidents.
- ACP searches.
- Checkpoint searches.
- Random gate searches.
- CID and military police investigations section.
- Health and welfare inspections.
- Building and area searches.
- Postal inspections.
- Aircraft and luggage searches.
- Predeployment and redeployment searches.
- Protective service missions.
- Other US government agency support.

1-33. A well–trained PEDD team can conduct a significantly more effective search of an area or facility in a much shorter time than a number of people can. Using PEDD teams helps reduce the potential risk to persons who would otherwise have to do the search without the benefit of the dog's superior sense of smell. More detailed information concerning the employment of PD, PNDD, and PEDD teams is discussed in *Chapters 5, 6*, and *7*.

Chapter 2

Administrative Requirements

This chapter contains information that is useful in planning an MWD program or adding MWD teams to an existing program. The guidelines in this section are not meant to be absolutes for utilization, training, or maintenance. They give general guidance and information that should be considered when planning a program. Leaders will find the following information helpful for writing authorization documentation statements and for planning the local MWD program. It is important to remember that all MWDs are trained as PDs in addition to the specific type of detector dog (narcotic or explosive).

DETERMINATION OF A NEED FOR MILITARY WORKING DOGS

2-1. The installation provost marshal (PM), military police unit commander, or task force commander must take the initiative to establish the local MWD program if it does not already exist. The specific needs must be determined and the costs of the program must be justified. The decision process involves a thorough risk and crime analysis, an accurate evaluation of the requirements of the entire military police mission, and the use of the working dog yardsticks located in this chapter.

2-2. Some of the factors to consider when determining the need for MWD teams include—

- The mission of the team.
- The size of the area of responsibility of the team.
- The size of the installation or the area population.
- The number of personnel to be served by the team.
- Incident rates for all crimes.
- The present capability and the commitment of a portion of the manpower resources as handlers.
- The types of terrain on the installation or in the probable areas of deployment.
- The types of combat support missions for which MWD teams can be used.
- The number of critical facilities or areas.
- The capabilities of each type of team.
- The number of HRP on the installation or in the area of responsibility.

VERIFICATION OF AUTHORIZATIONS

2-3. Initial MWD authorizations should be probationary using the following criteria:

- Adjustments must be made based on actual use data.
- A baseline ratio of 24 hours of MWD utilization to 8 hours training per week (4 hours patrol and 4 hours detection) must be maintained.
    - Utilization includes all missions (force protection/law enforcement, security, and combat support), whether patrolling, detecting drugs or explosives, or performing other functions (in combination), when a handler and dog are being employed together as a team.
    - Training includes all training activities to maintain, improve, regain, or develop dog skills such as obedience, controlled aggression, patrolling, scouting, detection of persons, and detection of drugs or explosives.
- An automatic review must be made by the respective MACOM after two years.

2-4. There is a trade-off when adjusting between the training and utilization necessary to reinforce basic skills or add new skills. For example, four hours per week may be adequate to maintain detector dog proficiency or to train in new skills, but additional training may be justified at times and training hours may be increased accordingly. This increase in training time should be accompanied by an equivalent reduction in utilization time. However, training hours should not be increased merely as a substitute for lack of utilization.

MISSION REQUIREMENTS

2-5. The mission requirements for the following support missions should be considered. These requirements should be used only as guidelines.

INSTALLATION FORCE PROTECTION AND LAW ENFORCEMENT SUPPORT

2-6. In order to provide continuous 24-hour, year-round MWD protection of the installation, nine MWDs are needed. Six of these dogs will be PEDD and three will be PNDD. This requirement is based on each MWD team being used a minimum of 24 hours per week. This does not include hours needed for training, grooming, or health care requirements. This does allow for temporary duty (TDY), handler leave, and illness.

2-7. The goal is to have one PEDD per installation available at all times for bomb threats or explosive type missions (not counted as part of the working force protection requirement). Installations with a three- or four-star headquarters require one additional PEDD authorization. This is due to the increased workload, facility criticality, facility vulnerability, location, past history of terrorist target selections, and present threat assessments.

HIGH-RISK PERSONNEL SECURITY MISSIONS

2-8. DOD tasks the Army to provide PEDD support for presidential, vice presidential, Secretary of State, and Secretary of Defense visits throughout the world. Currently, there is not a published formula to predetermine mission frequencies or requirements for these missions. On average, 40 teams inside the MACOM support the various missions above. These figures do not include travel or mission preparation time.

CUSTOMS SUPPORT

2-9. While no formula exists to quantify requirements for military customs support, this remains an ongoing mission that must be viewed as a requirement for additional MWD authorizations. MWDs are used to screen household goods shipments, privately owned vehicles (POVs), container express (CONEX) containers, military equipment, US mail, and port operations. Workload varies by installation population, location, unit missions and deployments, and the time of year.

CONTINGENCY OPERATIONS SUPPORT

2-10. Currently, approximately 50 MWD teams are supporting various contingency operations. This requirement will continue until mission completion. This figure represents 25 percent of Army MWD authorizations. Support for contingency operations has continued to rise since 1990 and will continue to do so. This increase must be factored into planning for MWD authorizations. A conservative average of 100 MWDs is needed for contingency operations on a yearly basis. This number is to ensure that MWD teams do not perform back-to-back deployments and that they receive proper proficiency training according to *AR 190-12* (4 hours per week in each specialty and 95 percent accuracy for PEDDs or 90 percent accuracy for PNDDs).

KENNEL CONSTRUCTION

2-11. When new MWD programs are started at locations where there are no facilities in place, kennel construction must be complete before the dogs arrive. The MWDs should be ordered through the military standard transportation and issue procedures (MILSTRIP) process approximately one year before the expected completion date of the kennel facility. In operations where MWD teams will be deployed for long periods of time and will be rotated, a permanent facility must be constructed within one year. Deployable field kennels may be used in the interim. The joint service MWD committee states that the DOD's design guide for MWD facilities must be referenced in planning the construction or renovation of a kennel facility.

2-12. Once the need and the justification for an MWD program are established, the PM, military police commander, or task force commander initiates a request to the installation facilities engineer to develop the kennel facility design. Before the kennel design has been finalized, the activity or installation veterinary corps officer (VCO), the physical security inspector, safety personnel, and the KM must be consulted to ensure that the minimum requirements are included in the final design. After kennel designs are completed and construction costs are estimated, the kennel project is submitted to the installation commander for approval and funding. The request for authorization of MWD teams must include the statement that the commitment has been made to fund and build kennels within one year prior to the arrival of the MWDs at the installation. Refer to *Appendix B* and *DA Pam 190-12* for details on kennel construction.

PROCUREMENT

2-13. All MWDs must be procured from the procurement section at the 341st Training Squadron, Lackland AFB, Texas. Local procurement of an MWD is prohibited. Requisition of an MWD team (one dog and one handler) requires two separate transactions. Handlers are obtained as a result of a personnel action requesting military police soldiers with the ASI Z6 in the required pay grades. MWDs are obtained as a result of a supply action. Commanders will assign an appropriately qualified handler to every MWD on hand. Assignment will be consistent with the policy of "one dog and one handler" as outlined in *AR 190-12*.

2-14. When the replacement of an MWD and not a handler is needed, requisitions are done according to *AR 700-81*. Initial requisitions for dogs should be submitted when positions for handlers are authorized. The MWDs in the Army supply system are—

- Dog, patrol: narcotic detector; national stock number (NSN) 8820-00-435-7542, line item number (LIN) D33800.
- Dog, patrol: explosives detector; NSN 8820-00-188-3880, LIN G33732.

2-15. The method of how to replace an MWD that has died, been adopted, or has undergone euthanasia is to submit a requisition in the MILSTRIP format that includes a statement that the unit or installation already has a trained handler on station. The MILSTRIP should also be submitted when the VCO has stated in writing that an MWD only has a 12-month life or work expectancy and/or is placed in deployment category IV (permanently nondeployable).

2-16. KMs should send requisitions for new or replacement dogs, in the MILSTRIP format, to their local property book office and forward a copy to their respective MACOM. The headquarters of the Air Force Security Police Agency will review all requisitions for dogs prior to processing by the 341st Training Squadron, Procurement Section.

TRAINING REQUIREMENTS

2-17. Training is needed at each installation to ensure that teams remain proficient. This realistic and challenging training should be scenario driven and relate to all of the possible operations in which MWD teams will operate according to the unit METL. Properly trained MWD teams are better prepared to accomplish all facets of MWD missions.

INITIAL TRAINING

2-18. MWD teams are initially trained at the 341st Training Squadron, Lackland AFB, Texas. This training is done at a basic level for both the handler and the dog and is not intended to prepare either the handler or the dog for immediate duty in an MWD team. The training is only intended to give a basic overview of the MWD program, the health and care of MWDs, and proper training techniques. It is the responsibility of the gaining unit to train a new handler or dog on all missions conducted by Army MWD teams. Proficiency training is continuously required to ensure that teams are reliable for use in patrol, narcotic detection, and explosive detection.

PROFICIENCY TRAINING

2-19. The effectiveness of MWDs trained in patrol, narcotic detection, or explosive detection depends on the continual reinforcement of the abilities of the dog team through proficiency training. This training is mandatory and if it is not conducted properly, an appointed certification authority can decertify the MWD team. Failure to meet the minimum training standard for at least the 35 days prior to a certification action will result in MWD teams not being evaluated for certification.

2-20. The minimum training standard is at least four hours of training each week in patrol and at least four hours of training each week in narcotic or explosive detection. The minimum standard of proficiency to maintain certification as a detector team is 90 percent or higher (with no more than a 10 percent false response rate) for narcotic detector dogs, and 95 percent or higher (with no more than a 10 percent false response rate) for explosive detector dogs.

2-21. Failure to maintain an average that meets or exceeds the minimum standard for three consecutive training events will result in the automatic decertification of the MWD team. Decertified teams may not be used in any way, and must undergo intensive retraining. These teams may be recertified only after retraining and a consistent demonstration to an appointed certification authority. Proficiency training should be conducted both on and off leash.

PROFICIENCY TRAINING OF PATROL DOG TEAMS

2-22. The effectiveness of PD teams depends on the continual reinforcement of mandatory proficiency training. PD teams must receive a minimum of 4 hours of patrol proficiency training each week. The training environment should approximate the military police working environment as closely as possible. Training should occur at varying times throughout the day and night and on varying days of the week, including weekends.

2-23. The emphasis in training is to develop the skills of both the handler and the dog so that they complement each other and the team becomes a working unit. As a proficiency evaluation criterion, the correct performance of the required task verifies that the handler understands how to control his PD and is able to do so.

2-24. Train the PD to protect with or without command based on any perceived threat to the handler. Because of the potential of serious bodily injury in this specialty, the certification of the handler and the dog as a team is very stringent. Ensure that the proficiency training of PDs includes an emphasis on critical tasks. There are no waivers for failure to accomplish a critical task by a PD team.

WEAPON FIRE TRAINING

2-25. Training on weapon firing ranges is essential for the PD to become proficient and not be deterred from attacking agitators during gunfire. The dog must not attack the handler during gunfire. The firing of all weapons assigned to the handler should be done with the MWD present whenever possible. PDs can be desensitized with the firing of many different types of weapons. This can often be accomplished by arranging for the handlers to take the dogs to weapon ranges of different units.

2-26. Determine the dog's reaction to the sound of gunfire. It may be necessary for the handler to use counterconditioning techniques until the desired proficiency is achieved. Counterconditioning techniques include starting at distances of 300 meters and slowly bringing the gunfire closer to the dog (or as safety allows, bringing the dog closer to the weapon). The goal is for the dog not to bark or show any signs of aggression when the handler fires all assigned weapons. This can be a slow process over the course of several days.

OBEDIENCE COURSE TRAINING

2-27. The obedience course exposes the PD to various obstacles (see *Figure 2-1*) that simulate walls, open windows, tunnels, ramps, or steps. The dog's exposure to these obstacles reduces the amount of time required to adapt dogs to different environments. The dog learns to negotiate each of the obstacles. When confronted with a similar obstacle in the working environment, the dog is not deterred from completing his mission. The obedience course also develops the handler's ability to control the dog's behavior both on and off leash. The obedience course is not a substitute for exercise. A dog should never be required to negotiate the obedience course until he has been warmed up by proper exercise.

Figure 2-1. Obedience Course Obstacles

PROFICIENCY TRAINING OF DETECTOR DOG TEAMS

2-28. Continuous training of detector dogs is required so they maintain proficiency in the ability to seek and find drugs or explosives, depending on their specialty. Ensure that the detector training is always documented properly. Documentation and training are critical to both maintaining team certification and to establishing probable cause for a search based on a detector dog's number of correct responses.

2-29. There are several differences in the training requirements for PNDD and PEDD. Proficiency training scenarios for both types of detector dogs must be varied to avoid conditioning the dogs to a repetitious training pattern. The following factors will be varied for each proficiency training exercise:

- The general training exercise area.
- The number of training aids planted.
- The specific locations of the training aids in the area being used for training.
- The amount of explosives or drugs used in the training aid plant.
- The type of explosives or drugs used.
- The time of day or night of the training.
- The type of training aid container.
- The type of distraction material planted with the training aid.
- The length of time the training aid is left in place before the detector dog team search.
- The person used to handle and/or plant the training aid.
- The height above or below the floor level of the training aid plants.
- The size of the room or area in which the training aids are planted.

DECOYS

2-30. The decoy is the person who role-plays the primary adversary for training and evaluating the MWD team. The decoy may be a suspect, a subject, an attacker, an agitator, a drunk, an escapee, an enemy, or any of a number of other persons an MWD team may expect to encounter while performing military police duties. A decoy may also be neutral or an ally, such as another military policeman, a supervisor, a lost juvenile, or an innocent person passing through an identification check. Use both males and females for decoys, agitators, and suspect role-players. Vary the clothing the role-players are dressed in to expose the dogs to civilian, military, and the ethnic dress of different countries, if possible.

2-31. Decoys must be taught how to handle the arm protector sleeve or wrap and other aggression tools such as the bite suit (*Figure 2-2*), the hidden wrap, and leg wraps for self–protection when the dog attacks. The sleeve or wrap used should be a hidden sleeve rather than the full, heavy padded sleeve. The hidden sleeve can be concealed under a sweater or field jacket to achieve more realistic conditions for training the MWD.

Figure 2-2. Bite Suit

2-32. KMs and plans NCOs teach decoys how to understand and use the different drives the MWD shows. Once understood, these drives or needs of the MWD can be exploited to cause the desired result in training. Decoys must learn the type of pattern to take to and from the MWD when training on aggression tasks or the bite. For safety and optimum training effect, decoys will learn how to target the MWD into the bite gear and what actions to take during a partial and full bite.

VALIDATION

2-33. The KM or plans NCO will conduct a validation prior to requesting certification. Validations will then be conducted quarterly to verify the patrol and detection accuracy rates recorded on *DA Form 2807-R (Military Working Dog Training and Utilization Record)* and/or *DA Form 3992–(R) (Narcotics or Explosives Detector Dog Training and Utilization Record).* Additional validations will be given to any patrol or detector MWD whose demonstrated proficiency appears to differ significantly from the rating recorded on his *DA Form 2807-R* and/or *DA Form 3992–R.* The validation will verify proficiency or identify the specific patrol/detector weaknesses that need corrective training.

2-34. Validations include at least 30 trials for a narcotic detector dog and at least 40 trials for an explosive detector dog. These trials will be divided equally with each substance the dog is trained to detect. Training aids will be distributed in at least five different areas over a five–day period. These areas should include (if available) vehicles, barracks, open areas, warehouses, theaters, planes, and luggage. A written record of the evaluation and validation will be maintained on record for a minimum of two years.

CERTIFICATION OF MILITARY WORKING DOG TEAMS

2-35. *AR 190-12* defines the standards and specific requirements for all certification of MWD teams. Certification of an MWD and handler is valid for one year and is immediately nullified when—

- The dog is assigned to another handler.
- The handler and MWD have been separated or have not conducted training for thirty-five or more consecutive days.
- The team fails to meet the minimum detector standard of 95 percent accuracy (explosives) and 90 percent accuracy (narcotics) with no more than a 10 percent false response rate for three consecutive months.

SCOUTING OR PATROLLING

2-36. The PD must be able to detect a person by scent, sound, and sight (on leash only). This task is rated critical. The following tasks must be accomplished to demonstrate proficiency in scouting or patrolling:

- Daytime standards.
  - Detect and respond to the scent of a person who is hidden within 75 yards, and follow the scent to the person's location.
  - Detect and respond to a sound made by a person 100 yards away, and follow this sound to the source.
  - Visually detect and respond to a person in view 100 yards away, and pursue the person on command.
  - Demonstrate proficiency in pursuit, attack, and release.
- Nighttime standards:
  - Detect and respond to the scent of a person who is hidden within 75 yards, and follow the scent to the person's location.
  - Detect and respond to a sound made by a person 100 feet away, and follow this sound to the source.
  - Demonstrate proficiency in pursuit, attack, and release.

Note. There is no nighttime standard for sight due to the dog's natural ability to revert to other senses in the dark.

VEHICLE PATROL

2-37. The PD must be able to ride quietly and calmly inside a vehicle with the handler until provoked or commanded. This is a semicritical task.

TRACKING

2-38. Tracking is the only noncritical task on which PDs are trained. Tracking is noncritical because not all PDs have tracking potential. The local KM should know which dogs have been trained to track. PDs that demonstrate a potential for tracking are identified by the KM or plans NCO at each unit, and the dogs are then further developed in this task. Although tracking is a valuable skill for the PD team, improper performance of the task does not degrade the overall effectiveness of the PD team to perform the military police mission.

2-39. PDs selected for tracking training should be trained and employed for tracking on a tracking harness. The harness serves as a useful cue to the PD that it should focus his senses on tracking. At the start of the track, the handler will give the command "Track" and begin casting for the scent. Initially, scent pads are identified and then an item with the tracklayer's scent on it will be used. The dog is required to track and alert on any items left by the tracklayer.

2-40. The minimum level of proficiency a PD trained for tracking must attain is to be able to follow a human scent at least one hour old, for 1 mile over varied terrain, and on a course with several turns. Given suitable tracking conditions, a skilled tracker dog team can follow the natural wanderings of individuals or groups of persons for at least 3.1 miles over rugged and varied terrain and on a scent track that is at least 12 hours old. While tracking, the dog is also capable of alerting his handler to the presence of tripwires and ambushes.

2-41. Evaluation of tracking tasks must include various ages and lengths of the track. The areas used to conduct tracking training and evaluations will be rotated frequently. Tracks will be at least one-third of a mile long but no longer than one mile. The age should vary from ten minutes to one hour. Tracks will include turns, but evaluators should ensure that the person laying the tracks is not trying to deceive the dog with undue tracklaying. Articles will be placed on the track for the dog to find.

INSPECTIONS

2-42. Inspections help ensure that the MWD section is in compliance with Army standards. *Appendix C* contains an example inspection memorandum to assist the inspection officials on what areas to inspect at an MWD section.

DEPARTMENT OF THE ARMY INSPECTIONS

2-43. Personnel from headquarters, Department of the Army (HQDA), conduct annual inspections of the MACOM MWD program. These inspections are used to assist the MACOM PM and program manager in moving the MACOM MWD program forward. This inspection will be on all aspects of the MACOM MWD program (see *Appendix C*).

MAJOR ARMY COMMAND INSPECTIONS

2-44. MACOM program managers conduct annual inspections of handlers, MWDs, training, team use, team proficiency, equipment, and kennel facilities. This ensures compliance with *AR 190-12*.

PROVOST MARSHAL AND COMMANDER INSPECTIONS

2-45. PMs and/or commanders conduct monthly inspections of handlers, MWDs, training, team use, team proficiency, equipment, and kennel facilities. These inspections are conducted at least monthly on one or more areas of the kennels and quarterly on all aspects of the installation MWD section to ensure compliance with *AR 190-12*.

2-46. A written record of the monthly and quarterly inspection is maintained on file with the appropriate corrective action (both in memorandum format), stating the corrective action taken for each deficiency. This written record is maintained for one year past the inspection date.

2-47. The responsible VCO will conduct an inspection of the kennel facility at least quarterly to ensure health and welfare for both the handlers and dogs assigned to the MWD section. A copy of this inspection is maintained on file for a minimum of two years. The responsible PM and/or commander takes all actions necessary to correct any deficiencies noted by the VCO. Veterinary inspections are to be conducted monthly, if possible.

TRAINING RECORDS

2-48. There are two forms that must be used to annotate MWD training, *DA Forms 2807-R* and *3992-R*. These forms are official military documents and must be maintained daily with factual information. Many times these documents are used as court documents, so the importance of the completion and accuracy of these documents cannot be stressed enough. Example forms and the procedures for filling these forms out are in *Appendix D*.

PROBABLE CAUSE FOLDERS

2-49. Probable cause folders are one of the most important methods of proving an MWD team's reliability. Legal requirements to conduct a search of private property based on the response of an MWD do not allow the probable cause folder a margin for error. All the items contained in the probable cause folder are copies. The following original records must be kept on file at the kennel facility:

- ***Department of Defense (DD) Form 1834 (Military Working Dog Service Record).*** *DD Form 1834* must be maintained on the MWD and kept as the initial form in the probable cause folder. This form must be accurate and signed on a change of a handler. When the MWD is trained on new substances, ensure that they are listed on the *DD Form 1834*. The 341st Training Squadron at Lackland AFB, Texas, must approve new substances.
- ***Lackland AFB Form 375 (Patrol Dog Certification).*** *Lackland AFB Form 375* follows the *DD Form 1834* in the probable cause folder and lists patrol certification information for the MWD.
- ***Lackland AFB Form 375a (Detector Dog Certification).*** *Lackland AFB Form 375a* follows the *Lackland AFB 375* (if dual certified), and lists postgraduation certification information for the MWD.
- ***Air Force (AF) Form 1256 (Certificate of Training).*** A copy of the handler's *AF Form 1256* follows *Lackland AFB Form 375a*. The handler provides a copy of the *AF Form 1256* for the probably cause folder.
- **Certification Letter**. This letter attests to the fact that the appointed certification authority has witnessed a demonstration of the capabilities of the dog team. This letter, at a minimum, should list the handler and dog, the whelp, the type of MWD (PEDD or PNDD), the social security number (SSN) of the handler, and the date of certification. This letter should be maintained in the probable cause folder after the handler's graduation certificate until the next certification of the team or a change of handler.
- **Quarterly Validation and Evaluation by the KM**. This document is used to show that the KM has conducted a quarterly evaluation on the MWD team. If the KM is not available, the plans NCO can also conduct this evaluation. This document follows the certification letter in the probable cause folder.
- **Training and Utilization Records**. The last items that should be maintained in the probable cause folder are the last three months of training and utilization records for the MWD team. These records include *DA Forms 2807-R* and *3992-R*. Once an additional month has passed, the latest month will be transferred to the MWD team folder in the filing system.

TRANSPORTATION REQUIREMENTS

2-50. Every unit, installation, and activity with MWD teams should assign enough support vehicles so that all missions and kennel support activities can be properly conducted and maintained. The minimum standard for transportation is one support vehicle per two MWD teams (two dogs and two handlers). Vehicle requirements should be determined by mission requirements and the overall number of MWD teams. Some modifications are necessary so that the MWD can be seen and can provide a visual deterrence during law and order or force protection and antiterrorism missions. These modifications will maximize the mission effectiveness of the dog while providing the dog with a comfortable, supportive, and protective area inside the vehicle. This type of modification is normally a kennel crate or cage secured in the rear of the vehicle; the crate or cage should have a nonslippery surface.

2-51. MWD handlers perform complex missions. Four-by-four type passenger vehicles should be used to ensure that MWD teams can get to any mission during inclement weather or on rough terrain.

2-52. Pickup trucks are essential for the transportation of explosive training aids. *AR 190-12* lists the requirements for explosives transported in government vehicles. However, pickup trucks should not be used for transporting MWDs for training or during road patrol type duties. This vehicle offers limited protection for the MWD from inclement weather and innocent bystanders.

2-53. Warning signs stating "CAUTION—MILITARY WORKING DOGS" should be placed on all sides and the rear of any and all vehicles used to transport MWDs. The signs should also be printed in the host nation language. The only exception to this is the use of unmarked vehicles for President of the United States and Vice-President of the United States missions.

NARCOTICS AND EXPLOSIVE TRAINING AIDS

2-54. Detector dog training and evaluations require the frequent use of narcotic and explosive training aids. Due to the potential safety hazards and risks involved with these aids, it is absolutely mandatory for all MWD personnel and commanders to know and follow the regulatory guidance. MWD leaders and handlers must participate in periodic review and training on the requirements noted in *AR 190-12*. These requirements are clearly stated in *AR 190-12*.

Chapter 3

Managing Military Working Dog Operations

Managing an MWD program requires attention from the program manager, the KM, and the plans NCO. This chapter will help these individuals better understand their specific requirements and concerns regarding the scheduling of various missions.

SCHEDULING AND EMPLOYMENT

3-1. The KM or plans NCO assists and advises the commander or operations officer in scheduling and employing MWD teams. The KM and plans NCO are the most qualified to determine the duty schedules, work cycles, and hours of operation to support each mission. All MWD teams should be utilized at least 24 hours per week to perform nontraining related duties.

3-2. Scheduling must allow for a minimum of four hours of patrol proficiency training per week and four hours of detector training per week. Details concerning proficiency training requirements are discussed in *Chapter 2*. Daily scheduling should include enough time for feeding, grooming, and exercising the MWDs as well as maintaining equipment and kennel facilities.

PREDEPLOYMENT OPERATIONS

3-3. MWD leaders and planners need to ensure that handlers are prepared for all possible missions before the moment of execution. Careful planning and concurrent training is an ongoing process that involves all soldiers in every kennel section. No MWD team should ever be sent on a mission without the proper preparation.

Program Managers Predeployment Responsibilities

3-4. MACOM program managers play a key role in MWD team readiness. The careful management of information concerning each kennel enables leaders and planners to keep the unit in a constant state of readiness for any possible deployment or no-notice mission.

3-5. Program managers keep track of the—

- Total number of personnel at each kennel in the MACOM.
- Certification status of all teams at each kennel in the MACOM.
- Deployability status of all MWDs and handlers (see *Chapter 8*).
- Total number and location of deployed MWD teams.
- Status of kennel facilities at deployment locations.
- Status of VCO support at deployment locations.

Kennel Master Predeployment Responsibilities

3-6. The KM's responsibility is to ensure the daily accomplishment of all missions. The readiness of all assigned MWD teams is a process that requires continuous effort from all leaders. KMs must maintain statistics and make the necessary corrections to have all personnel and equipment available for future deployments and missions.

3-7. The KM is responsible for—

- Managing the number of personnel assigned to each kennel.
- Verifying the status of certified teams and the dates of certification.
- Maintaining the deployability status of MWDs and handlers.
- Ensuring that soldiers have soldier readiness packets prepared according to *AR 600-8-101*.
- Conducting monthly and quarterly counseling.
- Maintaining accurate and detailed training records that will give the forward deployed KM an understanding of strengths and weaknesses.
- Ensuring that MWD handler equipment for short-notice deployments is prepared at all times.
- Conducting annual 9-millimeter handgun qualification. More than one qualification should be conducted (if possible) to better train MWD handlers on the use of the 9-millimeter handgun and to ensure that handlers are always qualified.
- Ensuring that handlers meet the standards outlined in *AR 600-9* and *FM 21-20* by conducting height and weight testing and physical fitness tests.
- Ensuring that the mandatory common military training and common task tests are conducted according to *AR 350-1*.
- Training soldiers in tasks associated with deployments (for example, kennel setup, utilization, and missions).
- Ensuring that MWD handlers have received any needed driver's licenses or transportation motor pool (TMP) licenses before a deployment order arrives.
- Ensuring that soldiers have received the needed individual readiness training in preparation for deployment in a combat zone.
- Conducting mine awareness training.

Plans Noncommissioned Officer Predeployment Responsibilities

3-8. Deployment planning must include coordination with the supporting military VCO. The plans NCO must consider the expected duration of the deployment and the types of facilities anticipated at the destination.

3-9. The plans NCO is responsible for ensuring that the handlers and MWDs in the MWD team or kennel section are technically proficient. The plans NCO is best able to evaluate and recommend to the KM which MWD team is most qualified and prepared for deployments and missions. Through continuous and realistic predeployment training, the plans NCO ensures that the MWD teams are proficient in—

- Crowd control training.
- Building searches.
- Tactical movements (through obstacles and from point to point).
- Gunfire training (handler and decoy).
- Defense training.
- Controlled aggression.
- Explosive detection (area searches, bunker searches, and tactical vehicle searches).
- Narcotics detection (area searches, bunker searches, and tactical vehicle searches).

Additional Predeployment Considerations

3-10. Army mobility requirements necessitate that each MWD has a shipping crate. Shipping crates are used to transport MWDs, simplify handling and loading, and give MWDs adequate space and ventilation during transport. Shipping crates may also be used as temporary kennels at intermediate sites or at the deployment destination.

3-11. The average water requirement for a single MWD is 10 gallons a day. Enough dog food must be taken to last 90 days or until resupply can be established. MWDs must be fed the standard high-performance dog food, contracted and supplied by the General Services Administration (GSA), unless otherwise directed by the supporting VCO. MWDs are fed twice daily unless the supporting VCO directs another meal frequency.

3-12. First aid kits must accompany the MWD team on every mission. Information on the deployment destination, climate, and terrain help determine the specific contents of the first aid kits. For example, if MWDs are employed on rough terrain, a pad toughener is included in the first aid kit.

3-13. All mission-related equipment must travel and arrive with MWD teams so that they can be fully operational as quickly as possible. Any equipment that is not sent with the MWD teams should be sent by other means to arrive at the location within 60 days of the MWD team's arrival. An example equipment list is in *Appendix E*.

DEPLOYMENT CONSIDERATIONS

3-14. There are many important factors to consider when leaders and MWD planners are preparing for deployments. To ensure that every deployment results in mission success, every aspect from notification to redeployment must be considered. These aspects include, but are not limited to, field kennels, training, and operations.

FIELD KENNELS AND SUPPORT FACILITIES

3-15. A suitable kennel site must be carefully selected. Before deployment, some possible locations may be selected on a map from a leader's reconnaissance. As soon as possible after arrival, the sites should be inspected to choose the most suitable location. When selecting a site, consult a VCO (whenever possible) about possible health hazards.

3-16. Locating the kennel in a congested area may be necessary to protect it from enemy attack. A temporary screen or physical barrier may be needed to block the dog's sight so he will be able to rest.

3-17. When shipping crates are used as temporary kennels, the interior, exterior, and ground beneath the crate must be cleaned daily to prevent the accumulation of moisture and waste or insect infestation. It must be raised four to six inches off the ground to allow for adequate drainage and to reduce parasite breeding places. In hot climates, place crates under trees, a tarpaulin, or plywood to provide shade and ventilation. Spread gravel (if available) around and under the crates to allow for drainage and the easy removal of solid waste. A large tent or prefabricated building can serve as a temporary storage location for equipment, rations, and other supplies.

3-18. A second tent or building erected at the kennel site to house handlers and the KM provides the kennel with sufficient manpower for security of the kennel area. Although conditions around a temporary kennel may be primitive, a high standard of sanitation is essential to prevent diseases and parasite infections. Permanent kennel facilities must be erected within one year of the arrival of MWD teams to the location of operational use. Modular expandable MWD transport containers (NSN 961120, 961130, or an equivalent) (*Figure 3-1*, page 3-4) with integrated air conditioning, heating, and equipment storage are recommended for use as temporary kennels in field environments.

DEPLOYMENT TRAINING

3-19. As soon as possible after arriving at the deployment destination and while making sure that site security is established, deployment training must begin. Deployment training allows the MWD teams to quickly develop familiarity with their new location. This is also known as right-seat ride time. The KM or plans NCO needs to evaluate or reevaluate the degree of skill of each of the MWD teams so that the KM and NCO are fully prepared to support the deployed force. Initial evaluation and training should consist primarily of field problems such as attack, search, re-attack, and scouting and basic obedience. Training should also include the handler firing his assigned weapon while maintaining control of the MWD. Some training should be done at dusk, and then in darkness, so that the handler may see (for safety) and the MWD may adjust to the noise and flash of weapons.

Figure 3-1. Modular, Expandable MWD Transport Container

Maneuver and Mobility Support Operations

3-20. MMSO consist of measures necessary to enhance combat movement and the ability to conduct movement of friendly resources in all environments. These measures ensure that commanders receive personnel, equipment, and supplies as needed. MMSO is conducted across the full spectrum of military operations. The primary focus of military police during MMSO is to ensure the swift and uninterrupted movement of combat power and logistical support. KM teams will coordinate MWD operations with the plans and operations officer or task force commander.

Area Security Operations

3-21. When setting up listening posts or conducting combat patrols, consider the availability and suitability of PDs for the particular mission. The KM or plans NCO must be briefed early regarding the mission. This will ensure that the most suitable PD team can be selected and that the handler can prepare for the mission. The handler should check the MWD for any limiting ailments and conduct training rehearsals to become familiar with the mission. The KM or plans NCO and selected handlers must be involved in planning conferences and briefings involving the utilization of an MWD team.

3-22. In defensive operations, combat patrols are used to provide early warning, to confirm intelligence information, and to detect or deter enemy action. PD teams greatly enhance the security of reconnaissance and combat patrols.

3-23. On combat patrols, a PD works at maximum efficiency for only two or three hours. During inclement or extreme hot or cold weather, this time will vary. The KM, plans NCO, and VCO should advise the commander on the MWDs' particular capabilities. The team is most effective in uninhabited areas. If an MWD frequently alerts on friendly forces and is continuously taken off his alert, he soon loses interest and reliability. Only the best teams should be selected.

Defiles and Route Reconnaissance/Surveillance Operations

3-24. MWD teams can be used to enhance the security at defile and route reconnaissance/surveillance operations. Consider the following to ensure mission success:

- The mission at hand (the possible threat scenario).
- The size of the area to be covered.

- The type and number of MWD teams.
- The duration of the mission.
- Sustainment supplies.
- Communications.

3-25. Operational planning must ensure that the proper numbers and types of teams are requested to perform the mission. Planning considerations must allow for proper rotations and rest-to-work ratios for the MWD teams.

Checkpoints

3-26. MWD teams are often called on to enhance the effectiveness of checkpoints. After receiving the mission to perform checkpoint duty, planners and handlers need to consider the following:

- The force protection condition (FPCON) (see *Appendix F* for details).
- The type and number of teams required versus those available.
- Sustainment equipment.
- Communications.
- Available support elements.

Area Defense

3-27. Areas that cannot be covered by static defensive posts because of vegetation, terrain, or some other peculiar feature can be secured using PD teams. Prior to positioning any MWD team in a defensive posture, ensure that intelligence preparation of the battlefield (IPB) details any possible placement of friendly or enemy mines. When relieving others from established defensive positions, use range cards to familiarize the team with the area.

3-28. A PD team can detect personnel using concealed avenues of approach and provide an early warning to defenders. Each MWD team's area of responsibility should be large enough so that the team can move to take advantage of the prevailing wind directions. A fixed position should be established to provide covering fire for each MWD team in case the team must withdraw closer to the defensive perimeter after the MWD has alerted on an attempted hostile probe or penetration.

WARNING AND RESPONSE PROCEDURES

3-29. Several procedures should be developed for warning friendly forces of a PD's response because circumstances may prevent the use of a radio. Also, different types of responses tailored to fit a variety of potential situations should be developed. For example, it may be desirable to have the MWD team follow the alert and locate the cause. At other times, combat intelligence may require the team to maintain the alert position until assistance arrives or to withdraw to a more advantageous position. Whatever action is taken, the handler should not release an MWD unless it is necessary to defend—

- Himself.
- Other personnel.
- Protected resources.

INTERNMENT AND RESETTLEMENT OPERATIONS

3-30. I/R operations consist of those measures necessary to guard, protect, and account for people that are captured, detained, confined, or evacuated by US forces. In any military operation involving US forces, accountability and the safe and humane treatment of detainees is essential. US policy demands that all persons who are captured, interned, evacuated, or held by US forces are treated humanely. This policy applies from the moment detainees become the responsibility of US forces and continues until the time they are released or repatriated. (See *ARs 190-8* and *190-47* and *FMs 3-19.40* and *27-10*.)

3-31. MWDs offer a psychological and actual deterrent against physical threats presented by personnel housed in an I/R facility. They cannot be used as security measures against US military prisoners. MWDs reinforce security measures against penetration and attack by small enemy forces that may be operating in the area. MWDs also provide a positive, effective alternative to using firearms when preventing disturbances or escapes by personnel.

3-32. MWD handlers will not use their MWDs to guard prisoners inside prisons or detainee holding facilities and confinement facilities. In addition, MWD handlers will not use their MWDs to degrade, torture, injure, or mistreat EPWs, detained personnel, civilian internees (CIs), or other detainees in US custody.

Note: MWDs are not used during interrogation operations.

Demonstrations

3-33. Conduct periodic demonstrations in full view of personnel housed in an I/R facility to increase the psychological deterrent of an MWD. Emphasize how easily and quickly an MWD can overtake a fleeing individual. Highlight the MWD's ability to attack and overcome physical resistance. Demonstrate an MWD's scouting ability. To ensure a successful demonstration, use only the most qualified MWD teams.

Perimeter Security

3-34. Use an MWD team as an addition to perimeter security by making periodic, unscheduled patrols around the perimeter fence during periods of darkness. During inclement weather, a temporary blackout or an electrical failure increases the number and frequency of patrols.

WARNING

Never set patterns or establish routines when conducting security patrols. Enemy observing security measures can exploit set times by attacking when the patrol ends or before it begins.

Narcotic and/or Explosives Detection

3-35. Walk an MWD team through living areas to search for contraband. PEDD teams can be utilized to detect explosive devices and components in the I/R facility and PNDD teams can be utilized to detect narcotics.

Work Details

3-36. Position an MWD team between the work detail and the area offering the greatest avenue of escape. MWD teams provide a valuable addition to work detail guards, particularly those employed in areas offering the greatest potential for escape. However, MWD teams are deterrents and must not be used as guards.

POLICE INTELLIGENCE OPERATIONS

3-37. *AR 525-13* directs all military law enforcement to develop foreign and domestic criminal intelligence. MWD teams are an integral part of the police intelligence operations (PIO) process. PIO are conducted in conjunction with all functions in which the MWD teams perform. For example, a PEDD team working at a checkpoint to detect explosives will also be gathering intelligence. This holds true for all MWD functions.

3-38. Leaders and planners of MWD operations must develop procedures to ensure that the MWD teams gather and submit all intelligence in a timely manner. Standardized reports are an effective means that allow for the accurate and detailed submission of intelligence. Contact the supporting intelligence section (S2), provost marshal office (PMO), or CID office to ensure that all PIO coordination is in place prior to performing MWD functions.

Law and Order Operations

3-39. Law and order operations are those that MWD teams perform to enforce laws, orders, and regulations and to maintain discipline and safety. Law and order operations can be performed in both combat and peacetime environments

Random Antiterrorism Measures Program

3-40. The random antiterrorism measures program (RAMP) is designed to be used at all FPCON levels to increase physical security awareness throughout an installation. RAMP involves the random use of measures, identified in *AR 525-13* and *TC 19-210*, for FPCON levels to make installation operations less predictable and therefore more secure.

3-41. The use of MWD teams during RAMP provides for more thorough detection of narcotics and/or explosives and for a psychological deterrent. An MWD presence can also increase confidence in installation security in general.

3-42. KMs coordinate with the PMO or S3 to ensure that RAMP checks are successful. Coordination will include—

- The reconnaissance of patrol areas (map reconnaissance as a minimum).
- Rehearsals and inspections before operations.
- Transportation.
- Equipment needed on site.
- A threat briefing.
- Rules of engagement.
- Use-of-force criteria.
- After-action review (AAR) requirements (see *Appendix D*).

Temporary Duty Missions (United States Customs Service/High-Risk Personnel Security Missions)

3-43. USCS safeguards the internal security of the United States through border searches. Because of this, they have broad powers in conducting searches. By utilizing an MWD team, all vehicles, ships, luggage, and baggage can be checked at border clearance crossings or ports. MWD teams are called upon to support local or federal authorities in detection operations. The Joint Enforcement Military Customs Operation places MWDs at border clearance ports to conduct searches of vehicles and cargo entering the United States. The MWD team takes no law enforcement action. The MWD team only searches and reports. *AR 90-12* gives a more detailed explanation of civil support by MWD teams.

3-44. The Air Force has been designated to serve as the primary service point of contact for all very important person (VIP) missions received from the Office of the Secretary of Defense, USSS, and/or Department of State. The VIP coordination officer who (under authority from the Secretary of Defense) tasks and coordinates all missions performed by DOD PEDD team personnel, regardless of service affiliation, issues all mission tasks.

3-45. MWD leaders and planners ensure that the certification of the team selected to perform the mission is well documented and current. Other considerations in planning for USCS support include—

- The travel arrangements of the MWD team.
- Government travel credit cards.
- Lodging for the handler.

- The kennel and boarding.
- VCO support.

Military Working Dog Public Demonstrations

3-46. Public demonstrations are conducted to show the abilities of MWD teams. These demonstrations are coordinated through the KM to ensure that the appropriate team is available for the target audience. The goal of public demonstrations is to deter criminal activity and to promote public acceptance of the MWD.

3-47. At no time should demonstrations make an MWD a form of entertainment. The KM or plans NCO should ensure that the details of sensitive police training techniques are not explained to the public. Avoid lengthy or frequent public demonstrations that allow MWDs to tire easily or become bored.

Health and Welfare Demonstrations

3-48. Initially, the MWD team should provide the responsible commander with a basic demonstration of the MWD's ability. The KM or plans NCO places a training aid in an undisclosed location, and the handler enters the search area with the MWD. The handler presents and conducts a systematic search of the area. When the MWD alerts, the appropriate reward is given. This demonstrates the MWD team's proficiency.

3-49. After the demonstration has been conducted, the unit should ensure that the area being searched is free of soldiers and that all rooms are unlocked. MWD teams are a tool the commander uses to find any illicit narcotic or explosive devices. It is the responsibility of the unit to properly search the room after the MWD has cleared the room. If the MWD team responds, military police personnel are called to continue the search. The MWD handler conducting the search should not search for the illicit narcotics or explosives. Unit commanders should also ensure that no personnel enter the building or area until the search is complete. This is usually accomplished by placing NCOs at various entranceways to stop personnel from entering the area.

Competitive Events

3-50. MWD teams should be encouraged and allowed to participate in competitive events, seminars, or conferences conducted by outside agencies or police canine organizations. Exposure to competitive events can challenge MWD teams to develop skills or learn advanced employment and training techniques. These events may require new tasks or methods that the handler can use to identify and apply to military missions. In this way, the team's proficiency and effectiveness are enhanced.

Recovery Operations

3-51. After every deployment or extended off-site mission, MWD leaders need to ensure that vital actions are conducted upon the return of the MWD team. These actions include—

- Conducting an AAR after every MWD deployment or extended mission to assist in capturing lessons learned. See *Appendix G* for detailed information on AARs.
- Conducting veterinary inspections.
- Returning any explosive or narcotic training aids.
- Updating training requirements and records.
- Recertifying teams.

Chapter 4
Legal Considerations

The use of MWDs has the same legal considerations as the use of a military police soldier. Each situation and location may be subject to various Status of Forces Agreements (SOFAs), general orders, and other legal restrictions. The following chapter addresses some fundamental areas. It is important to consult with the Staff Judge Advocate whenever legal questions pertaining to an MWD arise.

USE OF FORCE

4-1. Procedures for the use of force are found in *AR 190–14*. This regulation states that personnel engaged in law enforcement or security duties will use the minimum amount of force necessary to control the situation. This regulation further details the levels of force and encourages commanders to substitute nonlethal measures for firearms when they are considered adequate for safely performing force protection, antiterrorism, and law and order functions. The regulation states that in evaluating the degree of force required for specific situations, the following options should be considered in the order listed:

- Verbal persuasion.
- Unarmed defense techniques.
- Chemical aerosol irritant projectors (subject to host nation or local restrictions).
- The military police baton.
- MWDs.
- The presentation of a deadly force capability.
- Deadly force.

4-2. Although high in the order of measures, it is clear that the release of the MWD as a use of force is a nonlethal option to be used before deadly force is considered. It must be understood that the release of an MWD to pursue, bite, and hold a person is likely to result in an injury to the suspect. This injury can be very minor such as scratches or bruises; however, serious bodily harm is possible. For this reason, it is required by *AR 190-14* that all lesser means of force must be used first whenever possible.

4-3. The use of force differs between MWD handlers and regular military police because of the uniqueness of the MWD. Positive control of the MWD must be maintained at all times throughout all of the use-of-force measures. Often, just the presence or arrival of a handler with his MWD can control a situation. If the desired level of control is not achieved, verbal persuasion is the first degree of force to be used. If verbal persuasion is not enough force to maintain or gain control, the next measure may be considered. If the situation allows, military police may use unarmed defense techniques, spray irritants, and the baton before MWD handlers use the MWD as a measure of force.

4-4. A handler is not likely to use unarmed defense techniques when holding an MWD on leash. The handler must remain in positive control of the MWD; therefore he will not be able to get close enough to a suspect to employ unarmed defense techniques. The use of chemical aerosol irritant projectors or the baton would also hinder the handler's ability to keep positive control of the MWD. Therefore, the options are not reasonable measures of force available to the handler.

4-5. Military police and task force commanders, along with PMs, establish clear policies regarding the release of MWDs according to *ARs 190-12* and *190-14*. If the minimum amount of force necessary to control the situation requires the use of MWD force off leash, the handler must ensure that the following steps are followed:

- Ensure that the MWD has identified the same target as the handler before releasing him to bite.
- Give the warning order, "Halt, halt, halt, or I will release my dog." If assigned in a foreign country, give the order in the primary language of the host nation.
- Warn all bystanders to cease all movement.
- Call the MWD off the pursuit and regain control of him if the suspect indicates surrender or stops moving.

4-6. Use extreme caution when removing an MWD from a suspect. Regain and maintain leash control of the MWD until he has become calm enough to obey the commands "Heel" and "Stay." Do not release an MWD if the suspect is out of sight, unless an off-leash building search is authorized. Do not release MWDs in areas where children are present.

4-7. MWD teams are not used for crowd control or direct confrontation with demonstrators unless the responsible commander determines it to be absolutely necessary. When it is necessary, dogs are kept on a short leash to minimize the danger to innocent persons. Dogs are not released into a crowd. Civil disturbance contingency plans include specific criteria for the use of MWD teams that are consistent with *FM 3-19.15, AR 190-12,* and *AR 190–14*.

SEARCHES

4-8. Rules 311 and 313 through 316, *Manual for Courts-Martial (MCM),* prescribe HQDA policies concerning the conduct of searches and inspections and the disposition of property seized in conjunction with those activities. Property refers to property of the United States or of nonappropriated fund activities of an armed force of the United States wherever it is located.

4-9. Commanders and law enforcement personnel must be aware of the distinction between searches, seizures, inspections, and inventories and the provisions for the execution of these actions as outlined in the *MCM.* Knowledge of this information is necessary to ensure the legal use of a PNDD or PEDD.

4-10. Probable cause is a reasonable belief that a crime has been committed and that the person, property, or evidence sought in connection with the crime is located in the place or on the person to be searched. Probable cause must be established prior to searching any persons, property, or residence. Probable cause can be established in the following ways:

- Personal observation.
- Information obtained from a reliable informant or from the witness who observed the crime.
- Use of a detector dog team.
- Any combination of these three.

4-11. The detector dog team is considered a "reliable informant," according to Rule 315 (f)(2), *MCM.* The reliability of a team is established through the initial and annual certifications of the detector dog team by the certification authority and through proper documentation of the MWD's training records.

4-12. Certification establishes the foundation for an officer to grant a search authorization or warrant. Detector dogs are first certified after finishing their basic training. To meet the legal requirements permitting their operational use, each detector dog team must undergo an annual certification and a validation.

4-13. Each detector dog team undergoes a certification when first assigned to duty. A designated certification authority conducts the certification. The detector dog team must perform a demonstration for the person who authorizes any search. This may be any commander from company level up to the post commander. The guideline is that the person who witnesses the demonstration can only authorize a search of units he has command authority over. This person must also review the probable cause folder of the dog and must validate the team as reliable, in writing. This validation is good as long as the commander is assigned to that command.

4-14. The certification authority should witness a demonstration of the dog's capabilities. All substances the dog has been taught to respond on should be used. Additionally, the dog's capability to detect past/present odor should be demonstrated. At the completion of the demonstration, the results are documented in letter form, signed by the commander, and entered into the dog's probable cause folder. The MWD handler is responsible for preparing and maintaining the probable cause folder and the contents (see *paragraph 2-49*, page 2-9*)*. The probable cause folder accompanies the handler each time the dog is used on an actual search. Each document in the folder should be explained to the commander requesting a detector dog search. The commander with command authority should witness a simple demonstration of the dog's capabilities prior to a search. The above procedure should be followed whenever possible. This procedure establishes the dog's reliability on a continuing basis and is essential in establishing probable cause.

4-15. PEDD and PNDD teams must maintain high standards of accuracy to provide the desired level of confidence in their searches. PNDD teams must maintain a 90 percent accuracy rate for the number of trials conducted, while PEDD teams must maintain a 95 percent accuracy rate for the number of trials conducted. The installation commander authorizes the support of local and federal law enforcement agencies. *DOD Instruction 3025.12* governs the use of all PEDD teams. It governs the use of PNDD teams and ensures compliance with *Section 1385, Title 18, United States Code (USC) (The Posse Comitatus Act) (18 USC 1385 [The Posse Comitatus Act])*. Complete certification, recertification, and decertification details for narcotic and explosive detector MWD teams can be found in *AR 190-12*.

Chapter 5

Patrol Dogs and Characteristics

PDs can be used in traditional law and order functions, but commanders and planners should not overlook the tactical enhancements that a PD team offers to combat support operations. All aspects of military police functions can be enhanced by the employment of MWD teams. This chapter discusses many practical applications of the PD team.

EMPLOYMENT TECHNIQUES DURING COMBAT SUPPORT OPERATIONS

5-1. PD teams can perform many combat support operations. These teams enable the tactical commander to free up soldiers and employ their resources in other areas. This force multiplier is especially valuable in area security and force protection and antiterrorism operations.

PERIMETER SECURITY

5-2. PD teams are useful for perimeter (*Figure 5-1,* page 5-2) and distant support posts because they are usually located away from normal activity and large numbers of people. These posts also usually enclose large areas, and it would take a large number of single sentries to secure the area effectively. The large size of these areas allows the PD team to change positions to take advantage of the prevailing wind direction. These posts may be secured only during periods of advanced security and high threat. Occasional random posting of these areas is recommended, especially during periods of limited visibility (rain, snow, dust, smoke, and fog) or night conditions. Barriers and obstacles (such as fences, buildings, gullies, and streams) must also be identified and considered in security post planning. Place the PD team so that these obstacles offer the least interference to security.

5-3. Perimeter security, depending on the area to be covered, should use a two-team concept. The teams overlap each other while walking around the perimeter. PD teams should not come into direct contact with each other. Considerations include—

- The commander's intent.
- Priority intelligence requirements.
- Rules of engagement.
- Use-of-force criteria.
- The wind direction and distracters in the area (see *Chapter 8*).
- Friendly and enemy forces around the area.
- Rest areas within the perimeter to allow the MWD to relieve himself.
- Watchtowers or other security forces in the area.

5-4. The PD team should initiate the perimeter security operation by walking in a zigzag pattern around the exterior perimeter. The team should be allowed to go outside of the security area to ensure that no unwanted intruders are present.

Figure 5-1. Perimeter Security

Intrusion Detection System Augmentation

5-5. Continuing development and use of Intrusion Detection Systems (IDSs) for security make it necessary to reassess the role of PD teams in the security of priority resources where IDSs are used. Although IDSs are designed to greatly reduce the probability of a successful area penetration, they are not designed to replace PD teams. IDSs cannot see beyond the outer clear zones, counter intrusions, or scout the progress of an intruder force after penetration. They are also subject to malfunction and breakdown. PD teams cannot counter all potential problems, but they can be used selectively to counter some of the IDS limitations. With the advice of the KM, security planners should develop plans that set an effective balance between sensor systems and PD teams. Some suggested functions for integrating PD teams into IDS-augmented security systems are as follows:

- PD teams may be used inside or outside critical sites or locations of HRP type personnel areas to give immediate response to areas not clearly seen by closed circuit television (CCTV) or to support security response teams in their role of alarm assessment
- PD teams may be used to scout hidden intruders from the point of penetration to the hiding location inside protected areas.
- PD teams may be posted outside critical sites during periods of increased threat to expand the protected area beyond the range of the IDS. This use is particularly desirable when there is heavy vegetation beyond the perimeter clear zone or when protected areas are located on or near installation perimeters and may, therefore, be easily accessible to the public.
- PD teams may be used as an element of the security response team or as an element of other designated response forces.
- PD teams may be used to compensate for decreased IDS effectiveness when poor visibility or other environmental conditions adversely affect the IDS equipment or when the IDS is not functioning.

SECONDARY RESPONSE FORCES

5-6. Secondary response forces are typically used as blocking forces and for sweeps and counterattacks. PD teams can greatly improve the capabilities of the secondary response force. Some factors that must be considered are that—

- Blocking positions must be selected downwind from the PD team's position when PDs are used with blocking forces. In this type of mission, PDs should only be used to alert the forces of possible intruders.
- The secondary response force moves into the wind with the PD team in the lead position when the PDs are used for sweeps. It is imperative that security be placed on the PD teams to ensure that they are protected.
- A PD team can often pinpoint the exact location of individual intruders.

5-7. A PD team can help detect, alert, and catch hidden intruders while clearing a position that has been counterattacked successfully. PEDDs can be used to help clear the area of explosives. (See *Chapter 7*.)

LISTENING POSTS

5-8. On a listening post, the PD team should be positioned forward of the tactical area to reduce distractions to the PD, yet close enough to maintain contact with friendly forces. In selecting a location, consider the primary detection senses of the PD. Whenever possible, PD team listening posts should be located downwind from any potential enemy position or dismount avenue of approach. Other locations force the PD to rely only on his hearing and sight for detection, reducing his effective use of scent.

COMBAT PATROLS

5-9. In defensive operations, combat patrols are used to provide early warning, to confirm intelligence information, and to detect or deter enemy action. PD teams greatly enhance the security of reconnaissance and combat patrols.

5-10. On combat patrols, a PD works at maximum efficiency for only two or three hours. The team is most effective in uninhabited areas. If a PD frequently alerts on friendly forces and is continuously taken off his alert, he soon loses interest and reliability.

5-11. Patrol leaders and members must be briefed on actions to take when a dog handler is seriously wounded or killed. Dogs that have worked closely with patrol members and have developed a tolerance for one or more of them will usually allow one of the patrol members to return it to the kennel area. However, the dog may refuse to allow anyone near his handler, and other handlers may need to be called to help. The canine must be separated from the handler by—

- Coaxing the dog away with friendly words or food.
- Covering the dog with a poncho to immobilize it.
- Staking out the dog or leashing him.

5-12. A human will always take precedence over an animal. If no effort is effective, the dog may be destroyed. However, all efforts should be made to avoid this; an MWD is an expensive and valuable asset to replace.

5-13. If a situation arises where a canine is injured but the handler is not, the handler must be allowed to accompany the canine. If wounded or killed, an MWD should be evacuated using the same assets and should receive the same consideration as that given to a soldier under the same circumstances.

TACTICAL PATROLS

5-14. PD teams can be used in tactical patrols to detect enemy presence, to help the patrol avoid discovery, and to locate enemy outposts. When the PD alerts, the handler should signal the patrol to halt until the cause of the alert can be identified and the patrol can proceed safely. If a firefight develops while the PD team is at the point position, the PD team should respond to fire team's directions and act as a regular member of the patrol.

5-15. Generally, the best locations for PD teams are directly in front of the patrol or on the flanks to enhance the flexibility and security of the formation. Prevailing wind direction should be used to improve the patrol's chances for early warning of enemy forces. If the wind direction is from the rear of the patrol, the PD is forced to rely on sight and sound and may not be as effective. The handler must concentrate on the PD to read his alert and will not be able to use a weapon rapidly. Therefore, members of the patrol should be assigned to protect the PD team.

5-16. PD teams can be used to enhance the capabilities of an ambush patrol. A PD team should be placed in front of the patrol to minimize distractions, yet close enough to maintain contact with other patrol members. PDs on ambush patrols must remain silent and not respond aggressively to approaching enemy forces.

Internment and Resettlement Operations

5-17. PD teams are used during I/R operations for camp perimeter security and to guard against escapes. The PD's keen senses of sight, smell, and hearing assist the handler and camp authorities in detecting unauthorized activity. A PD team's presence also acts as a deterrent to escapes. In the event of an escape, the PD is a valuable asset that can assist in the search and recovery of internees.

PD Duties and an Internment and Resettlement Camp in Iraq

On 13 April 2003, SGT Carey A. Ford and his MWD Rex responded to a riot at Camp Bucca, Iraq. It was believed that an Iraqi internee had taken a pistol belt and some personal equipment from a guard. The prisoner had taken this equipment back to his tent and the 600 other internees refused to cooperate with the guard force. SGT Ford and MWD Rex went into the I/R camp unarmed. The MWD team moved through the disruptive and unruly crowd providing a secure environment for the quick-reaction force to extract the prisoner and retrieve the stolen pistol belt and personal equipment. SGT Ford related that EPWs would not come anywhere near his MWD Rex. "We work that fear to our advantage."

Four days later, SGT Ford and Rex were called on to scout for and apprehend an escaped EPW. The fugitive had a substantial lead on them, having made his escape two hours prior to SGT Ford's notification. SGT Ford entered a possible unexploded ordnance and/or antipersonnel minefield with Rex, who was also trained to detect explosives. The MWD team was determined to locate the dangerous fugitive and prevent access to any possible weapons caches since many had been located in the area. After hours of relentless searching, SGT Ford and Rex located and apprehended the escapee. Their actions provided the psychological deterrent necessary to prevent future escape attempts.

Riot/Crowd Control

5-18. Extreme caution should be applied before committing PDs to a potential crowd control situation. Although the presence of a PD can be a powerful statement to an enemy or terrorist, it can elicit a reverse and potentially adverse effect on individuals in a riot/crowd control situation.

5-19. It is difficult to employ PDs properly during demonstrations or riots due to the high levels of confusion and excitement and the large number of potential antagonists. PDs should be trained to remain undistracted by loud noises and crowds. A crowd's reaction to a PD in a riot environment may be dependent on the location. In some areas of the world, the appearance of PDs at an angry crowd scene results in an escalation of violence. The crowd will often challenge soldiers to use the PDs as a measure of force. This is particularly true if a situation can be manufactured or provoked so as to be interpreted as an unreasonable use of force by the military. The presence of PDs can cause extreme fear (deterrence) in some cultures and regions. This fear can be advantageous and should be exploited when needed.

5-20. Committed PD teams may be used as an alternative to the use of deadly force to gain control of a riot situation. If the commander located on the ground directs the employment of PD teams for direct confrontation with demonstrators, PD teams should be held in reserve, out of the sight of the crowd. As the situation escalates, PD teams may be moved forward to within the sight of the crowd, but well away from the front lines.

5-21. When committed to the front lines in direct confrontation, PDs are allowed to bite only under the specific circumstances authorized by the responsible commander. PDs should never be released freely into a crowd. Other riot control force personnel should be positioned approximately three meters from any PD team when conducting any form of riot control. This is to prevent unintentional bite injuries to friendly, riot control force personnel that may happen to get between a PD and an instigator participating in the riot or demonstration.

5-22. PD teams must move a safe distance from the crowd to ensure the safety of the PD. Use of PD teams for direct confrontation with demonstrators is not recommended. According to the Army's use-of-force policy, the application of physical force against a person who cannot reasonably be suspected of committing an offense is not appropriate. Presence in a crowd does not necessarily constitute an offense or a situation where it would be reasonable to apply the degree of physical force of a PD. A PD would not be used unless the person is engaged in some other readily apparent criminal behavior such as theft, destruction of property, or assault.

EMPLOYMENT TECHNIQUES DURING FORCE PROTECTION AND ANTITERRORISM OPERATIONS

5-23. PD teams contribute significantly to the military police force protection and antiterrorism effort. PDs are directly responsible for executing specific security-related measures and for providing support to other government agencies that contribute to force protection.

5-24. MWD support is accomplished under the leadership of the respective command PM. Force protection is accomplished through an active role in physical security, HRP security, law enforcement, and antiterrorism efforts. Force protection and antiterrorism operations consist of those actions that prevent or mitigate hostile actions against DOD personnel (to include family members), resources, facilities, and critical information. During force protection operations, offensive and defensive measures are coordinated and synchronized to enable the joint force to perform while degrading opportunities for the enemy.

5-25. In operations, PD teams are used to enhance the detection capabilities of the combat support force and to provide a psychological deterrent to hostile intrusions. If the PD teams are properly located in a hostile environment, the PD's alert provides an initial warning to the presence of a hostile force. In past combat operations, PD teams often provided warning of attacks early enough to allow a response force time to deploy and to prevent enemy forces from reaching their objectives. PD teams have also helped clear protected areas of hostile persons, explosives, and weapons after attacks.

Crowd Control in Kosovo

During Task Force Falcon (2A) in Kosovo, LTC Cloy (the battalion commander of 1-36 INF, 1 Bde, and 1AD) requested MWD support to assist the Alpha company clear a main supply route (MSR) that had been blocked by 10,000 Kosvar Albanians. LTC Cloy had already deployed both quick-reaction forces to assist Company A because the crowd would not stay off the MSR. LTC Cloy called the battalion operations center and directed that the MWD teams be sent to the scene.

SFC Barnes, SSG Rogers, SSG Barters, and SGT McClintock responded with their dogs Queeny, Danny, Sandor, and Fedor and linked up with an infantry squad. The MWD teams and infantry squad moved dismounted to the town of Gjilani. The crowd was starting to become violent. LTC Cloy was in the middle of the crowd, standing in the center of the city at the critical MSR intersection. LTC Cloy attached the MWD teams to Company A. CPT Curtain, the company commander, told SFC Barnes to move the crowd away from his area and to establish a perimeter. The MWD teams moved into the diamond formation with the dogs on guard. Splitting the crowd in half, the MWD teams moved through the crowd and reached the city center MSR intersection in the center of the city. The handlers then used the dogs to move the crowd back out of the road. Company A and the four MWD teams established a perimeter line to keep the MSR open and traffic moving. They also apprehended troublemakers in the crowd.

Later that evening when the dog handlers entered into the DFAC, LTC Cloy stood up, gained everyone's attention, and stated, "SFC Barnes, you and your handlers were like Moses. Your dogs parted the Red Sea!"

ACCESS CONTROL POINTS

5-26. A PD team may be used to enhance force protection and antiterrorism efforts at ACPs. The PD team's function at an ACP is to deter unauthorized access, crime, and terrorism. The handler that is assigned to the ACP should use the PD so that the public can observe and hear the dog. Care must be taken to ensure that the handler constantly keeps the PD under control to prevent incident or injury to innocent personnel.

BUILDING AND AREA SEARCHES

5-27. A PD team is especially effective in checking and searching buildings such as commissaries, post exchanges, finance offices, banks, and warehouses. The PD team can check buildings visually while patrolling and can stop and dismount so that the handler can physically check doors and windows. These checks should be made with the PD on leash. To take maximum advantage of the PD's scenting ability, the handler should approach buildings from downwind.

5-28. If a building is open or forced entry is evident, PD teams may be used to track hidden intruders from the point of penetration to their location. PDs may be released to apprehend an intruder suspected of committing a serious offense when the only alternatives are escape or the use of deadly force.

5-29. Prior to searching a building, the handler will determine whether to make the search with the PD on or off leash. Some of the factors that must be considered include the following:

- Friendly forces or innocent bystanders in the area.
- The time of day or night.
- Evidence of forced entry.
- The type and size of the building or area to be searched.
- The danger to any innocent persons in the area.
- The PD's ability to work off leash.
- The threat level and IPB.

5-30. Prior to releasing the PD inside a building or enclosed area, the handler will give a clear voice warning to any persons inside the building or in the area. They will be told to come out immediately and that failure to comply with these instructions will result in the release of a trained MWD. An interpreter may be needed to assist the PD team in giving the warning in the host nation language. All persons will be warned that the PD may attack without warning and that they could receive physical injuries. The handler maintains voice control of his PD throughout the search.

Note: A warning similar to the one for off-leash searches will precede on-leash searches. The searching team should be followed by at least one other military policeman to protect the search team.

APPREHENSION OF SUBJECTS

5-31. The use of PDs to attack and hold an individual enemy or terrorist should not be overlooked. A PD can be sent into a building, confined space, or cave to find, attack, and hold an individual until the handler and friendly forces can arrive to place the subject into custody. Although disadvantages do exist (injury or death of the PD), the PD is better-suited than a human to immediately find and subdue a subject in an area where the interior structure is unknown to friendly forces.

WALKING PATROLS

5-32. The PD team's contribution to MWD team law enforcement efforts is most effective when the team is on foot. Some of the law enforcement duties that a PD team can do as a walking patrol include checking or clearing buildings and patrolling parking lots, dependent housing areas, and troop billet areas.

5-33. Since it is easily seen, a PD used during daylight in congested areas provides a good psychological deterrent to certain crimes. PDs are tolerant of people, and the presence of a large number of people does not significantly reduce the PD's effectiveness except when conducting detection missions.

5-34. Using the PD team during darkness or periods of reduced visibility and in areas where there are fewer distractions can enhance the PD's detection ability. Military police may not detect a person fleeing a crime scene at night, but a PD will normally be able to detect and locate a fleeing suspect. When necessary and appropriate, the PD is also much more effective at pursuing and holding the suspect.

5-35. Periodic use of PD teams around on–post schools, especially when school is starting and dismissing, may help deter potential vandalism, child molestation, exhibition, and illegal drug activities. PD teams can also provide effective security for communications facilities, military equipment, and command posts (CPs).

ALARM RESPONSES

5-36. In responding to an alarm condition at facilities such as clubs, finance offices, or banks, the PD team should be among the first military police on the scene. The PD team may be used to search and clear the building and immediate area and help with any apprehensions. If the PD must be used to find a suspect, other persons should attempt to avoid contaminating the area with scents or tracks that may confuse him.

VEHICLE PARKING LOTS

5-37. PD teams are effective in detecting and apprehending thieves and vandals in vehicle parking lots. The presence of the PD team may deter potential acts of theft and vandalism. The PD team can be most effective by alternating between vehicle and foot patrols of the parking lots. During hours of darkness, when there is no activity in a parking lot, the PD team should approach the lot from the downwind side. If the PD alerts, the handler should locate and challenge the suspect for identification.

5-38. If a suspect tries to escape or avoid apprehension and the handler is reasonably certain the suspect has committed or is attempting to commit a serious offense, the handler should give a clear warning to the suspect to halt. When lesser means of force have failed to stop the fleeing suspect, the PD may be released to pursue, attack, and hold. If it is not safe to release the PD, the handler may attempt to follow the suspect if the PD has the ability to track.

Chapter 6

Narcotics Detector Dogs and Characteristics

MWDs are trained to perform dual roles. In addition to the uses and characteristics of the PD, the PNDD is trained to detect the odor and presence of specific drugs. This specialized ability to detect certain illegal drugs makes the PNDD a valuable tool to help commanders and other government agencies maintain law and order. The PNDD can be employed in combat and/or force protection and antiterrorism operations worldwide.

EMPLOYMENT FOR COMBAT SUPPORT OPERATIONS

6-1. The PNDD provides commanders with unique capabilities in combat support environments. The narcotic detection capabilities of these dogs can be used to maintain the order and discipline of US soldiers as well as combatants and other persons involved in operations (EPWs, CIs, refugees, or other detained/interred persons [during time of war/conflict]). In addition to narcotic detection functions, PNDD teams should not be overlooked for their PD functions as discussed in *Chapter 5*.

INTERNMENT AND RESETTLEMENT OPERATIONS

6-2. In addition to perimeter and other security duties, PNDD teams assigned to I/R camps can be used to ensure that no illegal drugs are at the I/R camps by conducting searches of all areas. Commanders and planners should consider having all inbound supplies or equipment searched by PNDD teams. Inbound and outbound mail and vehicles should also be considered as a means for the movement of illegal drugs and, as such, subject to a search by PNDD teams.

POSTAL OPERATIONS

6-3. Postal inspections, including mail storage areas and drop boxes, can be conducted in conjunction with a military customs inspector (MCI) and according to *DOD Directive 5030.49*. The postmaster or MCI can request PNDD support from the area PM or military police brigade S3 with coordination through the KM team.

6-4. A PNDD team can be used to search mail that is received or shipped. Searches can be conducted as the mail comes off or before it goes on the truck prior to being shipped. The PNDD team can search up and down the conveyer belt as the mail is removed from or placed on the truck. If a conveyer belt is not being used, the handler can instruct postal personnel how to arrange the mail in a secure area so that the PNDD team can search it more effectively. When a PNDD team is used in the close quarters of a mailroom or delivery area, all nonessential personnel should remain clear of the area while the team is actively working.

6-5. When the PNDD responds to a parcel indicating the possible presence of illegal drugs, the CID or the military police investigations section will be notified for further action. At this point the PNDD team's actions are finished.

AIRCRAFT AND LUGGAGE SEARCHES

6-6. Before beginning the search, ensure that all personnel have exited the aircraft and all compartments are opened. The PNDD team searches the aircraft before searching any luggage or cargo. When approaching the aircraft, conduct a fresh-air perimeter search from the downwind side. Once a perimeter search has been conducted, the team boards the vessel at either the front or rear. Once the area has been cleared to search, the handler ensures that only one door is open to limit draft and wind current. A systematic search of all areas within the aircraft is conducted from the top to the bottom. The handler advises other military police on the scene where the PNDD displayed a positive response, and the augmenting military police conduct a detailed search of the area surrounding the response.

6-7. When conducting inbound baggage inspections, the items should be aligned in a sterile area in rows with at least six feet between rows. The handler conducts an on-leash search, presenting each piece of baggage. When searching cargo, every effort should be made to search the cargo before it is palletized. The cargo should be aligned in rows, with at lest six feet between rows, in a sterile area (same as baggage). The handler conducts an on-leash search, presenting cargo as required. Once the PNDD team has searched the cargo, it may be palletized while remaining in the sterile area prior to embarkation.

6-8. Conducting debarkation operations is the same as inbound operations. The host unit (to which the cargo belongs or is being delivered) provides a representative (E6 or above) and a work detail to assist the PNDD team. When pallets are removed from the aircraft, they are placed in a sterile area. Once in the sterile area, the work detail unloads the pallets and arranges the cargo, as required by the PNDD handler.

FORCE PROTECTION AND ANTITERRORISM OPERATIONS

6-9. PNDD teams are extremely useful in force protection and antiterrorism operations. Commanders and other government agencies use PNDD teams to protect the community from the harmful effects of illegal drugs. PNDD teams protect the force and surrounding communities by deterring criminal activity and enhancing law enforcement capabilities.

HEALTH AND WELFARE INSPECTIONS

6-10. PNDD teams conduct health and welfare inspections when requested by the commander and coordinated through the KM, PMO, and S3. The commander contacts the KM directly to schedule a health and welfare inspection. The KM, in conjunction with the installation PM, the CID special agent in charge, and the military police investigations supervisor, briefs all levels of the command on the availability and capabilities of the PNDD teams available to them on the installation.

6-11. When scheduling the inspection, the commander should have the following information available:

- The requested time of inspection.
- The areas to be inspected.
- The number of living quarters.
- The number and size of common areas.
- Other pertinent information.

Appendix H provides a guide to assist handlers with health and welfare inspections.

6-12. Once the coordinating instructions have been established, the KM coordinates a time and location to brief the commander and demonstrate the PNDD team's capabilities. It is important that the unit requesting the health and welfare inspection practices operational security (OPSEC) and informs only those personnel who have an immediate need to know. This usually consists of the commander and first sergeant or sergeant major. The KM advises the PM and the military police investigation supervisor when the inspection is scheduled and if any additional support will be needed.

6-13. A systematic inspection of the living quarters and common areas is conducted. If the PNDD responds to a locked wall locker, the soldier assigned to that wall locker is summoned to open it. Once the wall locker has been opened, unit representatives conduct a detailed search of the wall locker and its contents.

DEPARTMENT OF DEFENSE SCHOOLS AND DRUG ABUSE RESISTANCE EDUCATION PROGRAM INSPECTIONS

6-14. KMs may receive requests from Drug Abuse Resistance Education (DARE) officers or DOD school principles to conduct a health and welfare inspection of a school. When conducting school inspections in conjunction with DARE, basic building inspection techniques apply. The primary concern when inspecting schools is the safety of the students and the secrecy of the search.

6-15. When possible, searches should be conducted in conjunction with a school activity, such as an assembly or fire drill. This will allow the handlers to conduct the search with a decreased risk of student involvement. Another option is to search the school after normal school hours or on a weekend or holiday. In any event, the use of the PNDD team on school grounds must be cleared by the school administration and the PM must be notified in advance.

ASSISTANCE TO MILITARY POLICE INVESTIGATIONS SECTION AND/OR THE DRUG SUPPRESSION TEAM

6-16. The military police investigation section or the drug suppression team (DST) may contact the KM to schedule an inspection for a number of reasons. PNDD teams may be requested to conduct an inspection of a residence or building where illegal drugs have already been found. In addition, unit commanders may request a PNDD team to inspect the quarters of a soldier who has shown a positive response on a urinalysis. The PNDD is not used to verify or test a substance that has already been found. PNDD handlers contact the military police investigations section or the DST in any circumstance that a PNDD has responded positively according to local standard operating procedures (SOPs).

MILITARY CUSTOMS MISSIONS

6-17. The US Army has been designated as the proponent service for the military customs program. According to current regulations, PNDD teams are used to inspect aircraft cargo areas and airframes when the military customs inspector program manager (MCIPM) directs these inspections. When conducting military customs missions, the handler should first inspect the vessel, then if deemed necessary by an MCI, continue with baggage.

6-18. When scheduling military customs missions, the MCIPM coordinates with the KM according to local SOPs. At a minimum, the MCIPM notifies the KM of—

- The requested time and date of the search.
- The number and types of vessels to be searched.
- The amount of cargo to be searched.
- The number of passengers and crew requiring the MCI's clearance.
- The determination on whether the search is a high-risk or selective search.

6-19. PNDD teams may inspect vessels such as military or civilian aircraft, boats or ships, vehicles, rail cars, or any other vehicle capable of carrying passengers and/or cargo according to the military customs program. These inspections are primarily conducted outside the continental United States. However, some missions in the continental US may be necessary for units returning from foreign theaters where the installation is their first stop within the customs territory of the United States. In addition, the military customs program may require preclearance for units or individuals deploying to areas outside the customs territory of the United States where an MCI has not yet been assigned or is not available. When conducting preclearance of mobilized or deploying soldiers' personal baggage while in formation, the MCI must ensure that the soldiers do not distract the PNDD team.

VESSELS, BAGGAGE, AND CARGO INSPECTIONS

6-20. When establishing sterile areas, the MCI coordinates with the PNDD handler and ensures that the area is of sufficient size with minimal distractions to ensure the optimal working environment for the PNDD. When scheduling working parties for embarkation or debarkation, the PNDD handler coordinates with the MCI to ensure that sufficient personnel will be present to accomplish the mission in a timely manner, but not so many soldiers that they might present a distraction to the PNDD. If an examination has been deemed necessary by the MCI, the handler presents each piece of baggage or equipment to the PNDD.

UNITED STATES CUSTOMS SERVICE MISSIONS

6-21. Missions in support of the USCS come from the MWD program manager via the KM. Handlers must have—

- Active government credit cards.
- Current passports.
- Appropriate civilian attire (or a civilian clothing allowance).
- Complete transportation requirements.

6-22. When temporarily assigned to the USCS, local USCS agents, policies, and procedures direct the PNDD team. Weapons are not authorized for handlers assigned to the USCS. The uniform of the day is determined by the USCS and is normally civilian attire.

6-23. Prior to deploying, every attempt should be made to conduct pretraining exercises with the USCS. These pretraining exercises will allow the MWD to acclimatize to larger quantities of contraband, which may be found while on the mission. The USCS will provide large training aids and determine the PNDD saturation point prior to deploying.

6-24. PNDD teams can expect to perform searches on aircraft, baggage, and tractor trailers as well as vehicle lines at US borders. When assigned to the USCS, the handler does not conduct physical searches of any kind. If the PNDD responds to an area, the handler advises a USCS agent and continues with the search, leaving the agent to conduct the actual physical search of the area.

6-25. Tractor trailers are searched in a USCS sterile area. Once the cargo has been offloaded and placed in standard lines, the PNDD team performs an exterior inspection of the vehicle before moving to all interior compartments and concluding with a systematic search of the cargo.

6-26. While assigned to a US border, the PNDD may be employed in vehicle lines. Vehicle lines are generally 10 to 20 vehicles in a row (placed bumper to bumper) that are inside a sterile area. The PNDD team travels up one side of the vehicle line and back down the other. The handler presents all productive areas. In the event that a USCS agent selects a vehicle for an individual vehicle inspection, that vehicle will be placed in a sterile area and the driver will open all the compartments of that vehicle. Once all occupants are clear of the vehicle, the PNDD will conduct a systematic inspection of the entire vehicle.

6-27. The PNDD will only be used in his narcotics detector role while deployed in support of the USCS or other federal agency. The PD capabilities of the PNDD will not be used for any reason. If the handler is instructed to use his PNDD in any method other than that of a narcotics detector dog, he must immediately notify his KM.

PNDD Support to US Customs Counterdrug Missions

In the summer of 2003, SGTs Payne and Mason from Fort Knox, KY, and SPC Johnson from Fort Sill, OK, were tasked to support a US Customs counterdrug mission along the southwest border between Mexico and the United States. After an overview and train up, the PNDD handlers and their assigned dogs went to work with the customs agents to detect illegal drugs at various points of entry into the country.

These teams became very proficient in detecting large quantities of illegal drugs. The PNDD teams detected narcotics in false floors, dashboards, gas tanks, seats, and hidden panels of many vehicles.

After three months of searching vehicles at checkpoints, the PNDD teams were responsible for detecting over 4,800 pounds of marijuana, over 95 pounds of cocaine, and other undisclosed contraband.

RANDOM VEHICLE SEARCHES

6-28. The use of a PNDD at ACPs and installation gates can increase installation antiterrorism capabilities by providing a visual deterrent to terrorist groups and other criminal elements. The installation commander, task force commander, or base camp commander may task the PM to conduct random vehicle inspections according to local SOPs and the FPCON. When this occurs, the PM submits a schedule of random vehicle inspections to be conducted within a time frame that includes times, dates, and locations of inspections to be conducted. The schedule will be approved or disapproved and the inspection parameters will be dictated (will every vehicle be inspected or will vehicles be inspected at random according to the guidance provided) by the commander and returned to the PM. The PM then notifies the KM accordingly and provides him with a copy of the schedule with the commander's signature and inspection parameters. The PNDD's patrol capabilities may be used at ACPs to search for stowaways and personnel trying to gain access to the installation or a controlled area by hiding in delivery vehicles or other large cargo type vehicles. Extreme caution must be used when using the team in this manner. The commander must be aware that once the PNDD is given the command to search for personnel the MWDs aggression level will increase, thus increasing the chances of the PNDD biting personnel.

ON-CALL DRUG RESPONSE

6-29. PNDD handlers will normally be placed in an on-call status. On call means the handler must remain within response time and distance according to the local SOP. When on call, PNDD handlers do not consume alcoholic beverages; however, they are in a duty position that allows them to wear appropriate civilian attire as well as leave the post according to the local SOPs.

6-30. When on call, PNDD handlers maintain positive control of a secondary communication device issued by the command. The secondary communication device may be a digital pager or cellular telephone as determined by the command. When recalled, the PNDD handler returns to the kennel in the most expedient legal manner to retrieve his PNDD. The PNDD will only be transported in an approved government owned vehicle (GOV). At no time will an MWD be transported in a POV. The handler may respond in appropriate civilian attire according to the local SOP.

6-31. KMs provide insight and guidance to the PM on technical matters of the SOP. The KM should provide recommendations as to acceptable distances which may be traveled while on call, required response times, the required secondary mode of communication, and any installation-specific items of interest, such as an alternate uniform based upon the customary civilian attire that is prominent in the area.

Chapter 7

Explosive Detector Dogs and Characteristics

In addition to the uses and characteristics of the PD, the PEDD is also trained to detect the odor and presence of explosives. This specialized ability to detect explosives makes the PEDD a valuable asset to help commanders and other government agencies maintain law and order, while enhancing safety procedures. PEDD teams can be employed in combat and/or force protection and antiterrorism operations worldwide. PEDD teams are valuable assets. The task force commander, the PM, and military police planners should consider the use of PEDD teams during combat support operations. This chapter details sound employment techniques and the capabilities of the PEDD teams that enable leaders to develop plans.

INTERNMENT AND RESETTLEMENT OPERATIONS

7-1. The Army is the DOD executive agent for all EPW/CI operations. The military police is tasked with coordinating shelter, protection, accountability, and sustainment for I/R operations. The I/R function addresses military police roles with US military prisoners, EPWs, CIs, refugees, or other detained persons (during time of war/conflict). Although the combat support military police unit initially handles EPWs/CIs, modular military police (I/R) battalions with assigned military police guard companies and supporting MWD teams are equipped and trained to perform I/R operations for sustained amounts of time (see *FM 3-19.40*).

7-2. PEDD teams may be used for searching internees' belongings or confinement cells. PEDD teams should not be used to search a person. A PEDD is trained to attack with or without command; any sudden movements the internee may make could be perceived by the PEDD as an aggressive act. Safety must remain a priority for both the handler and military personnel.

7-3. The commander can also use PEDD assets to search the property of attached military personnel when needed. This is especially helpful when new units are rotating or replacing old units. These types of searches ensure that no explosives are taken out or brought in by US forces.

POSTAL OPERATIONS

7-4. Postal inspections, including mail storage areas and drop boxes, can be conducted in conjunction with an MCI and according to *DOD Directive 5030.49*. The postmaster or MCI can request PEDD support from the area PM or military police brigade S3 with coordination through the KM team.

7-5. The PEDD team can be used to search mail that is received or shipped. Searches can be conducted as the mail comes off or before it goes on the truck prior to being shipped. The PEDD can search up and down the conveyer belt as the mail is removed from or placed on the truck. If a conveyer belt is not being used, the handler can instruct postal personnel how to arrange the mail in a secure area so that the dog can search the mail more effectively. When a PEDD team is used in the close quarters of a mailroom or delivery area, all nonessential personnel should remain clear of the area while the team is actively working.

7-6. When the dog responds to a parcel, indicating the possible presence of explosives, the area is evacuated immediately. Military police secure the area and set up a temporary CP a safe distance from the suspected explosives. Use the stand-off distance according to *Appendix I*, as applicable. Use the PEDD team to search the safe area for any possible secondary explosives. Explosive ordnance disposal (EOD) personnel are requested to dispose of the explosives or render them safe. EOD personnel may give explosives and improvised explosive device (IED) components to federal agencies for further investigation or turn over the scene to federal agencies after explosive hazards have been neutralized. The PEDD team remains at the temporary CP and resumes detection operations once EOD personnel remove the explosives.

CHECKPOINTS AND ROADBLOCKS

7-7. PEDD teams can be used at checkpoints and roadblocks to supplement the control and movement of vehicles, personnel, and materiel (*Figure 7-1*). PEDD teams are added to security forces to detect explosives and to prevent illegal actions that may aid the enemy. Using these control measures deters terrorist activities, saboteurs, and other threats.

Figure 7-1. Checkpoint Search

7-8. When using PEDD teams at hasty or deliberate checkpoints and roadblocks, a search area must be designated for vehicles and a separate area must be designated for the vehicle occupants or any dismounted personnel. The PEDD team conducts a search of the vehicle only after security forces at the checkpoint ensure that all of the occupants have exited the vehicle and that a threat to friendly forces does not exist. Security forces at the checkpoint provide overwatch while the PEDD team searches for explosives.

Note: At no time should a PEDD handler be used to search the interior of a vehicle. PEDD handlers are trained on searching techniques with a PEDD and should not be used for security force tasks.

7-9. In the event that the dog indicates the presence of explosives, the PEDD team withdraws immediately from the vehicle. The area is evacuated and EOD personnel and the chain of command are notified. The PEDD team stays behind a barrier at the safe-distance location in case the EOD team leader requests further detector assistance. (See *Appendix I* for stand-off distances.)

Note: Be sure that radio operations are stopped until the PEDD team and security forces reach a safe distance from the suspected explosives.

CORDON AND SEARCH OPERATIONS

7-10. Military police leaders and other task force commanders should consider using PEDD teams when conducting cordon and search operations. PEDDs are trained and able to greatly improve the combat force's success in uncovering hidden explosives and people.

7-11. Once the military police or other security forces have established the cordon, the PEDD teams should move in and operate with the search elements. During the search, the PEDD teams provide warning of hidden explosives or persons while the rest of the search element employs overwatch and search techniques.

7-12. The search element leader directs the actions of the team if people or explosives are detected. The PEDD handler should be prepared to advise the commander on the capabilities of the dog and actions required by the rest of the team to ensure the PEDD team's safety. It is important that all members of cordon and search teams rehearse together before conducting operations.

FORCE PROTECTION AND ANTITERRORISM OPERATIONS

7-13. PEDD support for force protection and antiterrorism operations can be requested anywhere in the world and in different circumstances. In forward deployed areas or hostile environments, PEDD teams are often requested to ensure that any suspected explosives are detected before harm can come to US Army, joint, and allied forces. In peacetime environments around the world, PEDD teams perform the detection operations much the same as in combat support operations. The environment surrounding explosive detection operations does not interfere with the explosive detection capabilities of the PEDD team.

SUSPICIOUS/UNATTENDED PACKAGE RESPONSES

7-14. Once a suspicious/unattended item that may contain explosives has been discovered, the military police or security forces on the scene request a PEDD team. Normally, military police evacuate the area and establish a CP a safe distance away from the threat using the stand-off distance listed in *Appendix I* of this manual. The PEDD team checks the perimeter and secures the area for secondary explosives. EOD personnel should respond to the scene and make contact with the PEDD team to be briefed on the location of the package. The PEDD team moves to within 50 meters of the package and begins an area search up to package. The PEDD team also scans the immediate area for secondary devices, when applicable.

> ***Note:*** At no time should a PEDD team be used to search the package itself. EOD personnel have received special training in handling suspicious/unattended packages and determining if a suspicious/unattended package contains explosives.

7-15. A PEDD handler must not present, touch, open, or physically disturb a package that is suspected of concealing explosives. Once the MWD team has discovered a suspicious item, EOD personnel assume responsibility for the area and take control of the situation in order to render the package safe or dispose of it as necessary.

PEDD Operations in Support of Operation Iraqi Freedom (OIF)

In April of 2003, SGT Raymond Rivera was performing explosives detection missions with his PEDD Kevin in support of OIF near Baghdad, Iraq. SGT Rivera was requested to sweep an area that a support element was in the process of clearing. There was so much debris that a forklift was being used to move large amounts of furniture and building supplies.

The PEDD team would search a portion of the area and then the forklift would come in and haul off the debris searched. After the third load was hauled off, the dog responded to a mattress on the ground with an office table on it. Expecting an explosive devise, SGT Rivera notified the NCOIC of the detail and the area was cleared of all personnel (to a safe distance) and secured. EOD personnel were notified of the suspected explosives and responded to the scene.

SGT Rivera was standing by at the safe area while EOD personnel uncovered two hand grenades that were buried in the sand just below the surface. SGT Rivera said, "Even though it was only two grenades, I was amazed that Kevin found them under all the debris. I'm glad we found them and not our soldiers."

BOMB THREATS

7-16. A PEDD team should be requested for assistance in detecting explosives whenever bomb threats are made against persons or property for which the Army is responsible. The PEDD team's search for explosives begins only after the area or building has been evacuated of all personnel and security has been established. EOD personnel only respond when a suspect item is found; they do not conduct or participate in searches except in support of the USSS. EOD units place a team on standby once the unit is alerted of a bomb threat. PEDD teams coordinate this with the local EOD unit prior to any operation.

7-17. The PEDD team, EOD personnel, and site leadership collaborate to resolve the situation at a nearby CP (see *Appendix I* for stand-off distances). When PEDD teams are located on an installation, their use must be included in the bomb threat planning guidance.

7-18. A systematic search by the PEDD team should include the assistance of a spotter to ensure safety and accuracy. PEDD teams are to be used only to detect the possible presence of explosives.

WARNING:

At no time will the handler touch, open, or otherwise disturb a possible explosive device. IEDs are easily disguised and can be triggered to explode by motion, sound, light, heat, or remote devices.

7-19. Handlers and spotters must not change or otherwise disturb anything in the search. If the lights or any electric device are off or on, do not touch the switch until a thorough search is conducted.

7-20. Ensure that the handler and spotter are provided with available safety equipment including—

- A flack vest (in addition to body armor).
- Flashlights or helmet-mounted lights.
- Gloves.
- Safety goggles.

> **WARNING:**
>
> **Always keep track of how much time has elapsed when using a PEDD team at bomb threat incidents. Sufficient time must be maintained to ensure that the team exits the building or area well before the announced or suspected detonation time.**

7-21. If the PEDD responds to the presence of explosives, the PEDD handler and spotter note the location and leave the area. EOD personnel then begin measures to render the device safe or remove it. On the request of EOD personnel, the PEDD team may then continue the search until another suspected explosive is found or until the area is cleared by EOD personnel.

7-22. If the search is 30 minutes or longer in duration, explosive training aids should be planted periodically so that the dog can find them. This helps to keep the dog interested in the search. Training aids should only be planted in areas that have already been searched by the PEDD team.

HIGH-RISK PERSONNEL SEARCHES

7-23. Military police provide the task force commander with valuable area security capabilities to protect C2 headquarters, equipment, and services essential for mission success. HRP security is an important task within the area security function. PEDD teams may be employed to enhance explosive detection and protection capabilities.

7-24. The KM or plans NCO determines which PEDD teams will perform the mission. The number of PEDD teams for any mission is determined by the threat, the number of PEDD teams available, any conflicting missions, and other environmental considerations noted in *Chapter 8*.

7-25. PEDD teams may be used as part of a combined arms operation or as an element of a larger security force. When used as part of a combined arms team, the PEDD teams need to be involved will all troop-leading procedures, including rehearsals. All personnel must have a complete understanding of the mission in order to create a safe and secure environment for the HRP. The PEDD teams provide explosive detection and personnel detection capabilities. PEDDs are also a deterrent to hostile forces or terrorists operating within the area of operations.

7-26. Prior to an HRP visit to an area, the PEDD team assists with force protection actions by searching for explosives and IEDs. PEDD also conduct searches of and provide security for outer perimeter areas. During HRP operations, the handler—

- Briefs the on-scene commander of the dog's capabilities.
- Communicates with EOD personnel.
- Obtains the HRP's itinerary and determines locations to be searched.
- Requests a spotter or helper to assist.
- Ensures that areas are secure after searches are done.
- Informs the commander of results and continues follow-on missions.

ACCESS CONTROL POINTS

7-27. Using PEDD teams at ACPs can increase an installation's or base camp's force protection and antiterrorism capabilities by providing a visual deterrent to terrorist groups and other criminal elements. The PM notifies the KM of the requirement to perform vehicle searches at ACPs.

7-28. In the event that the PEDD detects the presence of suspected explosives, the military police forces at the ACP secure and evacuate the area (see *Appendix I* for stand-off distances). EOD personnel are then notified to render the item safe or remove it. PEDD handlers do not attempt to remove or otherwise disturb any suspected explosives.

7-29. PEDD team patrol capabilities may be used at ACPs to search for stowaways and personnel trying to gain access to the installation or a controlled area by hiding in delivery vehicles or other large cargo type vehicles. Extreme caution must be used when utilizing the team in this manner. The commander must be aware that once the PEDD is given the command to search for personnel the potential for injury exists.

HEALTH AND WELFARE INSPECTIONS

7-30. PEDD teams conduct health and welfare inspections when requested by the commander and coordinated through the KM, PMO, and S3. The commander contacts the KM directly to schedule a health and welfare inspection. The KM, in conjunction with the installation PM, the CID special agent in charge, and the military police investigations supervisor, briefs all levels of the installation command on the availability and capabilities of the PEDD teams available to them.

7-31. When scheduling the inspection, the commander should have the following information available:

- The requested time of the inspection.
- The areas to be inspected.
- The number of living quarters.
- The number and size of common areas.
- Other pertinent information.

Appendix H provides a guide to assist handlers with health and welfare inspections.

7-32. Once the coordinating instructions have been established, the KM coordinates a time and location to brief the commander and demonstrate the PEDD's capabilities. It is important that the unit requesting the health and welfare inspection practice OPSEC and inform only those personnel who have an immediate need to know. This usually consists of the commander and 1SG or sergeant major. The KM advises the PM and the military police investigations supervisor when the inspection is scheduled and if any additional support is needed.

7-33. A systematic inspection of the living quarters and common areas is conducted. If the PEDD responds to a locked wall locker, the soldier assigned to that wall locker is summoned to open it. Once the wall locker has been opened, unit representatives conduct a detailed search of the wall locker and its contents.

AIRCRAFT AND LUGGAGE SEARCHES

7-34. Before beginning the search, ensure that all personnel have exited the aircraft and that all compartments are opened. The PEDD team should search the aircraft before searching any luggage or cargo. The PEDD team conducts a fresh-air perimeter search from the downwind side when approaching the aircraft. Once a perimeter search has been conducted, the team boards the aircraft at either the front or rear. Once the area has been deemed clear, the handler ensures that only one door is open to limit draft and wind current. A systematic search of all areas within the aircraft is conducted from the top to the bottom.

7-35. If the dog alerts the handler to a possible presence of explosives, military police or security forces evacuate and secure the scene. The PEDD team moves a safe distance away from the suspected explosives but remains available to further assist EOD personnel, if requested. Once EOD personnel clear the area, the PEDD team may resume the search until another suspected explosive is found or until the area is cleared by EOD personnel.

7-36. When conducting inbound baggage inspections, the items should be aligned in a sterile area in rows with at least six feet between them. The handler should conduct an on-leash search, presenting each piece of baggage. When searching cargo, every effort should be made to search the cargo before it is palletized. The cargo should be aligned in a sterile area in rows with at least six feet between them (the same as baggage). The handler should conduct an on-leash search, presenting cargo as required. Once the PEDD team has searched the cargo, it may be palletized while remaining in the sterile area prior to embarkation.

7-37. Conducting debarkation operations is the same as inbound operations. The host unit (to which the cargo belongs or is being delivered) provides a representative (E6 or above) and work detail to assist the PEDD team. When pallets are removed from the aircraft, they are placed in a sterile area. Once in the sterile area, the work detail unloads the pallets and arranges the cargo as required by the PEDD handler.

Chapter 8
Veterinary Support

At most locations, veterinary support is provided by an assigned Army VCO. When there is no assigned VCO, an attending VCO from a nearby military installation is assigned responsibility for providing veterinary support. The Army provides veterinary support for all MWDs as authorized by *DOD Directive 6400.4* and *ARs 40-1, 40-3*, and *40-905*. Emergency, civilian veterinary support is authorized for MWDs when there is not a VCO available and/or care is beyond the capability of the veterinary treatment facility (VTF). The responsible VCO and Army Veterinary Service chain of command are responsible for establishing standard procedures for approval, provision, and payment for emergency, civilian veterinary support.

VETERINARY RESPONSIBILITIES

8-1. The US Army Surgeon General provides professional veterinary support for the entire MWD program through the Army Veterinary Corps. This responsibility includes—

- Medical and surgical care at training facilities, bases, and installations.
- Inspections to ensure that the kennel facilities are sanitary.
- A professional review of plans for new construction and modification of kennels, support buildings, and sites.
- Prescribing an adequate feeding program.
- Instructing handlers and supervisors in all matters related to the health of MWDs.
- Conducting research to improve the MWD program.

8-2. The responsible VCO provides treatment for the diseased or injured MWD at the kennel site or at the installation VTF. The Army Veterinary Corps is responsible for equipping this facility and for providing medical and surgical supplies. The VCO is responsible for the MWD veterinary treatment record (VTR) and related information about examinations, immunizations, and treatment.

8-3. The VCO instructs handlers about dog health, care, feeding, and first aid. This instruction helps the handler to develop a better understanding of the MWD's health needs and improves the handler's ability to care for the MWD. The VCO prescribes an appropriate feeding program based on the MWD's health, the climate, and working conditions.

8-4. The VCO is involved in the review of all plans for new kennel construction or kennel modifications. This ensures that potential health and safety hazards can be corrected before construction begins.

VETERINARY INSPECTIONS

8-5. At least quarterly, the VCO inspects the MWD kennel facility to make sure that proper sanitary standards are being maintained. It is recommended that an animal care specialist conduct a monthly check of the kennels to ensure that no major safety violations are present. The VCO ensures that insect and rodent control efforts are adequate and that the handlers are properly maintaining the general health of the MWD. The quarterly inspection also includes the following elements:

- An examination of the kennel facilities for safety hazards and distractions that may interfere with the rest and relaxation of the MWDs.

- A review of the adequacy of the kennel structure (particularly for environmental conditions) and feeding and watering schedules. The VCO should make recommendations to help prevent disease and injury.
- A review of kennel sanitation and cleaning procedures and MWD food storage procedures to minimize the risk of disease transmission to the MWD.

MANDATORY TRAINING

8-6. The local VCO is required to provide training to the MWD handlers upon initial assignment and at least annually thereafter. More frequent training of handlers is recommended but not required. The Department of Defense Military Working Dog Veterinary Services (DODMWDVS) recommends that handler training be conducted at least quarterly. Topics include the following:

- Procedures on when and how to obtain veterinary support (an SOP is appropriate for this).
- General animal health.
- Grooming.
- Prevention of environmental injury (such as heatstroke or cold injuries).
- Emergency first aid (for bloat [gastric dilatation-volvulus], heat stroke, snake bites, and cold injuries).
- Administration or application of routine medications.
- Proper MWD feeding.
- Feed sanitation (proper storage, preparation, and cleanup).
- MWD chemical, biological, radiological, nuclear, and high-yield explosives protective measures.

8-7. It is best to schedule mandatory training so that seasonal problems are addressed prior to their occurrence and appropriate refresher training is provided prior to a deployment. This training is not meant to turn handlers into animal health technicians, but to give them the tools needed to provide emergency care until they can acquire veterinary support.

WEIGHT CHECKS

8-8. Each month the MWDs body weight must be recorded and sent to the responsible VCO. The KM may communicate this to the VTF. Body weight is also measured and recorded monthly in order to monitor body condition and detect important trends in weight loss or gain that may indicate illness. The KM and/or handler documents data such as food intake, stool consistency, and vomiting in the monthly records.

RECORDS/HEALTH CERTIFICATES

8-9. VTRs are maintained by the local VTF. It is very important that MWD handlers ensure that all health information is annotated in these records.

8-10. An examination is given just before shipping an MWD from one installation to another. The MWDs are examined to detect injury or disease and to support the safe shipment and continued good health and performance of the MWD. A health certificate is issued for the state or country into which the MWD is being shipped for TDY and must accompany the MWD. When the TDY period exceeds 10 days, a health certificate may be required for the MWD's return to his home station. A VCO signs this certificate no more than 7 days prior to reembarkation. If the TDY location is in a different Army VCO's area of operation, the gaining VCO is notified prior to the MWD's departure from the home station. An examination is given as soon as possible, but not more than 10 days after the arrival of an MWD at a new assignment, TDY, or deployed location and within 3 days of the MWD's return to his home station. It is important that the handler who deployed with the MWD is present for the examination so that the VCO can get accurate and complete information on any health issues while the MWD was deployed.

Feeding and Food Control

8-11. MWDs require a diet that is significantly different from that of pet dogs. Their work demands much higher levels of energy and larger quantities of essential nutrients. The standard high-performance diet, contracted and supplied by the GSA for MWDs, meets these nutritional needs and is the only approved feed unless otherwise directed by the attending veterinarian.

8-12. Special diets may be procured and fed to individual MWDs when the VCO determines that other than the standard diet is required. Normally, these diets are available from the VTF, but the MWD unit is responsible for procurement costs.

8-13. Some MWDs have been trained using a food reward schedule that also requires special food. Food reward is used only if an MWD fails to respond to any other type of reward. Procure the special food through normal supply channels or authorized local purchase. The amount of food each MWD should be fed depends on his weight, the amount of activity, and the climate. Local VCOs determine the proper amounts to feed. The responsible VCO must be told when an MWD is trained using food.

8-14. The VCO also prescribes the time of day each MWD is to be fed. This depends on the MWD's duty schedule and the schedule of other kennel activities. MWDs should be allowed 2 hours to eat; leftover food is disposed of within another two hours and feeding pans are cleaned and put away. Never leave uneaten food in the kennel past the authorized 2-4 hour feeding period. If the MWD finishes his meal prior to the end of the feeding period the feed pan may be removed.

Parasite Control

8-15. MWDs are at risk of becoming infested with internal parasites (intestinal worms and heartworms), and external parasites (fleas, ticks, and mange). Some of these parasites may also transmit infectious diseases to the MWD. In order to prevent infestation and disease transmission, all MWDs must be on a routine parasite prevention program supervised by the responsible veterinarian. At a minimum, this must include the administration of a monthly oral (pill or chewable) heartworm preventative and internal parasite preventive and the administration of a topical, veterinary-prescribed external parasite preventative against fleas and ticks.

8-16. When deployed to areas with high parasite risk, additional preventive measures such as prescription flea and tick collars and the oral antibiotic doxycycline will be prescribed by the supporting veterinarian. Doxycycline must be given with food. The handler must ensure that adequate supplies of all preventives are issued by the VTF prior to departing for TDY or deploying. Flea and tick collars will only be used when prescribed by the veterinarian and must be removed when the MWD is not under direct physical control (when off the choke chain and leash) of the handler to prevent accidental ingestion.

Food Storage

8-17. Food should be stored in a sealed container that will not allow contamination or rodents to enter the food (for example, pail, utility 7 gallon, NSN 7240-01-411-0581). Bulk storage of food must be in an area separate from the storage of other equipment, cleaning supplies, or materials that may contaminate the food. Bags should be elevated 4-6 inches above the ground, and neatly stacked with a 4-inch space between the stacks and the walls. The sanitation standards for the storage of MWD rations are identical to those required for the storage of bulk rations for personnel in a troop issue subsistence activity, exchange, or commissary.

Deployability Categories

8-18. The purpose of reporting MWD deployability status is to assist the MWD program managers, PM, commanders, and KMs with the selection of MWD teams for deployment and TDY and to rapidly identify MWDs with limited duty requirements. These guidelines take medical and physical fitness factors into consideration including—

- The anticipated MWD duty stress at deployed locations compared to the home location.
- The availability of veterinary support at the deployed location.

- The climate and environment of the deployed location.
- Any diagnosed or suspected medical conditions that may affect the MWD's performance.
- The MWD's physical conditioning and stamina as observed by the KM and handler.

8-19. It should be noted that some determinations are and will remain subjective; thus the need for prudent judgment by all individuals involved in these assessments remains. This is particularly true regarding the effect of age on an individual MWD. Old age is not a disease and MWDs are not downgraded just due to age; however, with increased age comes increased incidence of illness or disability. These problems may warrant a change in deployability status.

8-20. Deployability status is initially determined after the VCO examines the MWD at the receipt physical or semiannual physical exam. The KM determines training and proficiency status. The VCO and KM should discuss their findings and then categorize each MWD's status and forward the listing through the KM to the PM and MWD program manager. Updates to deployability status lists are performed at least monthly or as needed as changes occur in an individual MWD's status.

Category 1—Unrestricted Deployment (Both Outside the Continental United States and in the Continental United States)

8-21. The MWD is medically fit for any contingency or exercise. There are no limiting or compromising factors. Medical factors may exist but will not limit performance. Special diets, controlled drugs, or special medications can be sent with the MWD (or be carried by the handler) in a quantity sufficient to last the full duration of the anticipated deployment.

Category 2—Limited Deployment (Continental United States or Home Theater of Operation)

8-22. The MWD is medically fit for regions and missions with a minimal requirement for acclimation to heat or physical stress where complete veterinary support is available. The MWD is fit for short duration deployments. There are no significant limiting or compromising factors. Medical problems may exist which slightly limit performances but are controllable. Special diets or medications are required; however they are available in the deployed location or can be sent with the MWD. The reason for limited deployability must be reported.

Category 3—Temporarily Nondeployable

8-23. A medical condition exists that impedes daily duty performance and is under diagnosis, observation, or treatment. The MWD may still be allowed to work at the home station but in a limited capacity.

Category 4—Permanently Nondeployable

8-24. An unresolved medical or physical problem exists that frequently or regularly impedes daily duty performance and an early retirement date cannot be given. The reason for nondeployability must be reported. Medical or physical conditions warrant replacement within one year. The responsible veterinarian and KM should initiate an MWD disposition process at the time an MWD is declared category 4.

Kennel Sanitation Procedures

8-25. Cleanliness is one of the most important factors for the good health of an MWD. The KM must enforce sanitary measures in and around the kennel area. A good standard of sanitation is the result of the cooperation of the handlers, KMs, supervisors, and the VCO.

8-26. The VCO and the PM or military police commander should set the standard of sanitation. This standard must be maintained by each of the handlers and the KM.

8-27. Kennels must be sanitary, in a good state of repair, and thoroughly cleaned every day. Kennels should be disinfected at least once every week using only those disinfecting products approved by the VCO. Kennels should also be disinfected whenever an animal is removed from a kennel so that the kennel will be ready to be occupied by another animal.

8-28. Stool must be removed from the runs as often as necessary as but no less than twice daily to prevent the MWD from becoming dirty. The method of disposing of stool depends on local conditions and the type of sewage system present. If stool must be carried from the area in cans, the cans must be cleaned and disinfected after each use.

8-29. The entire kennel area must be free of refuse and garbage that could attract rats and insects. Mosquito control measures must be used in ditches and swampy areas in the vicinity of the kennels. Disinfectants and disinfectant procedures should be used only with the approval of the VCO.

MEDICATION

8-30. At times, a VCO prescribes special medication (to be given separately or mixed with food) for a sick or injured MWD. The handler must know how to administer these medications in both pill and liquid form.

8-31. When any foreign substance is placed directly into a dog's mouth, his first reflex is to spit it out. The handler must learn how to administer medication properly so that the MWD is forced to swallow. With the VCO's approval, pills or liquids can be mixed with a small amount of canned dog food or other soft food and fed to the MWD. To administer capsules or tablets—

- Place the fingers of your left hand over the MWD's muzzle and insert your left thumb under the lip and between his upper and lower teeth right behind the canine tooth, pressing your left thumb against the roof of his mouth to open it.
- Place the capsule or tablet into the MWD's throat at the extreme rear of the tongue to prevent him from spitting it out. Quickly remove your hand, close the MWD's mouth, and gently massage his throat until he swallows. The entire procedure must be quick and smooth to ease the MWD's apprehension and resentment.

8-32. Liquid medication is best administered with the help of another person. The handler—

- Holds the MWD's upper and lower jaws together with his left hand. The assistant pulls the MWD's lips away from the teeth at one corner of the mouth with his right hand. The MWD's nose is then pointed slightly upward, forming a natural funnel by the lip, and the assistant pours the liquid into this funnel.
- Elevates the MWD's head only slightly above the horizontal when giving liquid medicine. If the MWD's head is raised higher, he has difficulty in swallowing. Give the MWD adequate time for swallowing to prevent the liquid from getting into the trachea, nose, or lungs. Use extreme caution in giving oily liquids.
- Allows the MWD to lower his head and rest before proceeding if any signs of distress, such as coughing or struggling, appear. Do not give oral medications or any liquids if the MWD is unconscious or cannot swallow.

FIRST AID

8-33. Normally, the handler's early recognition of symptoms of illness or injury allows sufficient time to get assistance from a VCO. However, situations may arise when medical help is not immediately available. The seriousness of the incident may require that the handler take emergency actions to protect the life or health of the MWD. The first aid instructions that follow are commonly used to save life, prevent further injury, and reduce suffering from pain. In all emergency situations, notify the VCO as soon as possible so that the MWD can receive professional medical attention.

8-34. First aid kits should always be available in the kennel area and at training sites. The handler should also carry a first aid kit as part of his or her equipment for the MWD on all operational missions. The VCO will determine the contents of first aid kits. When items are used from the first aid kit, they should be replaced with new items immediately.

MUZZLES

8-35. When an MWD has been injured, the first consideration is to calm and immobilize the animal. Pain and distress, however, may cause the MWD to respond to the handler in an unpredictable manner. The MWD may not respond to verbal commands and may attempt to bite the handler and anyone helping the handler. Whether to apply a muzzle (*Figure 8-1*) or not depends on the nature of the emergency. If the MWD is unconscious, shows any difficulty in breathing, or has suspected head injuries, a muzzle should not be used. Otherwise, a muzzle should be used for safety.

Figure 8-1. Muzzle Use

8-36. There are several types of muzzles, but the leather basket muzzle is the best and most comfortable. It allows for free breathing and causes the least alarm and apprehension. The MWD can still inflict a wound while wearing this type of muzzle, so caution is required.

8-37. An improvised muzzle can be made using the MWD's leash and is called a leash muzzle. To apply the leash muzzle, tighten the choke chain on the MWD's neck by pulling the leash tightly with the right hand. Place the left hand, palm up, under the choke chain on the neck. Grasp the leash tightly as it passes through the palm of the left hand, wrap the leash once around the MWD's neck, and bring it up and across the left side of his head. Finally, wrap the leash twice around the muzzle and grab it tightly with the left hand.

CAUTION

The leash muzzle may be used when the leather muzzle is not available or when it would not provide adequate safety. Do not use the leash muzzle when the MWD is overheated, is having difficulty breathing, or is indicating that he may vomit. Do not leave it on for long periods in hot weather.

FRACTURES

8-38. If a fracture occurs, restrain the MWD and encourage it to lie down and remain quiet. If the face and jaw are not injured, put the MWD in a loose muzzle. Notify the veterinarian immediately when a fracture occurs or is suspected. If there is any bleeding, treat the MWD according to *paragraph 8-40,* then proceed with splinting and moving the MWD to the veterinarian if he is not immediately available.

8-39. Prior to moving the MWD, the fracture area needs to be immobilized with a splint. This should be possible for most leg fractures. The leg should be splinted in the same position it is placed by the MWD when it is hurt. Do not try to straighten or reduce the fracture. Splinting an MWD's leg is similar to splinting a soldier's leg and may be done with the same materials: bandages, torn cloth, splints from the medic's bag, or other available material including sticks, boards, and rolled newspaper. After the leg is splinted, or if the fracture cannot be splinted, gently move the MWD onto a litter or improvised litter, such as a piece of plywood, or into a kennel and carry it to a vehicle for transport. It is best to slide the litter under the MWD rather than lifting the MWD onto the stretcher. The MWD may need to be strapped to the litter to allow safe movement. Make sure to support the injured leg or area during movement.

Bleeding Wounds

8-40. Bleeding must be quickly controlled, particularly wounds in the foot or leg that are prone to bleed freely. Applying pressure directly on the wound may control bleeding; use a sterile bandage or clean handkerchief or pinch the edges of the wound with your fingers. As soon as possible, apply a pressure bandage.

Burns

8-41. Most burns occur when an animal comes into contact with hot water, grease, tar, or other scalding liquids. MWDs may also get electrical burns by chewing electrical wires. If an MWD is trapped in a burning building, it may suffer from smoke inhalation in addition to surface burns. The handler can prevent all of these situations through positive control of the MWD. If a burn does occur, applying cold-water soaks or ice packs to the burn for approximately 20 minutes may treat minor burns. The cold helps to reduce the pain. The hair around the burn should be clipped away and the burn should be washed gently with a surgical soap. Blot it dry with sterile or clean soft gauze or cloth. Protect the burned area from rubbing by applying antibiotic ointment and a loose fitting gauze dressing.

Shock

8-42. An animal may go into shock after injuries to internal organs, excessive bleeding, or trauma. Shock can be recognized by the following symptoms:

- A glassy look to the eyes.
- A rapid or weak pulse or rapid shallow breathing.
- A dropping body temperature. The lips and feet may feel cold.
- A paleness of the membranes of the mouth and eyes.
- Slow capillary refill time. To determine capillary refill time, press firmly against the MWD's gums until they turn white. Release the pressure and count the number of seconds until the gums return to their normal color. If it is more than two to three seconds, the MWD may be going into shock. Failure to return to the red-pink color at all indicates that the MWD may be in serious trouble and needs immediate assistance.

8-43. If a handler suspects that an MWD has internal injuries or is going into shock, request help from a VCO immediately. Keep the MWD warm and quiet, and lower his head to prevent possible brain damage. If it is necessary to move the MWD, use a litter. Administration of intravenous fluids by appropriately trained personnel will help to prevent permanent injury from shock.

Artificial Respiration

8-44. There are conditions that may cause an MWD to stop breathing. When this happens, do not panic. Follow these procedures:

- Open the MWD's mouth and check for any obstructions. Extend the MWD's tongue and examine his throat.
- Clear the MWD's mouth of any obstructions then, close his mouth, and hold it gently closed.
- Inhale, and then cover the MWD's nose and mouth with your mouth.
- Exhale gently. Do not blow hard. Carefully force air into the MWD's lungs and watch for his chest to expand. Repeat every five to six seconds or at a rate of 10 to 12 breathes per minute.

SNAKEBITES

8-45. The bite of a poisonous snake can cause serious illness and death if not treated immediately. Keep the MWD quiet and calm, and request veterinary assistance or move the MWD as quickly as possible to a treatment facility. Panic or exertion causes snake venom to move more rapidly through the bloodstream. If possible, kill the snake so it can be shown to the VCO. The handler should be careful not to be bitten by the snake or come into contact with the snake venom. Bites may occur on the face or neck of the MWD. When this happens, remove the choke chain and loosen or remove the collar and muzzle. Swelling occurs rapidly after a snakebite, and this equipment may restrict breathing. Position the MWD's head extended from his body to allow him to get maximum airflow. An ice pack applied to the bite area helps to slow the flow of blood and helps to keep the venom from spreading.

CAUTION

A tourniquet should not be applied. Constricting the blood flow could increase tissue death.

FOREIGN OBJECTS IN THE MOUTH

8-46. An MWD may get a stick or other foreign object stuck in his mouth or throat. The most common foreign object that becomes stuck is a Kong® or ball reward that is too small for the MWD. The MWD may cough, gag, have difficulty swallowing or breathing, paw at the mouth, and drool. If the MWD is having difficulty breathing, cautiously and gently open his mouth. Look for any abnormal object in the throat, under the tongue, between the teeth, in the gums, or in the roof of the mouth. If the MWD is having trouble breathing, removing the object is an emergency. The ball, Kong, or other object can be removed using pliers or clamps. If the object is not causing breathing problems, it is best to leave it in place and get veterinary assistance.

POISONOUS SUBSTANCES

8-47. There are many toxic chemicals an MWD may come in contact with and/or ingest. These include insecticides, herbicides, rodenticides, and antifreeze. The symptoms of poisoning vary. Unless the handler is certain the MWD has eaten a poison, do not treat for poison. Occasionally, MWDs accidentally swallow narcotics or explosives training aids. Exposure to poisons and accidental ingestion of training aides should be prevented by positive control of the MWD by the handler. If the handler knows that the MWD has eaten a poison, narcotic, or explosive, take the following actions:

- Request veterinary assistance immediately.
- Determine the type and quantity of poison, chemical, narcotic, or explosive that has been swallowed. If any part of the substance or container is available, keep it for the VCO to examine.
- If the Army VCO is not available, immediately make contact with a civilian veterinary to determine if vomiting is appropriate. Causing an animal to vomit is not recommended in all types of poisoning as it may cause more harm to the animal. When vomiting is appropriate, the animal may be induced to vomit by giving him one-quarter to one-half of a cup of hydrogen peroxide.
- Keep the MWD quiet and warm until the VCO arrives.

8-48. For future reference, keep a copy of the material safety data sheet for any training aide that may be ingested available. MWD first aid should be according to the material safety data sheet for people.

OVERHEATING

8-49. Overheating results when an MWD is unable to eliminate body heat rapidly enough. This condition requires immediate action by the handler to save the MWD's life. During hot, humid weather, an MWD may easily become overheated during training, during operations, or while being transported. A body temperature of 105° Fahrenheit or more, combined with poor response to commands, weakness, unsteady movement, vomiting, difficult or labored breathing, convulsions, and collapse are all symptoms of overheating. It is helpful to take the MWD's temperature in order to monitor the severity of overheating and progress in cooling, but temperature alone does not determine the need for or method of treatment.

8-50. For mild to moderate heat injury, the MWD's temperature is generally lower than 104° Fahrenheit. The MWD may have mild symptoms, to include uncontrolled panting, a rapid but somewhat weak pulse, mild to moderate difficulty breathing (but not cyanotic [blue]), and exercise intolerance or loss of stamina. This is equivalent to heat exhaustion in people.

8-51. When the symptoms of overheating occur, loosen the leash, collar, and muzzle as much as possible without loosing control of the MWD. This allows for more efficient panting and cooling. Follow these guidelines at the onset of symptoms:

- Carry the MWD to the nearest shade and fan him if no wind is present.
- Try to quickly lower the MWD's body temperature by running and sponging cool water over the MWD's head, body, and legs. If a body of water is available, allow the MWD to stand in the water but make sure the MWD's head remains above water. Running water works better than standing water.
- Place the MWD in an air-conditioned kennel, vehicle, or shelter.
- Apply ice packs to the MWD's groin, armpits, and head to increase speed of cooling.
- Dampen, but don't soak, the MWD's feet, ears, and abdomen with rubbing alcohol. Allow it to evaporate and then reapply more.
- Do not let the MWD over drink. He should not consume more than one cup of water until he is calm and his temperature returns to normal.
- Monitor the MWD's temperature every 5-10 minutes, and stop cooling efforts when it reaches 100-103° Fahrenheit. Keep monitoring his temperature for an additional 60 minutes.

8-52. If the MWD's temperature remains normal (about 100-103° Fahrenheit), treatment is complete; however the MWD must be examined by a VCO and a heat injury must be reported in his VTR. If his temperature climbs again, restart cooling efforts.

8-53. If the MWD's temperature drops below 100° Fahrenheit, begin warming by drying and wrapping him with sheets and blankets. Remove the heat when his temperature goes above 101° Fahrenheit.

8-54. For severe heat injuries, the MWD's temperature is generally over 104° Fahrenheit combined with poor response to commands, weakness, unsteady movement, vomiting, difficult or labored breathing, uncontrolled panting, convulsions, collapse, a dull or depressed attitude and behavior, a rapid but somewhat weak pulse, exercise intolerance or loss of stamina, and cyanosis (blueness) or dark red mucous membranes. This is equivalent to heatstroke in people. If this occurs take the following actions:

- Initiate all cooling and monitoring efforts as noted in *paragraph 8-51*.
- Administer 1 liter of intravenous fluids to the MWD as quickly as possible if appropriately trained personnel are available.
- Continue cooling and monitoring until a veterinarian can evaluate the MWD.

Bloat (Gastric Dilation-Volvulus)

8-55. Bloat is an acute stomach enlargement and twisting that may be due to gas, food, or water. It may occur if the MWD is fed immediately before or after hard exercise or when he is returned to his kennel after work or exercise and allowed to drink too much water.

8-56. An enlargement may be seen just behind the ribs, primarily on the left side. The MWD may also be restless and show signs of pain in the abdominal region. He may attempt to vomit or have a bowel movement. His breathing may be difficult or labored due to pressure from the enlarged stomach.

8-57. Handlers should notify the VCO and stop all watering and feeding. Walking may enable the MWD to relieve himself through bowel movements or passing gas. Most cases require extensive treatment by the VCO.

8-58. To prevent bloat, MWDs should not be fed within a two-hour period before or after hard work or vigorous exercise. It is also safer to provide multiple small meals each day, feeding twice daily, rather than one large meal. In hot weather, give small amounts of water during training or working MWDs to prevent excessive thirst. For the first hour after working or training, only three inches of water in the bucket should be available. After this cooling-off period, more water may be given.

First Aid Kit

8-59. The responsible VCO determines the contents of MWD first aid kits. The VCO should consider the skill of the handlers, access to veterinary support, and emergency situations that may occur. The DODMWDVS does not have a standard first aid kit supply list because needs vary. The local VCO and KM are responsible for developing and putting together an appropriate kit and training the handlers how to use it. The kit should contain supplies for cleaning and dressing minor wounds, cleaning and protecting an eye injury, bandaging or splinting minor breaks, and removing mild superficial foreign bodies (splinters or glass). It should also contain peroxide, antibacterial solutions, and a small supply of any prescription medications needed for individual MWDs.

ENVIRONMENTAL EFFECTS ON THE MILITARY WORKING DOG

8-60. Just as the environment affects soldiers performing their duties, the MWD is affected by his surroundings. It is important for planners and leaders to become aware of the many different conditions that can affect the performance of MWDs. Not all MWDs are affected in the same manner. Acclimation times will vary between conditions and MWDs. Planning considerations should always include the VCO when the MWD will be used or trained in environments that he is not accustomed to.

Wind

8-61. Always consider the direction and velocity of the wind when using an MWD for any detection operations. This consideration holds true even indoors. There may be a slight breeze that will move a scent away from the location of its source. Fans, air conditioning units, and objects in motion can create winds that will affect the performance of a detector dog.

8-62. In addition to these operational concerns, wind may create medical issues (such as blowing sand) that are negative. Wind also creates increased cooling, which is positive in hot weather but negative in cold weather.

Inclement Weather

8-63. Precipitation can degrade the MWD's performance of detection operations; however, do not assume that an MWD will not be able to perform satisfactorily. The amount of moisture accumulated and the amount of time elapsed since the scent was left are variables. MWDs have been proven to be effective even in inclement weather. MWDs that become wet and are returned to their kennels or crates while wet are at increased risk of skin injury and infection.

Heat

8-64. When operating in locations of extreme heat, the MWD should be allowed sufficient time to acclimate. Consult with the VCO at the location to determine how long an MWD should be allowed to acclimate. Care should be given to ensure that the MWD does not sustain injuries to the pads of his paws due to excessive walking on concrete or pavement in extreme heat conditions. Always ensure that the MWD has a fresh and constant source of drinking water.

Cold

8-65. Extreme cold temperatures should not deter MWD operations. Special care should be taken to enable an MWD to perform in cold locations. The accumulation of moisture on the MWD's paw pads can freeze and injure him. Using paw pad protectors to aid in the prevention of injury due to cold weather may be helpful, but the MWD must be trained to wear them. The MWD should be allowed to rest and warm up on the same cycle as the handler.

TERRAIN

8-66. The safety of the handler and MWD must be given the highest consideration when operating in hazardous terrain. Care should be taken when employment will take place in rocky or mountainous areas. Unfamiliar locations can pose threats in the form of caves, holes, loose footing, and other hazards. Whenever possible, conduct a reconnaissance of the area before employing an MWD.

DISTRACTIONS

8-67. There are numerous distractions that can affect MWD operations. These are conditions that can cause the MWD to focus his attention away from the desired task at hand. Planning and consideration should be given to eliminate or reduce distractions that hinder operations. Distractions are natural and man-made. Some examples of distracters are:

- Other dogs (both MWD and indigenous).
- Other animals.
- Human presence and activities.
- Noise.
- Smells.

Appendix A

Metric Conversion Table

This appendix complies with current Army directives, which state that the metric system will be incorporated into all new publications. *Table A-1* is a conversion chart.

Table A-1. Metric Conversion Chart

US Units	***Multiplied By***	***Equals Metric Units***
	Length	
Feet	0.30480	Meters
Inches	2.54000	Centimeters
Inches	0.02540	Meters
Inches	25.40010	Millimeters
Miles (nautical)	1.85320	Kilometers
Miles (statute)	1.60930	Kilometers
	Area	
Square inches	6.45160	Square centimeters
Square feet	0.09290	Square meters
	Volume	
Cubic feet	0.02830	Cubic meters
Fluid ounces	29.57300	Milliliters
	Temperature	
Degrees Fahrenheit	Subtract 32, multiply by 5/9	Degrees Celsius
	Weight	
Ounces	28.34900	Grams
Pounds	0.45359	Kilograms
Metric Units	***Multiplied By***	***Equals US Units***
	Length	
Centimeters	0.39370	Inches
Kilometers	0.53960	Miles (nautical)
Kilometers	0.62137	Miles (statute)
Meters	3.28080	Feet
Meters	39.37000	Inches
Millimeters	0.03937	Inches
	Area	
Square centimeters	0.15500	Square inches
Square meters	10.76400	Square feet
	Volume	
Cubic meters	35.31440	Cubic feet
Milliliters	0.03380	Fluid ounces

Table A-1. Metric Conversion Chart (Continued)

Metric Units	*Multiplied By*	*Equals US Units*
	Temperature	
Degrees Celsius	Multiply by 9/5, add 32	Degrees Fahrenheit
	Weight	
Grams	0.03527	Ounces
Kilograms	2.20460	Pounds

Appendix B

Kennel Construction

This appendix is intended to be a starting point in the design process with the identification of requirements and spaces. It should serve as a supplement to the installation design guide and not as a replacement. This appendix provides commanders and planners with guidelines and specifications to consider for the construction and equipment of kennel facilities. The floor plan (*Figure B-1*) is an example of how an MWD facility works, how areas relate to each other, and what is required for each space. It is not intended to serve as a rigid model for duplication.

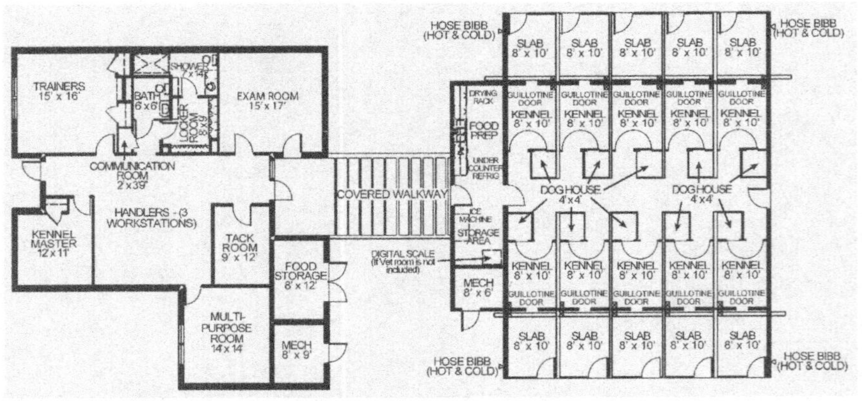

Figure B-1. Example of an MWD Kennel Floor Plan

TEMPORARY KENNELS

B-1. There are two major types of kennels, temporary (*Figure B-2*, page B-2) and permanent. According to *AR 190-12*, temporary kennels are to be used during deployments for no longer than one year, to allow sufficient time for permanent kennels to be built. Temporary kennels may be constructed out of any available resources. The design is not standard and varies from site to site depending on the location, mission, and duration of operations. Using a concept similar to constructing fighting positions, the temporary kennels will be improved continuously as needed or until a permanent facility is completed.

B-2. Using tents and shipping crates may be a hasty method of setting up temporary kennels. Although this method is not optimal for sustained operations, it may be useful until other accommodations can be coordinated. Erect the tents at a location that allows the MWDs a safe place to rest away from other troop activities after operations. Whenever possible, use terrain or buildings to place a barrier between troop activity and the temporary kennel. Never place the temporary kennel site near motor pools, dining facilities, or high noise areas.

Figure B-2. Temporary Kennel

B-3. In some areas of operation, it may be possible to obtain the use of a building or other structures for temporary kennels. In these cases, leaders must ensure that the building is well ventilated, safe, and structurally sound. Ensure that there are no hazards (such as chemicals, bare wires, holes, or debris) that can cause injury to the MWDs or handlers.

B-4. Engineers are often used to assist in the improvement of temporary kennels. In addition, fences should be erected and dog runs created.

***Note*:** From the start of operations, sanitation measures must be employed in all temporary kennel configurations.

PERMANENT KENNELS

B-5. The following information and figures give specific details for the construction of a permanent kennel facility and adjacent areas. These guidelines may be altered to allow for special needs or uses unique to each location.

EXTERIOR AREAS

B-6. An MWD kennel complex (*Figure B-3*) consists of kennels, a support building, an obedience course, an exercise area, an MWD break area, exterior storage, parking areas, drives, and walks. Often, as a result of the location, utility services are limited at the site. Construction should provide the following utilities:

- Water (potable with backflow prevention).
- A sanitary sewer.
- Electric.
- Telephone.
- Network and communication service to the site.

Other systems should be incorporated according to local procedures.

B-7. The entire complex should be enclosed, at the minimum, with a heavy-duty, 8-foot-high chain-link fence with three strands of straight wire (no barbed) at the top to prevent an MWD from climbing or jumping out. A 10-foot (minimum) vehicle gate that can be padlocked is to be installed to allow for food deliveries to the kennel and other access requirements. A personnel entry gate that is visible from the KM's office is also required. The personnel gate should have a cipher lock and a notification buzzer that sounds in the KM's office, the obedience course, and the kennel. The buzzer in the kennel should have the capability to be turned on when personnel are in the area and turned off to allow the MWDs to rest when personnel are absent. All gates should be self-closing and self-latching.

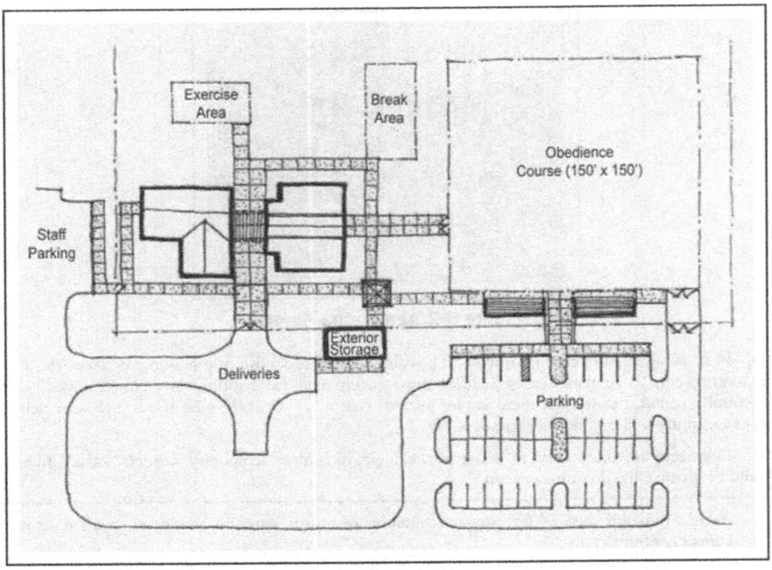

Figure B-3. MWD Kennel Complex

B-8. Security lighting for the area should be according to local policy. At a minimum, the lighting should be on from dusk to dawn and placed on all corners of the fencing area. Security lights should be placed in areas that the other lights do not cover.

B-9. Warning signs will be posted on the exterior fencing and buildings of the MWD kennel and exercise area. Signs should contain the following words: "DANGER—OFF LIMITS–MILITARY DOG AREA." Personnel approaching the kennel area should be able to see and read the warning signs under normal daylight conditions from a distance of 50 meters.

B-10. Locate a trash dumpster for the complex outside of the perimeter fence. Screen the dumpster and provide low-maintenance landscaping for the complex according to local policy.

Exterior Storage, Parking Areas, Drives, and Walks

B-11. An enclosed, 400-square-foot exterior storage building is required to store maintenance equipment, portable kennels, and obstacles. Interior/exterior lighting, power, and a water faucet should be included.

B-12. Staff parking should be a paved parking lot large enough for eight POVs, three GOVs, two visitor vehicles, and one handicap parking space. Pave access drives to accommodate parking, exterior storage, food deliveries, emergency vehicles, and the transportation of MWDs.

B-13. Provide paved walks to all areas. When designing, give careful consideration to creating a one-way MWD traffic system. To prevent confrontations between the MWDs, avoid situations that allow MWDs to meet head on.

Obedience Course (150 Feet by 150 Feet)

B-14. The obedience course plays an important role in maintaining the MWD's agility and stamina as well as in reinforcing obedience and proficiency training. The course should be grassed and free of hazards (trees, large rocks, holes, and burrs) that may be harmful to MWDs and handlers. The site should be graded for drainage but minimally sloped to provide a level field for training. The area should be enclosed with an 8-foot chain-link fence. Gates should be self-closing and self-latching with a minimum width of 5 feet to allow for lawn maintenance equipment. The area should be well lighted to eliminate shadows. Light poles should be located on the exterior of the fence. On and off switches should be manual. Weatherproof outlets should be provided at each light to provide power at the site. Speaker units connected to the chosen public address system should also be provided. Water should be at the course site with water faucets installed to provide water for MWDs and lawn watering. An irrigation system is desirable for long-term lawn maintenance.

B-15. Obstacles for the course (*Figure B-4*) include barrels, tunnels, steps, jumps, framed windows, and dog walks. Each obstacle is 15 to 20 feet from the previous obstacle; the course is run in a sequence. Obstacles should be constructed to the following specifications:

- Barrel 1 is 35 inches long with a 23-inch opening.
- Barrel 2 is 70 inches long with a 23-inch opening.
- Barrel 3 is 105 inches long with a 23-inch opening.
- The tunnel is 146 inches long with a 19-inch opening.
- The steps are constructed out of wood and are 94 inches high with 5 steps on each side. Each step is 43 inches wide by 16 inches high by 24 inches long. The top platform is 48 inches long by 43 inches wide.
- Jumps 1, 2, and 3 are constructed of wood and are 36 inches high at the maximum raised level. There are 6 removable boards that are 51 inches long by 6 inches high and 1 inch thick.
- The framed window is constructed of wood and is 8 feet tall. The maximum height for the MWD to negotiate is 36 inches. The window opening is 48 inches wide. Removable boards are 5 1/7 inches wide by 51 inches long and 1 inch thick.
- The A-frame is constructed of wood. Its two sides are 8 feet long by 4 feet wide by 1 inch thick with 1-inch by 2-inch boards every 6 inches to secure footing. It has a maximum raised height of 6 feet with a base spread of 10 feet.
- The dog walk is constructed of wood and is 26 inches high by 224 inches long and 11 inches wide with 1-inch by 2-inch boards every 4 inches to secure footing. Each end has a 45° incline ramp that is 11 inches wide by 49 inches long (included in the 224 overall length measurement).

Note: See *DA Pam 190-12* for more complete construction guidelines including the bracing necessary for the various obstacles.

Military Working Dog Break Area (10 Feet by 20 Feet)

B-16. An MWD break area should be located near the kennels. A break area allows the handler to release the MWD immediately after exiting the kennels or before entering the kennels so that the MWD can relieve himself. The area should be enclosed with an 8-foot chain-link fence. Gates should be self-closing and self-latching and a minimum of 5 feet wide to allow for lawn maintenance equipment. The ground can be a sandy area for cleaning ease. A water faucet is required to maintain the area.

Exercise Area (20 Feet by 40 Feet)

B-17. The exercise area is a space where the MWD can be released without the handler being present. The exercise area should not be a shared area with the obedience course, as it would conflict with training objectives. The exercise area should be visually separated from the obedience course to prevent MWD from being distracted during training. The exercise area should be grassed, hazard free, and graded slightly for drainage purposes.

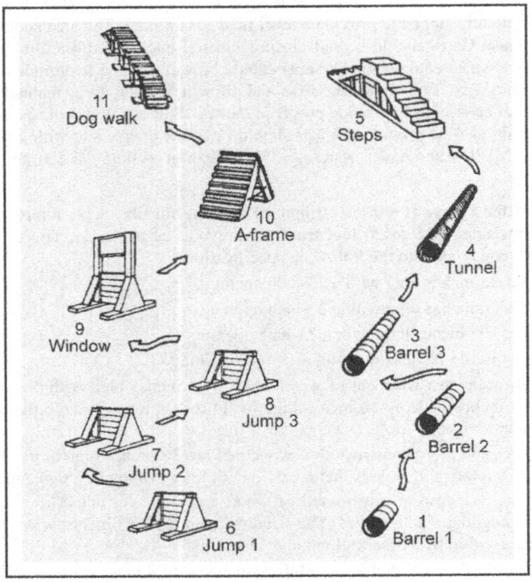

Figure B-4. Obstacle Course Layout

INTERIOR AREAS

B-18. Interior areas have heating, ventilation, and exhaust systems; general-purpose lighting; and power outlets in all spaces. Provide phone, fax, and local area network (LAN) connections to support each area and hot and cold water to all latrines, sinks, and showers. Construct the facility to meet all fire protection and life safety feature requirements.

Administration Areas

B-19. Since MWDs are likely to visit the administration area (*Figure B-5*, page B-6) at some point, finishes should be of durable material, require minimal maintenance, and be easily repaired or replaced in case of damage. (For example, flooring should be covered by carpet.)

Kennel Master's Office (132 Square Feet)

B-20. The KM's office serves as the nerve center for the complex. Ideally, the office should be located at the front of the administrative area with exterior windows that view the entry gate to the complex. It includes a lockable closet for supply storage. Furnishings include a desk, a chair, two visitor chairs, a four-drawer file cabinet, and a bookcase. Equipment includes a computer, a printer, a fax/copier, and a public address system or central intercom console.

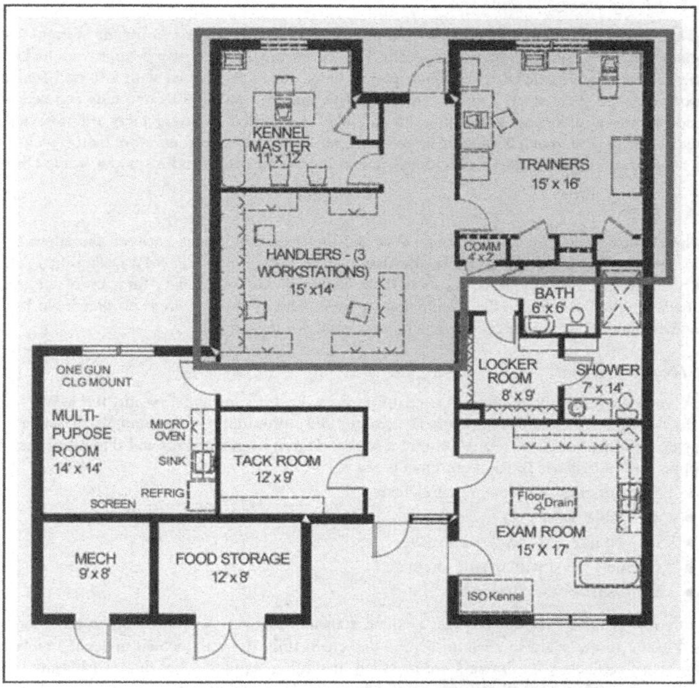

Figure B-5. Kennel Administrative Areas

Trainer's and Plans Noncommissioned Officer's Office (240 Square Feet)

B-21. The trainer office should provide workspace for two people. Lockable closets are required to store training aides for each trainer. Furnishings include a desk, a chair, and a four-drawer file cabinet for each trainer and a common bookcase. Equipment includes a computer, a printer, a copier, and a 4- by 8-foot dry erase board.

Handlers' Area (210 Square Feet)

B-22. The area for the handlers is an open area for general office functions. It includes—

- Three common workstations, depending on the size of the kennels, for handlers to update records and complete daily reports.
- Space for document storage.
- Furnishings (three workstations, file cabinets [one four-drawer vertical file per workstation or the equivalent], and a distribution counter.
- Equipment (computers and a printer).

As a general rule, one workstation should be available for every three handlers assigned to an MWD section.

Controlled-Substance Storage Room

B-23. This room is inside the administrative area and allows sufficient room for the storage safe in which controlled-substance training aids are stored. The doors are secured using approved locking devices. Lightweight doors are replaced with solid metal or wood doors or covered with a 9- to 12-gauge security screen or 16-gauge sheet steel. They are fastened with smooth-headed bolts and nuts and peened in place. All windows providing access to a storage room that is not staffed 24 hours a day are protected by a 9- to 12-gauge security screen or 3/8-inch or larger diameter steel bars spaced no more than 6 inches apart. The frames holding the screen or bars must be fastened to the window frame with smooth-headed bolts.

Special-Use Areas

B-24. Special-use areas (*Figure B-6,* page B-8) are the transition spaces between the administration area and the kennels. They are important in the day-to-day operations of the MWD facilities. They are not normally occupied on a daily basis. As such, these areas provide a secondary function of serving as a noise buffer between the kennels and the administrative areas. The special-use areas provide a workspace for the VCO and accommodate the unique storage requirements for the kennels.

Veterinary Treatment Room

B-25. A veterinary treatment room is used to perform health care and first aid for MWDs since many locations do not have veterinary treatment areas in close proximity to the kennels. In order to maintain sanitary conditions, treatment rooms require a higher degree of cleanliness and durability than the rest of the support building. Basic finish requirements are—

- Seamless floors with an integral base.
- Washable walls.
- Painted gypsum board ceilings.
- Cabinets faced with plastic laminate.
- Solid-surface countertops.

B-26. Interior partitions should provide a sound transmission class rating of 50 to 55. Floor drains are centered in the room to aid in cleaning, and a dedicated floor drain is located under the isolation kennel. Wall and base cabinets with drawers are installed, including stainless steel dual washbasins with hot and cold water. One section of wall cabinets should be lockable so that prescription medicines for MWDs may be stored in the room. In addition to the normal electrical outlets required by code, a 220-volt outlet is required to support other portable equipment that may be used. Equipment for the room includes the following:

- **An MWD isolation kennel.** The entire structure should be made of uniform, nonglare type 304 stainless steel. The frame, cross members, and vertical spokes should be welded together at every intersection. Latches and hinges are made of 12-gauge stainless steel. The floor of the pen is a raised, removable grate. The floor grate is a heavy-gauge stainless steel, plastic-coated section installed 1 3/4 inches off the floor. A dedicated floor drain is required for this kennel.
- **Standard table tub**. The tub is made of type 304 stainless steel with a minimum depth of 15 inches. The tub slopes to the drain end. The design includes a removable, bathing rack made of plastic-coated stainless steel that fits into the bottom of the tub to give MWDs better footing during treatments. Faucets include a flexible, stainless steel hose that will extend the length of the tub and spray powerfully to reach all areas of the tub.
- **Exam room light.** A ceiling-mounted exam room light with an arm extension and swivel capacity of 360° is recommended.
- **Walk-on platform scale.** The scale is a type 304 stainless steel platform with a wall- or post-mounted light-emitting diode (LED) display. The scale is able to perform calibration and weigh animals to the nearest 0.1 of a pound or kilogram.
- **Stationary exam table.** The table top is type 304 stainless steel die-formed from one piece of 20-gauge steel. It should have raised edges to prevent fluid runoff and be permanently attached

to a heavy base. There should be a minimum 35-inch work height from the floor to the top of the table. The heavy base should have adjustable leveling screws.

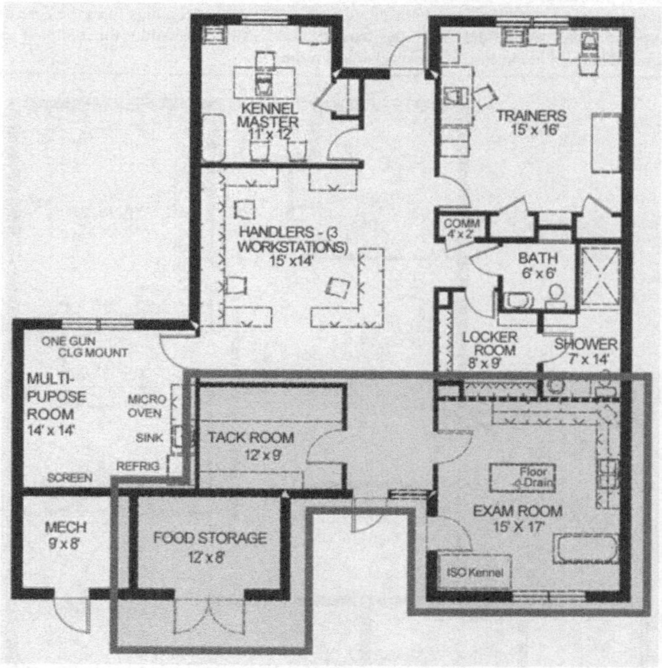

Figure B-6. Kennel Special-Use Areas

Tack Room (108 Square Feet)

B-27. The tack room is for the storage of extra MWD equipment, such as bite suits and portable kennels. The tack room is located near the entrance to the kennels but not the main building entrance. Requirements include 12 linear feet of shelves that are 18-inches wide and 36 inches deep and allow for 36 inches above each shelf; a 24-inch space with hooks to hang items on; and 6 linear feet of shelves that are 18 inches wide and 12 inches deep and allow for 18 inches above each shelf.

Food Storage Room (96 Square Feet)

B-28. The food storage room is normally for the bulk storage of a 30-day supply of food for the MWDs. The room is adjacent to the kennels with an exterior entry. The entry is a pair of 36-inch doors to allow for palletized food delivery and storage. Doors require weather stripping and thresholds to aid in insect and rodent control. Temperature and humidity control equivalent to the office area is required to control food spoilage. A minimum separation of 2 inches is required between the walls and the stored food to allow for air circulation. Shelving should be installed to meet local delivery operations.

SUPPORT AREAS

B-29. The support areas (*Figure B-7*) provide space for the common-use areas of the building. Areas for the multipurpose room, latrines, communications, and mechanical necessities form the core of the support areas. Support areas should be centrally located and easily accessible for building occupants or maintenance personnel in the case of the mechanical room.

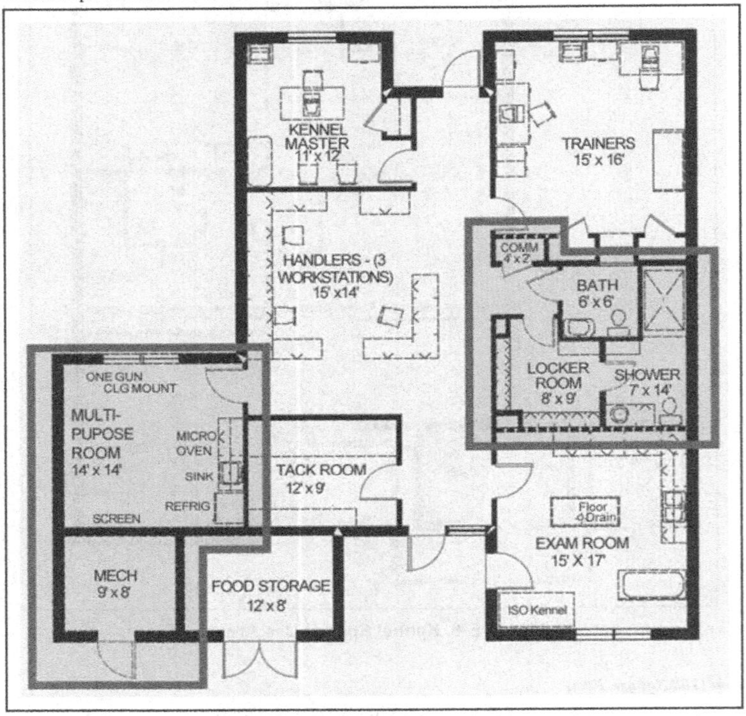

Figure B-7. Kennel Support Areas

Latrine, Shower, and Lockers (206 Square Feet)

B-30. A unisex area is divided into three rooms for the latrine, shower, and lockers. One room is a handicapped accessible latrine with a water closet and lavatory for general use. The locker room includes 10 full-height metal lockers and a bench for MWD handlers' use. The shower room includes a shower, bench, water closet, and lavatory. Do not neglect to include other accessories such as mirrors, soap dispensers, paper towel dispensers, disposal units, grab bars, and hooks.

Heating, Ventilation, Air Conditioning, Mechanical, Electrical, and Communications Systems

B-31. Heating, ventilation, and air conditioning (HVAC) is a process to add hot or cold air in a facility. It also introduces fresh air in to the facility to maintain indoor air quality and eliminate sick building syndrome. Design HVAC systems to optimize energy use and meet the design criteria for the area in which you are located. Place the hot water heater in the mechanical room and with a capacity that will meet the demands of the latrine and shower; multipurpose room; and veterinary treatment room. Locate electrical panels in this area as well. Although a recommended size is shown on the plans, the room should be big enough to support the equipment required for the location.

B-32. While the number of instruments and computers is not large, the communications closet is vital to the success of the organization. Provide a minimum of one prewired telephone and LAN outlet for each workstation, one in each common area, and one in the kennel food preparation area. Include telephone and LAN outlets for fax machines and printers. Conceal conduit in the walls.

Multipurpose Room (196 Square Feet)

B-33. The multipurpose room serves as a combination conference room, classroom, and break room. Food preparation requirements are a light, a countertop, and base and upper wall cabinets for storage. Lighting should have a variable-level control for use as a conference room and/or classroom. Furnishings include either a conference table for fifteen or three five-person tables; chairs; a bulletin board; and a 4- by 8-foot dry erase board. Equipment includes a video projection system, a projector screen, a sink, a microwave oven, and a refrigerator.

KENNEL AREA

B-34. The kennel area provides the daily living environment for the MWDs, giving each MWD a private place for eating and resting. Kennel areas should be built to accommodate large-breed dogs and designed as modular structures to allow for the future expansion of runs.

B-35. There are three types of kennel areas; indoor kennels, outdoor kennels, and indoor/outdoor kennels. The preferred standard is the combination indoor/outdoor kennel; however, there are factors that must be taken into consideration when selecting the kennel type. Cost is always a consideration, but climate also influences the selection. In cold weather climates, the benefit-to-cost ratio will drive the selection to an indoor kennel while the opposite is true for hot weather climates. The more temperate the climate the more suited the indoor/outdoor kennel is for use. Each type of kennel includes the same space requirements. A kennel includes kennel runs with doghouses, a food preparation area, a storage area, and a mechanical room.

B-36. The kennel should be separated from the administrative building by a minimum of 20 feet. This provides separation between different functions and aids in noise control. As a result of the constant interaction between the two facilities, a covered walkway is required to connect the facilities and provide protection from the elements. Care should be taken in the layout of the kennel in relation to the administrative building; a one-way system should be maintained in the movement of MWDs. The objective is to avoid situations where MWDs meet face to face and to prevent confrontations. Consider installing mirrors at corners, intersections, and blind spots to alert handlers of potential wrong-way traffic.

B-37. All entries and exits from the kennel area are required to be self-closing and self-latching. Doors exiting the kennel runs should open inward to aid in preventing MWD escape. Proper ventilation is important in the kennels to prevent the spread of diseases and to control odors. It is recommended that the ventilation standard be to cycle out the room air 10 to 15 times per hour. Kennel temperature should range from 45° to 85° Fahrenheit with humidity in the range of 40 percent to 70 percent. Temperature should be maintained within 10° of the exterior temperature. MWDs work more effectively and are more alert when the kennel temperature is close to the temperature of their working environment.

B-38. Potable water is required at the kennels. Hot and cold water lines are installed at ceiling level down the length of the 5-foot wide center corridor connecting to hose reels at each end of the corridor. A high-pressure washer is required to assist in sanitizing and cleaning kennel runs.

B-39. Since the kennel is basically a wet environment, all electrical receptacles should be provided with ground fault circuit interrupters and all-weather covers. Receptacles mounted at 48 inches above the finished floor are required at each end of the central corridor.

B-40. Each kennel run should have a waterproof light fixture that has an individual switch at the corridor entrance. Other areas require general-purpose lighting according to industry standards. All drain lines in the kennel should be a minimum of 6 inches in diameter and be designed to sustain flow velocities that will maintain a self-cleansing action. Due to the nature of the waste, cleanouts should be numerous and easily accessible. Floor drains should be included in the central corridor to aid in the cleaning of kennels.

Indoor/Outdoor Kennel

B-41. A combination indoor/outdoor kennel (*Figure B-8*, page B-12) simply implies that there is an interior run and an exterior run linked with each MWD living area. A guillotine type door will connect the interior run and the exterior run. Other door types may be used, but the idea is to permit isolation of the MWD during cleaning operations. Ideally, operation of the door is from the central corridor and from the outside of the exterior run. At 6 feet above the finished floor, use translucent wall panels in the wall separating the runs to provide natural light inside the kennel.

B-42. The roof should extend over the exterior runs and include gutters and/or downspouts to prevent excessive water in the runs. Exterior framing should be enclosed with a soffit to prevent birds from roosting. Exterior water faucets (with freeze protection where required) with both hot and cold water are required at each corner of the building. These are necessary for cleaning exterior runs. Floor drains (*Figure B-9*, page B-12) should be located in the corner of each run adjacent to the common wall separating exterior and interior runs. Drains should tie-in to a common waste line. Drain covers should be flush with the floor. Kennel floors should slope (at least 1/4- inch per foot) toward the drains to allow for quick water drainage and drying.

Indoor or Outdoor Kennel

B-43. The plan for the indoor or outdoor kennel (*Figure B-10*, page B-13) is the same for either style. The obvious difference is the wall that encloses the kennels. For indoor kennels, the exterior walls along the runs should have windows beginning at 6 feet above the finished floor to allow for natural light. Windows should be operable and hinged at the sill to tilt inward to prevent the escape of MWDs. For outdoor kennels, the exterior wall is simply an 8-foot chain-link fence. Water faucets with hot and cold water are required at each corner of the building. These are required to clean the gutter and drain system (*Figure B-11*, page B-13) for the kennel. A gutter and drainage system is provided at the back of the kennel run for either type of kennel for sluicing wastewaters from cleaning operations. Kennel run floors should slope to the gutter and the gutter should slope to the drain. Gutters should be sufficient to allow for easy cleaning. A minimum of 3 feet should be allowed between the end of the kennel and the wall or fence.

Kennel Run

B-44. Runs should be arranged so that openings are staggered in order to avoid MWDs facing each other across the corridor. Full-height partitions are required between runs for all kennel types. The first 6 feet of wall above the floor should be sealed concrete. For interior runs, steel clad acoustical panels should be considered from the top of the concrete partition to the ceiling. Steel clad acoustical ceiling panels should also be considered above the kennels.

B-45. Heavy-gauge chain-link fence is used for exterior runs and is also an option for interior runs. Because of the noise generated in the enclosed area of an interior run, it is important that noise-reducing materials be introduced to prevent stress to the MWD and hearing loss to personnel.

B-46. As a minimum, end walls should be full-height, galvanized chain-link fencing. Take care to prevent sharp edges from bolts, hinges, and tie wires on the interior of the runs where MWDs may be cut or harmed. Self-latching entry gates should swing 180° and be able to cover the opening of the MWD house. This allows the MWD to be penned in the house during run-cleaning operations. A stainless steel water bucket is required for each kennel, including a holder that prevents the MWD from overturning the bucket.

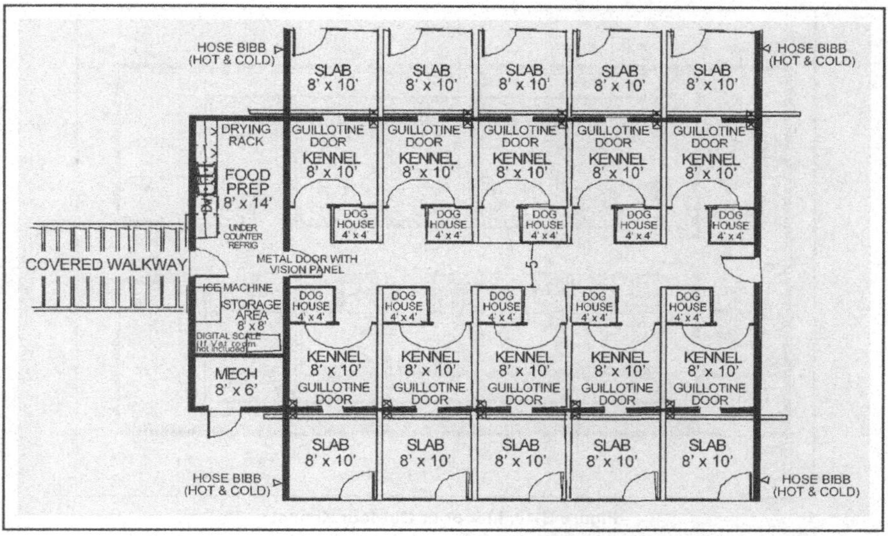

Figure B-8. Combination Indoor/Outdoor Kennel

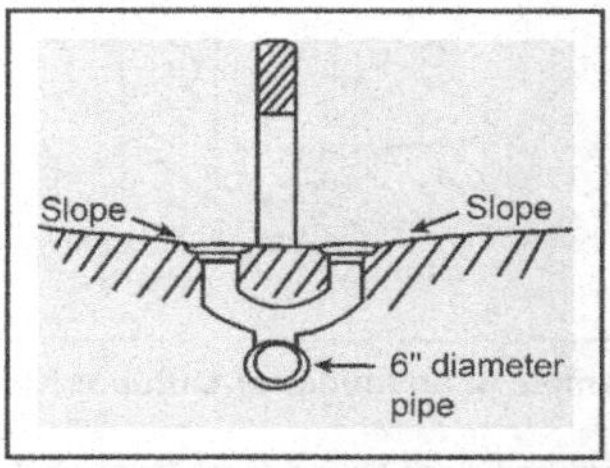

Figure B-9. Example of a Combination Indoor/Outdoor Kennel Floor Drain

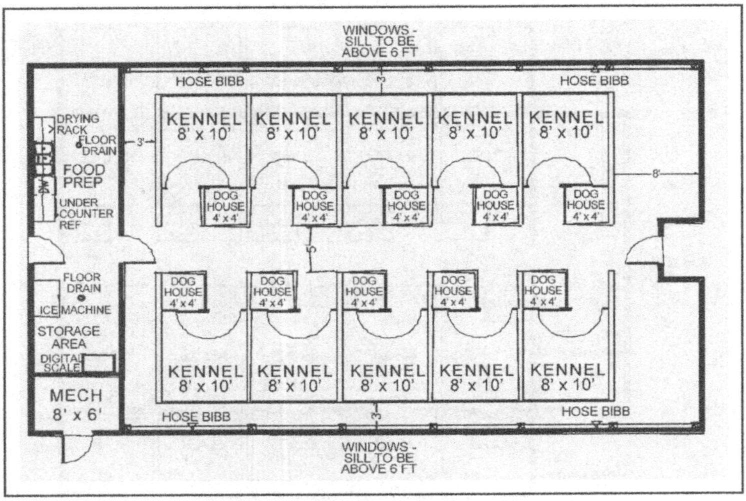

Figure B-10. Indoor or Outdoor Kennel

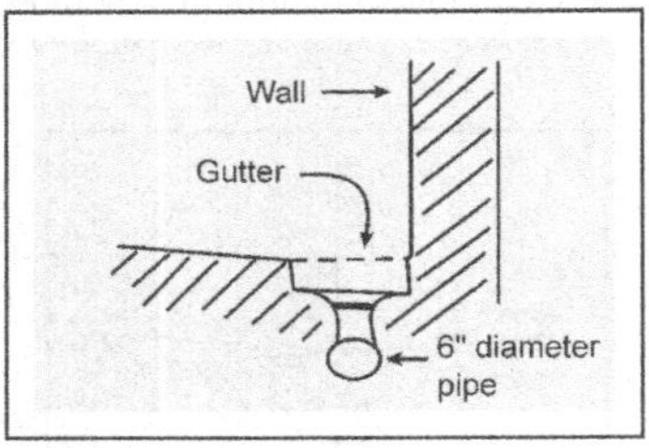

Figure B-11. Example of an Indoor or Outdoor Kennel Floor Drain

B-47. Flooring should be concrete and slope toward floor drains and/or gutters to provide rapid runoff of water and drying. Since runs are subject to frequent wash downs to clean and prevent disease, a smooth epoxy-type finish is needed. Temperature-controlled floors should be emplaced for locations that have inclement weather.

B-48. All materials in areas accessible to MWDs should be resistant to damage by scratching, biting, and chewing. Materials should be durable and easy to sanitize. Avoid using material such as angle iron that rusts over a period of time and generates sharp edges that pose a danger to MWDs.

B-49. MWD houses are 4 feet by 4 feet by 4 feet sealed, concrete boxes with a metal top. Tops should hinge toward the end wall in order to be raised to clean the house or treat an MWD. Once lifted, a holding device should be available to keep the top open. Wall opening dimensions are shown in *Figure B-12*, page B-14. The house should be a minimum of 8 inches and a maximum of 12 inches above the run floor. The house floor should have positive drainage toward the wall opening.

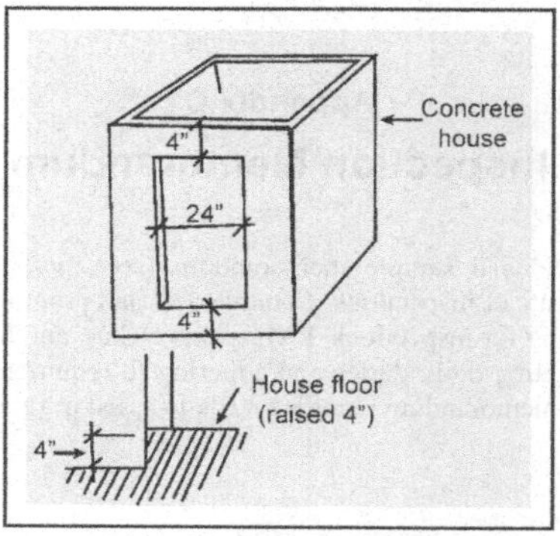

Figure B-12. MWD House

Storage and Mechanical Area

B-50. Additional storage for the kennel is provided. A commercial quality ice machine is required to aid in cooling down MWDs. Include a floor drain for the area. If a veterinary treatment room is not included, a digital scale should be located in this area. Design mechanical systems to optimize energy use and meet the design criteria for the area where the kennel is located. Include the hot water heater in the mechanical room and with a capacity to meet the demands of the kennel. Allow for the location of electrical panels in this area as well.

Food Preparation Room (112 Square Feet)

B-51. The food preparation area is where feedings are prepared and feeding pans/bowls are stored and cleaned. Food preparation for MWDs requires a large countertop with a deep, three-compartment stainless steel sink. Upper wall cabinets are required for general storage. An under the counter refrigerator, a dishwasher, and space to store food for daily use are necessary for operation. A 2-foot space at the end of the counter should accommodate a storing and drying rack for pans and bowls. Lighting and receptacles and a floor drain should be installed per the code. A telephone, bulletin board, feeding chart, and large garbage can should also be provided. Optional equipment includes a garbage disposal unit and a stackable washer and dryer.

Appendix C
Inspection Memorandum

This appendix provides a sample memorandum (see *Table C-1*, page C-2) for documenting the results of inspections. Commanders and program managers will find this useful in preparing for inspections. KMs, plans NCOs, and handlers will find this list helpful in conducting daily duties and adhering to requirements for inspections. Maintain completed memorandum records on file to assist in future inspections.

Table C-1. Sample Inspection Memorandum

DEPARTMENT OF THE ARMY
UNIT
ADDRESS OF UNIT

OFFICE SYMBOL 15 May 2004

MEMORANDUM FOR RECORD

SUBJECT: MACOM Technical Assistance Visit or Provost Marshal Inspection

1. **SECTION I. GENERAL INFORMATION.**

UNIT: 529th Military Police Company LOCATION: Kaiserslautern
DATE: 3-14-05 TIME: 0900

INSPECTOR (S): LTC Jeffery S. Captain ESCORT: SFC James Majors

TYPE OF VISIT: __X__ ANNOUNCED______UNANNOUNCED

TYPE	AUTHORIZED	ON HAND
Patrol		
Patrol Explosive	6	4
Patrol Narcotic	3	2
Explosive Detector	0	0
Narcotic Detector	0	0

2. **SECTION II. PERSONNEL.** **YES N/A NO**

a) Are all authorized MWDs on hand (unit TDA/MTOE document)? ___ ___ ___

b) Are all MWD handlers certified by the DOD Dog Center? ___ ___ ___

c) Are ASI trained handlers being used as specified in AR 614-200? ___ ___ ___

d) Is the assignment policy of one dog and one handler being followed? ___ ___ ___

e) Are all MWDs assigned to a handler? ___ ___ ___

f) Has an excess of dogs been reported to the MACOM? ___ ___ ___

g) Does the KM meet the minimum criteria to be appointed as a KM? ___ ___ ___

h) Are handlers being assigned duties that take away from their handling responsibilities (excluding PT, staff duty, field duty, and unit training)? ___ ___ ___

3. **RESPONSIBILITIES OF THE PROVOST MARSHAL AND COMMANDER.**

a) Is there an updated SOP signed by the PM explaining policies, procedures, and responsibilities for the direction, management, and control of the kennels? ___ ___ ___

b) Has the PM initiated an MWD program based on the evaluation of the installation support, customs, HRP support, and combat support missions? ___ ___ ___

Table C-1. Sample Inspection Memorandum (Continued)

OFFICE SYMBOL SUBJECT: MACOM Technical Assistance Visit or Provost Marshal Inspection		YES	N/A	NO
c)	Has the PM issued written guidelines governing the release of an MWD to apprehend a suspect?	____	____	____
d)	Are there specific criteria in contingency plans governing the use of dog teams during civil disturbance situations?	____	____	____
e)	Does the PM provide guidance on procedures for explosive detector dog support to civil authorities?	____	____	____
f)	Is the proficiency of each MWD team evaluated periodically by the PM?	____	____	____
g)	Has the PM and/or unit commander checked one or more of these areas monthly: kennels, kennel support building, training and exercise areas, equipment, handlers, dogs, training, utilization and team proficiency?	____	____	____
h)	Has the PM and/or unit commander inspected all areas of the MWD kennels quarterly, and is a written record on file?	____	____	____
i)	Are MWD requisitions submitted according to *AR 190-12* and *AR 700-81*?	____	____	____
j)	Are MWDs placed on the unit property book by NSN, name, and tattoo number?	____	____	____
k)	Are MWDs assigned to the KM on a *DA Form 3161* (*Request for Issue or Turn-In)* hand receipt (according to *DA Pam 710-2-1*)?	____	____	____
4.	**VCO RESPONSIBILITIES.**			
a)	Are the dogs examined by the VCO every six months (according to *AR 40-905*)?	____	____	____
b)	Does the VCO inspect the kennel facility and area for proper sanitary standards on a quarterly basis?	____	____	____
c)	Is the issued dog food and diet checked periodically?	____	____	____
d)	Is there a program in effect for insect and rodent inspection?	____	____	____
e)	Does the VCO concur with the kennel standards/facilities?	____	____	____
	If not what course of action has been taken? ____________________			
f)	Does the VCO provide regular training to handlers on MWD health care, feeding, and first aid?	____	____	____
g)	Does the VCO provide a list to the KM concerning the deployable status of all MWD?	____	____	____
h)	Does the VCO issue health certificates for dogs pending shipment?	____	____	____

Table C-1. Sample Inspection Memorandum (Continued)

OFFICE SYMBOL
SUBJECT: MACOM Technical Assistance Visit or Provost Marshal Inspection

		YES	N/A	NO
5.	**WARNING SIGNS.**			
a)	Are there warning signs with the following words posted at each instillation entrance, **"CAUTION—THIS AREA PATROLLED BY MILITARY WORKING DOGS**?"	___	___	___
b)	Are kennels located away from disturbing influences and adequately posted with **"DANGER—OFF LIMITS—MILITARY DOG AREA**?"	___	___	___
c)	Are the shipping crates stenciled with **"CAUTION—MILITARY WORKING DOG—DO NOT TAMPER WITH ANIMAL?"**	___	___	___
d)	Are shipping crates stenciled with **"LIVE ANIMAL**?"	___	___	___
e)	Are removable or permanent markings placed on the sides and rear of vehicles with the following phrase **"CAUTION—MILITARY WORKING DOGS**?"	___	___	___
6.	**KENNEL AREA.**			
a)	Has a risk analysis been conducted on the kennels?	___	___	___
b)	Is the kennel manned both day and night?	___	___	___
c)	Does the kennel have an IDS/CCTV system if not manned both day and night?	___	___	___
d)	Are approved locks and padlocks used to secure government equipment?	___	___	___
e)	Have primary and alternate key custodians been appointed, in writing, by the PM or commander?	___	___	___
f)	Are personnel authorized to issue and receive keys listed on an access roster?	___	___	___
g)	Are all keys controlled and accounted for on a *DA Form 5513-R* (*Key Control Register and Inventory)?*	___	___	___
h)	Is the *DA Form 5513-R* secured in an approved, lockable container?	___	___	___
i)	Is the key depository box permanently affixed to a wall?	___	___	___
j)	Are keys inventoried semiannually, and is the last report on file?	___	___	___
k)	Are all keys inscribed with a serial number?	___	___	___
l)	Are kennels used to house or care for other animals?	___	___	___

Table C-1. Sample Inspection Memorandum (Continued)

OFFICE SYMBOL SUBJECT: MACOM Technical Assistance Visit or Provost Marshal Inspection	YES	N/A	NO
m) Are adequate training areas adjacent to the kennel areas?	____	____	____
1) Is the training area fence 8 feet high?	____	____	____
2) Does the overhanging fence face inward?	____	____	____
3) Is the fence staked to the ground?	____	____	____
n) Does the kennel area have an adequate water supply?	____	____	____
o) Are the kennels of a standard design?	____	____	____
7. EQUIPMENT.			
a) Is all authorized equipment on hand?	____	____	____
b) Are Air Force shipping crates returned within 10 days after receiving a new or replacement MWD?	____	____	____
c) Is there one shipping crate available for each dog authorized?	____	____	____
d) Is leather equipment properly maintained and cleaned?	____	____	____
e) Is all metal equipment free of rust?	____	____	____
f) Is the web equipment cleaned with proper cleaning material to prevent deterioration?	____	____	____
g) Are there adequate medical supplies on hand for first aid treatment and training?	____	____	____
8. USE OF MWDs.			
a) Are the dog teams being employed a minimum of 24 hours per week?	____	____	____
b) Are PDs employed in all areas of the installation during both day and night?	____	____	____
9. NARCOTICS.			
a) Is the MWD section registered with the Drug Enforcement Administration?	____	____	____
b) Is a *DEA 223 (Controlled Substance Registration Certificate)* on file and current?	____	____	____
c) Have all personnel been designated, in writing, by the PM to handle training aids?	____	____	____
d) Has primary and alternate narcotic training aid custodians been appointed, in writing, by the PM?	____	____	____

Table C-1. Sample Inspection Memorandum (Continued)

OFFICE SYMBOL
SUBJECT: MACOM Technical Assistance Visit or Provost Marshal Inspection

		YES	N/A	NO
e)	Have primary and alternate narcotic custodians been cleared by a favorable Crime Records Center name check?	____	____	____
f)	Is the Crime Records Center name check maintained on file?	____	____	____
g)	Have all personnel authorized to handle or use training aids been briefed at least annually?	____	____	____
h)	Is a written record of the briefing being maintained on file?	____	____	____
i)	Are the monthly/quarterly inventories being conducted by a disinterested E7 or above?	____	____	____
j)	Are the monthly/quarterly inventories being maintained with the controlled substances?	____	____	____
k)	Are the monthly/quarterly inventories being maintained for at least two years?	____	____	____
l)	Are joint inventories being conducted on the change of primary or alternate custodians and is the inventory maintained on file?	____	____	____
m)	Is a daily issue/return green logbook on hand?	____	____	____
n)	Is the property book properly maintained to account for all authorized explosive training aids?	____	____	____
o)	Is an *SF 702 (Security Container Check Sheet)* posted on the safe?	____	____	____
p)	Are *SFs 702* maintained for 30 days?	____	____	____
q)	Is a controlled-substance accountability folder on hand?	____	____	____
r)	Is the controlled-substance accountability folder properly maintained?	____	____	____
10.	**EXPLOSIVES.**			
a)	Have primary and alternate explosive custodians been appointed, in writing, by the PM?	____	____	____
b)	Have primary and alternate custodians been cleared by a favorable Crime Records Center name check?	____	____	____
c)	Are records of the Crime Records Center checks on file?	____	____	____
d)	Are all personnel authorized to receive and handle explosive training aids designated, in writing, by PM?	____	____	____
e)	Is annual explosive safety training being conducted by EOD personnel or other qualified personnel?	____	____	____

Table C-1. Sample Inspection Memorandum (Continued)

OFFICE SYMBOL
SUBJECT: MACOM Technical Assistance Visit or Provost Marshal Inspection

		YES	N/A	NO
f)	Is a written record of the safety training being maintained on file for a minimum of two years?			
g)	Is there an SOP that outlines safety procedures in training areas, transportation of explosives, severe weather conditions, and accidents?			
h)	Are the monthly/quarterly inventories being conducted by a disinterested E7 or above?			
i)	Are the monthly/quarterly inventories maintained on file for at least two years?			
j)	Are all required explosive training aids available to ensure that proficiency standards are met?			
k)	Are chlorates being replaced quarterly?			
l)	Is each type of explosive being stored in a separate vapor proof container?			
m)	Is each vapor proof container properly marked?			
n)	Has the vehicle that is used to transport explosive training aids been approved by safety personnel?			
o)	Does the installation have an explosive detector training plan that has been approved by the installation commander which lists buildings and facilities authorized for training the PEDD teams?			
p)	Is a daily issue/return green logbook on hand?			
q)	Is the logbook properly maintained to account for all authorized explosive training aids?			
r)	Is a *DA Form 7281-R (Command Oriented Arms, Ammunition, & Explosives Security Screening and Evaluation Record)* on hand for each handler that is authorized to handle or receive explosive training aids?			
11.	**TRAINING.**			
a)	Does the KM (or handlers) give training to military police on MWD team functions and operational considerations?			
b)	Are the MWDs receiving a minimum of 4 hours of patrol and 4 hours of narcotic/explosives proficiency training each week?			
c)	Have all certified narcotic/explosive detector teams received quarterly validation testing by the KM?			
d)	Is the minimum proficiency standard for assigned patrol/narcotic detector dogs 90 percent and patrol/explosive detector dogs 95 percent?			
e)	Was the dog decertified when his proficiency standard was less than 90 percent and less than 95 percent for more than three consecutive months?			

Table C-1. Sample Inspection Memorandum (Continued)

OFFICE SYMBOL
SUBJECT: MACOM Technical Assistance Visit or Provost Marshal Inspection **YES** **N/A** **NO**

12. RECORDS.

		YES	N/A	NO
a)	Is the handler's information being recorded on the back of *DD Form 1834*?	___	___	___
b)	Have all *DA Forms 2807-R* been completed properly?	___	___	___
c)	Have all deficiencies for the MWDs been annotated and explained on DA Forms 2807-R?	___	___	___
d)	Is a *DA 3992-R* being maintained to record the MWD's reliability?	___	___	___
e)	Does each MWD team have a probable cause folder?	___	___	___

13. THE RESULTS OF THIS INSPECTION ARE—

____ EXCELLENT

____ SATISFACTORY

____ UNSATISFACTORY

14. REMARKS. All areas noted with a deficiency will be annotated in the remarks section. A copy of this technical assistance visit will be provided to OPMG, the MACOM program manager, and the KM. All records of technical assistance visits will be maintained on file at the installation for a minimum of two years after the inspection date with the corrective action memorandum.

15. POINT OF CONTACT. The point of contact for this action is the undersigned at COMM: 555-555-1511 or e-mail address.

(Original Signed)
JEFFREY S. CAPTAIN
LTC, USA
MACOM PM

Appendix D

Training Records

This appendix provides samples and details of how to properly use MWD training records. Records must be maintained for the life of the MWD. At least one year of historical documents must be maintained on file at each kennel for every MWD. Records can be discarded two years after the death, adoption, or euthanasia of the MWD. Nearly all MWDs have a dual purpose (PD and PEDD or PNDD); therefore, every MWD should have a PD and a PEDD/PNDD form on file. Kennels that have internet access can obtain records via the Military Working Dog Management System (MWDMS).

DEPARTMENT OF THE ARMY *FORM 2807-R*

D-1. *DA Form 2807-R (Military Working Dog Training and Utilization Record)* (*Figure D-1,* page D-3) is a standard form used by all MWD personnel. It is used to record the training and utilization of PDs. Use the following instructions when preparing a *DA Form 2807-R*:

ADMINISTRATIVE DATA

D-2. Enter the following information in the appropriate blocks:

- The date.
- The MWD's name and type, tattoo number, and age.
- The handler's name and grade.
- The organization and the location.

TRAINING

D-3. The training section of the form provides complete information on all of the training tasks that must be accomplished for a PD to maintain proficiency. It is not necessary to train on all of the controlled-aggression tasks during the same day; however, all of the controlled-aggression tasks must be trained every week.

D-4. Entries for the amount of time spent training on specific tasks during a particular day of the month are made in minutes. The monthly total for each specific task will be in hours and will be automatically entered in the far right-hand column titled "Total Hours." The handler's or trainer's evaluation of the MWD's daily rating is entered as S or U. Remarks are mandatory for any area the MWD fails to perform to standard. These remarks will be annotated on the reverse side of the form using the following format:

- Deficiency (DF). The MWD's shortcoming.
- Corrective action (CA). The handler's response to the MWD's deficiency.
- Corrective response (CR). The MWD's response to the CA.

D-5. Enter the appropriate information on—

- **Line 1.** Record the number of minutes per day that the MWD is trained on obedience on leash.
- **Line 2.** Record the number of minutes per day that the MWD is trained on obedience off leash.
- **Line 3.** Record the number of minutes per day that the MWD is trained on the obedience course.

- **Line 4.** Record the daily rating of the MWD on controlled aggression (either S or U). Write an explanation of any U rating on the reverse side of the form. In 4a to 4e, record the number of minutes per day that the MWD trained on each controlled-aggression task.
- **Line 5.** Record the number of minutes per day that the MWD trained on building searches on line 5.
- **Line 6.** Record the number of minutes per day that the MWD trained on gunfire with the handler firing the weapon. The number of rounds will be annotated on the reverse side of the form.
- **Line 7.** Record the number of minutes per day that the MWD is trained on gunfire with a third party firing a weapon. The number of rounds will be annotated on the reverse side of the form.
- **Line 8.** Record the number of minutes per day that the MWD is trained on scouting/patrolling. Record the distance (in feet) for scent, sight, and sound detection on lines 8a, b, or c.
- **Line 9.** Record the number of minutes per day that the MWD is trained on vehicle patrol.
- **Line 10.** Record the number of minutes per day that the MWD is trained on tracking.
- **Line 11.** Record the daily training rating (S or U). Explain deficiencies on the reveres side of the form. Any failure of a critical task requires a U daily training rating. See *DA Pam 190-12* for critical tasks.

UTILIZATION

D-6. The utilization section of the form provides a daily record of the time spent performing military police duties in the three general categories of combat support operations, patrol–law enforcement, and patrol–security.

D-7. The combat support operations category may be used to record time spent performing military police missions (actual or training) in support of combat units. Such operations include field training exercises, CP exercises, mobilization exercises, and other activities related to the tactical and strategic missions of the Army, provided that these missions are accomplished with a handler and an MWD together as a certified team.

D-8. The patrol–law enforcement category may be used to record time spent performing law enforcement or force protection type missions. These missions include but are not limited to—

- Random vehicle searches.
- Law enforcement patrols.
- Other duties performed while working in a law enforcement type mission.

D-9. The patrol–security category may be used to record time spent performing security type missions such as—

- Dignitary security.
- Critical asset patrolling in a peacetime environment.
- USSS, Department of State, or other explosive type searches.

D-10. The daily rating of the MWD's performance on military police duties is recorded as S or U. All U performance annotations will have an explanation for the DF and have the CA taken recorded on the reverse side of the form and continuation sheets, if necessary. The total monthly utilization hours are recorded in the last column, titled "Total Hours."

D-11. When training is conducted during the actual use and employment of an MWD, the time used for training may also be reported in the training section. This training is not an automatic entry every time the team has utilization hours. Time annotated as training during utilization hours will be actual time training on patrol or detector tasks. It is unethical to count all or even half of an MWD team's utilization hours as both utilization and training time for every shift, and unrealistic numbers will not be accepted. When time is recorded as utilization and training, a notation is made on the reverse side of the form explaining the double time entry and the type of training conducted. This will make it possible to differentiate between training time, training time while on duty, and duty (utilization) time. Although additional training during utilization hours is acceptable, it will not be accepted as the primary amount of mandatory training hours each week.

MILITARY WORKING DOG TRAINING AND UTILIZATION RECORD

For use of this form, see AR 190-12; proponent agency is ODCSOPS

NAME/TYPE OF DOG	TATTOO/NUMBER	AGE	NAME OF HANDLER	GRADE	ORGANIZATION AND LOCATION	MONTH AND YEAR
NATZ/PEDD	E309	2	BOOKSTEIN, DAVID B.	SGT	HHC 98TH ASG, BAMBERG, GE	SEP 2003

DAILY RATINGS: S - SATISFACTORY, U - UNSATISFACTORY (Explain deficiency and corrective action on reverse)
(Use reverse side of this form for any remarks or notes)

TRAINING (DAY OF MONTH)	1	2	3	4	5	6	7	8	9	10	11	12	13	14	15	16	17	18	19	20	21	22	23	24	25	26	27	28	29	30	31	TOTAL HOURS
1. ON LEASH OBEDIENCE	15	15	15	15	30		30	40	10	15			20	15		15	20	30	15		15	15	15	15					15	15		6.50
2. OFF LEASH OBEDIENCE	20	20	15	15	40		30	40	10	15			20	30		15	30	30	15		15	20	20	20					15	15		7.50
3. OBEDIENCE COURSE					20		20			15																						0.92
4. CONTROLLED AGGRESSION (S OR U)		S	S					S	S				S	S																		
a. FALSE RUN		5	5					15	5				5	5																		0.67
b. ATTACK		10	10					20	10				10	10																		1.17
c. SEARCH AND ATTACK		5	5					15	10				10	10																		0.92
d. STAND OFF		10	15					20	5				10	10																		1.17
e. ESCORT		5	5					15	5				5	5																		0.67
5. BUILDING SEARCH		15						15																20								0.83
6. GUNFIRE - HANDLER								10																								0.17
7. GUNFIRE - DECOY/AGITATOR					10														30													0.67
8. SCOUTING/PATROLLING (TIME)					15																			15					15			0.75
a. SCENT DETECTION (DISTANCE)					40																			30					30			
b. SIGHT DETECTION (DISTANCE)																																
c. SOUND DETECTION (DISTANCE)																																
9. VEHICLE PATROL																																0.00
10. TRACKING																																0.00
11. DAILY TRAINING RATING (S OR U)	S	S	S	S	S		S	S	S	S						S	S	S	S		S	S	S	S					S	S		21.92

UTILIZATION (HOURS PER DAY)	1	2	3	4	5	6	7	8	9	10	11	12	13	14	15	16	17	18	19	20	21	22	23	24	25	26	27	28	29	30	31	TOTAL HOURS
1. COMBAT SUPPORT OPERATIONS																																0
2. PATROL - LAW ENFORCEMENT	8	8	8	8	8			8	8	8			8	8	8	8				8	8	8	8						8	8		144
3. PATROL - SECURITY																																0
4. DAILY UTILIZATION RATING (S OR U)	S	S	S	S	S			S	S	S			S	S	S	S				S	S	S	S						S	S		144.0

DAILY FEEDING (QUANTITY OF FOOD (BY WEIGHT))	1	2	3	4	5	6	7	8	9	10	11	12	13	14	15	16	17	18	19	20	21	22	23	24	25	26	27	28	29	30	31	WT OF DOG DATE	LBS
2 CU S/D	A	A	A	A	A	A	A	A	A	A	A	A	A	A	A	A	A	A	A	A	A	A	A	A	A	A	A	A	A	A	A	1ST	75
2 CU S/D	A	A	A	A	A	A	A	A	A	A	A	A	A	A	A	A	A	A	A	A	A	A	A	A	A	A	A	A	A	A	A	15TH	73.5

DA FORM 2807-R, Oct 84 EDITION DEC 72 IS OBSOLETE, AND REPLACES DA FORMS 2810-R AND 2815-R, DEC 72

Figure D-1. Sample DA Form 2807-R

DA Form 2807-R
Continuation Sheet

DATE	REMARKS
2-Sep-03	**DF1:** MWD WAS DISTRACTED BY OTHER HANDLER; WOULD NOT FOCUS ON COMMANDS.
	CA1: CONDUCTED PROPER AVOIDANCE TRAINING TECHNIQUE.
	CR1: AFTER MWD DID NOT RESPOND TO INITIAL AVOIDANCE TRAINING TECHNIQUE, A PINCH COLLAR WAS USED, RESULTING IN REGAINED FOCUS.
	DF2: MWD BIT SLEEVE ON STAND OFF.
	CA2: CONDUCTED PROPER AVOIDANCE TRAINING TECHNIQUE.
	CR2: MWD PERFORMED STAND OFF CORRECTLY.
5-Sep-03	GUNFIRE TRAINING: 5 ROUNDS FIRED BY THIRD PARTY.
	DF1: MWD REPEATEDLY BROKE POSITION (SIT TO DOWN AND DOWN TO SIT).
	CA1: CONDUCTED PROPER AVOIDANCE TRAINING TECHNIQUE.
	CR1: MWD REMAINED IN POSITION.
	DF2: MWD WAS HESITANT TO OUT REWARD (BITE SLEEVE).
	CA2: CONDUCTED PROPER AVOIDANCE TRAINING TECHNIQUE.
	CR: MWD OUTED REWARD ON COMMAND.
7-Sep-03	MWD HAS BEEN BREAKING POSITION AFTER SHORT TIME DURING OFF/L OB WHEN HANDLER IS POSITIONED AWAY FROM HIM. TODAY MWD REMAINED IN THE SIT FOR 1 MIN 50 SEC, BROKE SIT, WAS VERBALLY CORRECTED AND REMAINED IN SIT FOR AN ADDITIONAL 25 SEC.
	DF: MWD BROKE FROM DOWN SEVERAL TIMES DURING PARADE REST.
	CA: CONDUCTED PROPER AVOIDANCE TRAINING TECHNIQUE.
8-Sep-03	GUNFIRE TRAINING: 10 ROUNDS FIRED BY HANDLER.
	CR: AFTER MWD DID NOT RESPOND TO AVOIDANCE TRAINING, ESCAPE TRAINING WAS CONDUCTED. EVENTUALLY MWD REMAINED IN DOWN. MWD IS SO EXCITABLE THAT HE HAS A TENDENCY TO CUE OFF HANDLER'S MOVEMENTS TO BREAK FROM POSITION. HANDLER IS CONCENTRATING ON TEACHING MWD NOT TO BE SO DISTRACTED BY OUTSIDE STIMULUS.
	DF1: MWD WAS GETTING DISTRACTED AND NOT HEELING IN THE PROPER POSITION.
	CA1: CONDUCTED PROPER AVOIDANCE TRAINING TECHNIQUE (HEEL-A-WAYS).
	CR1: MWD HEELS CORRECTLY.
	DF2: MWD WAS BITING SLEEVE DURING STAND OFF.
11-Sep-03	**CA2:** CONDUCTED PROPER AVOIDANCE TRAINING TECHNIQUE.
	CR2: MWD PERFORMED STAND OFF CORRECTLY.
	NO TRAINING DUE TO MWD GOING TO VET. MWD HAS BEEN DIAGNOSED WITH A POSSIBLE PULLED/STRAINED MUSCLE IN HIS RIGHT LEG, WHICH HE IS FAVORING AS HE WALKS. X-RAYS SHOW NO BREAKS OR FRACTURES. MWD CANNOT DO ANYTHING STRENUOUS FOR THE NEXT WEEK (NO RUNNING OR JUMPING UNTIL 18 SEP) INCLUDING BITE WORK OR OB COURSE. MWD CAN DO BASIC OB STARTING TUES, SEP 16.
19-Sep-03	GUNFIRE TRAINING: 15 ROUNDS FIRED BY DECOY.

Figure D-1. Sample DA Form 2807-R (Continued)

D-12. Record the appropriate utilization information on—

- **Line 1.** Record the hours per day that the MWD is used for combat support operations.
- **Line 2.** Record the hours per day that the MWD is used for patrol–law enforcement.
- **Line 3.** Record the hours per day that the MWD is used for patrol–security.
- **Line 4.** Record the daily utilization rating (S or U). Explain any U on the reverse side of the form.

TOTAL HOURS

D-13. Record the total hours of training and use in the last column on the right side of the form. The time should be recorded in hours and minutes. The MWDMS makes this entry automatically.

Daily Feeding

D-14. Record the quantity of food that the MWD eats every day in this section. If the MWD is fed once daily, the type and amount of food will be noted in the first block to the left. Under quantity of food, the amount eaten for each feeding will be annotated as follows:

- "A" if all food was eaten.
- "3/4" if three-quarters was eaten.
- "1/2" if half was eaten.
- "1/4" if one-quarter was eaten.
- "0" if none of the food was eaten.

D-15. If the MWD is fed twice daily, the second block is used to record the second daily feeding (which could be a different type of food). If the MWD is fed once daily, the second block may be used to record a change in the diet of the MWD, directed by the veterinarian. Block 2 may also be used to indicate the type and amount of food used daily for MWDs on a food reward system. Monitoring MWD food consumption is a critical portion of the required daily health checks.

D-16. The last column, titled "Wt of Dog," is used to record a semimonthly weight for the MWD. The weight checks occur on or around the first and the fifteenth day of each month. There is sufficient room under the headings "Date" and "Lbs" to record both the date when the MWD is weighed and the weight in pounds. Any drastic changes in food consumption or weight gain or loss are reported to the veterinarian.

D-17. Enter the following information in the appropriate blocks:

- Enter the VCO's prescribed amount of food to be given daily directly under "Daily Feeding." Both blocks are used when an MWD is fed twice daily.
- Record the amount of food eaten per day, in the appropriate day's block under "Quantity of Food (by Weight)."
- Record the date and weight of the MWD in the respective blocks in the bottom right-hand corner.

Remarks

D-18. The reverse side of *DA Form 2807-R* (*Figure D-1*) is used to annotate advanced techniques used during training and DFs with CAs and CRs listed. This section is very important for probable cause and shows the time line of events for training. Each day is listed with the action taken on that day. If no training is conducted due to company type training, the words "soldier development" are written on the line followed by the type of training conducted. "Soldier development" is only used when MWD handlers are conducting training in which the MWD is not needed or not taken. If no training is conducted due to leave, passes, or scheduled days off, no entry is needed; *DA Form 6 (Duty Roster)* can be used to verify these days.

DEPARTMENT OF THE ARMY *FORM 3992-R* (FIRST PAGE)

D-19. *DA Form 3992-R (Narcotics or Explosives Detector Dog Training and Utilization Record)* (*Figure D-2*, page D-9) is a standard form used by all MWD personnel. It is used to record the use and training of detector dogs. Use the following instructions when preparing a *DA Form 3992-R:*

Administrative Data

D-20. Enter the following information in the appropriate blocks:

- The date.
- The MWD's name and type, tattoo number, and age.
- The handler's name and grade.
- The organization and location.

TRAINING

D-21. The training section of the form is used for recording all training of the detector dog team in the particular detection skill. The detection areas of buildings, containers, and vehicles are listed. There is one blank space for detector training in other areas. Examples of other areas may include—

- Mail facilities.
- Open or wooded areas.
- Warehouses.
- Aircraft.
- Areas considered significant or unique to the command to which the MWD team is assigned.

D-22. Although some of the examples may be included under one of the three prelisted categories, the others should be listed separately. For areas covered under the three general headings, the handler or trainer annotates the specific location under the appropriate day in the remarks (for example, Smith High School, NCO Club, or Corps Auditorium).

D-23. The top line of each training category is divided diagonally into two blocks. The top of each block is used to record the number of training aids planted for that day's detector training. The bottom of each block is used for the number of training aids found by the proper response of the detector dog. If the MWD responds on the same training aid more than once in the same training scenario, it will not be counted as a find unless the number of plants is changed as well. At no time will there be more finds annotated than there are plants. Continually allowing the MWD to return to the last aid found creates false response problems in the MWD and should not be encouraged. It is the handler's responsibility to keep the MWD from returning to the scent cone of the last found training aid, whether the MWD is being worked on or off leash. Once the MWD has entered the scent cone, he must be allowed to work to the source and final respond. At no time will an MWD be yanked off an aid once he is working the scent cone.

D-24. The bottom line of each training category is used to record the search time. Search time is the amount of time devoted to detector training that day. Detector training is the time from the moment the handler enters a problem and commands the MWD to "Seek" until the MWD has completed the search of that training problem, excluding handler or MWD breaks. Add the daily time entries in each training category, and enter the results in the last column, titled "Total Hours."

D-25. Record the following information beside the appropriate training location (building, containers, vehicles, or other):

- **Plants/Finds.** Record the number of plants in the top half of the block and the number of finds in the lower half.
- **Search Time.** Record the amount of time devoted for detector training that day.
- **Training Total Hours.** Record the sum of hours performing detection missions in the far right

UTILIZATION

D-26. All operational detection missions are recorded in the utilization section. The detection missions are listed. Detection missions that are significantly different from these are listed separately to give an accurate record of the types of missions for which the detector dog team is used. There are two blocks under each day of the month for each utilization category. The top block is for recording the number of times the MWD detected a substance. The bottom block is for recording the total search or inspection time for the detection mission. The last column is the sum of the daily times. It gives a record of the total amount of time spent performing the detection missions for that month. The number of finds for the month can also be totaled and entered in the last column, above the time entry.

D-27. Record the following information beside the appropriate location (building, containers, vehicles, or other):

- **Finds.** Record the number of times the MWD finds the substance on a detection mission.
- **Search Time.** Record the total search time.

DETECTOR DOG SEARCH DATA

D-28. All relevant information about the productivity and success of each detection mission is recorded using the "Detector Dog Search Data" section. The location may be a building number, a unit designation, map grid coordinates, and/or any other information that helps identify where the detection mission was performed.

D-29. The MPR number is the number assigned to the case by military police to account for the custody and disposition of the substances found. The substance is the identification of the found material by a common name, such as heroin, marijuana, dynamite, or detonating cord.

D-30. Quantity is the measure of the amount of the substance found. Weight, volume, overall dimensions, length, or any other appropriate measure may be used.

D-31. The "Remarks" section may be used for adding any other relevant information about the substance found. This would include a field or laboratory verification of the type of substance, an EOD evaluation of an explosive device or explosive substance as live or inert, or the presence of other hazardous material in or around the substance found (such as razor blades, trip wires, or poisons). Record any information that may be useful in preparing the team for future searches or that may be applicable or useful to other detector dog teams.

D-32. Record the following information in the appropriate boxes:

- **Time.** Record the time the detector dog performed the mission.
- **Date.** Record the date the detector dog performed the mission.
- **Location.** Record the exact location of the detector mission (such as front gate, Fort Sill, Oklahoma; National Airport, Gabon, Africa; or building 442, Fort Bragg, North Carolina).
- **Military Police Report (MPR) Number.** Record the MPR number in this column, if applicable.
- **Substance.** Record the common name of the substance found, if applicable.
- **Quantity.** Record the quantity of the substance found, if applicable.
- **Remarks.** Use this column to record any other relevant information about the substance found (such as a positive response, no response, or a lab verification).

DETECTOR DOG PROFICIENCY (SECOND PAGE)

D-33. A correct response occurs when the MWD detects the substance and responds with the proper final response, and the training aid or substance is found in the general location where the MWD has responded. A not-at-source response is a training tool used to help an MWD team move closer to the source location of a training aid plant. The distance away from the source in order for a not-at-source response to be given is six to ten feet.

D-34. A missed response is when the MWD fails to detect and respond to the presence of a training aid with a final response. A handler miss or error is when—

- The handler fails to search or clear an area with the MWD and a training aid plant is missed.
- The handler fails to read an MWD's change of behavior and pulls the MWD away from the area where a training aid is planted and a training aid plant is missed.
- The handler presents an area but fails to ensure that the MWD is sniffing and a training aid plant is missed.

Handler misses or errors are considered missed explosives or narcotics and are counted against the find rate percentages of the MWD team.

D-35. False responses are noted in the second to last section of *DA Form 3992-R*. A false response is when the MWD responds as if he has detected a substance and the handler accepts that response, but no training aid or substance can be found within a reasonable distance of the MWD's final response.

D-36. The detector dog team's proficiency is computed monthly by adding the total number of correct responses ("a" for the month) and adding the total number of false/missed (MWD or handler) responses ("b" for the month). These two numbers are used in the following formula to obtain the detector dog team's proficiency rating:

$$\left(\frac{a}{a+b}\right)100 = \textit{proficiency}$$

D-37. For example, during the month the detector dog team made 93 correct responses on training aids, the MWD team also had 4 false responses, 2 MWD and 1 handler missed responses for a total of 7 false/missed responses. Applying these two numbers (a=93 and b=7) to the formula, the following is obtained:

$$\left(\frac{93}{93+7}\right)100 = 93\%$$

D-38. The computation above shows that the MWD team is working at a 93 percent proficiency rate. This rate is above the minimum for a PNDD team (90 percent), but it is below the minimum standard for a PEDD team (95 percent). A PNDD handler should continue training to maintain and possibly increase the team's proficiency. A PEDD handler needs to identify the causes of the team's substandard performance and immediately begin corrective training to bring the team up to and over the minimum 95 percent detection proficiency standard.

Note: Responses on actual substances during actual searches are not included in this computation. Search data cannot be included because it is impossible to determine the number of false or missed responses during an actual search.

D-39. Record the following information on—

- **Line a.** Record the total correct alerts. The MWDMS makes these entries automatically.
- **Line b.** Record the total false or missed alerts. The MWDMS makes these entries automatically.

Narcotics or Explosive Detector Dog Training and Utilization Record (Third and Fourth Pages)

D-40. The first part of section of the Narcotics or Explosive Detector Dog Training and Utilization Record section is used to record the number of each training aid used in training on each day of the month. The entry for each training aid type is divided into three rows to record the day of the month, correct alerts, and missed alerts, respectively. Record the appropriate information in the following sections.

- **Administrative Data.** Record the MWD's name and tattoo number, the month, and the year in the appropriate blocks.
- **Aid Type.** Enter the specific aid type in the upper left-hand corner of each "Detector Dog Proficiency" section.
- **Total Correct Alerts.** Record the number of correct alerts the MWD made during training.
- **Missed Alerts.** Record the number of missed and false alerts the MWD made during training.
- **Proficiency.** Record the actual proficiency in this block. The MWDMS makes this entry automatically.

Remarks

D-41. Remarks for *DA Form 3992-R* are recorded on a continuation sheet. Each DF (for example, a missed aid or false sit) is listed by the date it occurred. The CA and CR are also listed. If more than one DF is noted per day, number the DF, CA, and CR series to avoid confusion.

NARCOTICS OR EXPLOSIVES DETECTOR DOG TRAINING AND UTILIZATION RECORD

For use of this form see AR 190-12. The proponent is DCSPER.

MONTH AND YEAR	NAME/TYPE OF DOG	TATTOO NUMBER	AGE	NAME OF HANDLER	GRADE	ORGANIZATION AND LOCATION
SEP 2003	NATZ/PEDD	E309	2	BOOKSTEIN, DAVID B.	SGT	HHC 98TH ASG, BAMBERG, GE

TRAINING		1	2	3	4	5	6	7	8	9	10	11	12	13	14	15	16	17	18	19	20	21	22	23	24	25	26	27	28	29	30	31	TRAINING TOTAL HOURS 24.25
BUILDINGS	PLANTS /			7	8	4				4	6						11		5						4	8	6				7		70
	FINDS			7	5	4				4	6						10		4						4	8	6				7		65
	SEARCH TIME			120	80	60				60	60						130		70						60	120	120				120		16.67 HRS
CONTAINERS	PLANTS /									19	57						19		19														114
	FINDS									19	57						19		19														114
	SEARCH TIME									45	120						45		45														4.25 HRS
VEHICLES	PLANTS /					4													5														9
	FINDS					4													5														9
	SEARCH TIME					60													80														2.33 HRS
OTHER	PLANTS /									4																							4
	FINDS									4																							4
	SEARCH TIME									60																							1.00 HRS

UTILIZATION		1	2	3	4	5	6	7	8	9	10	11	12	13	14	15	16	17	18	19	20	21	22	23	24	25	26	27	28	29	30	31	UTILIZATION TOTAL HOURS 0.00
BUILDINGS	FINDS																																0
	SEARCH TIME																																0.00 HRS
CONTAINERS	FINDS																																0
	SEARCH TIME																																0.00 HRS
VEHICLES	FINDS																																0
	SEARCH TIME																																0.00 HRS
OTHER	FINDS																																0
	SEARCH TIME																																0.00 HRS

DA FORM 3992-R, Oct 84 EDITION OF 1 DEC 72 IS OBSOLETE

Sample

Figure D-2. Sample DA Form 3992-R

NO.	TIME	DATE	LOCATION	MPR NUMBER	SUBSTANCE	QUANTITY	REMARKS
				DETECTOR DOG SEARCH DATA			
1							
2							
3							
4							
5							
6							
7							
8							
9							
10							
11							
12							
13							
14							
15							
16							
17							
18							
19							
20							
21							
22							
23							
24							
25							
26							
27							
28							
29							
30							
31							
32							
33							

Sample

DETECTOR DOG PROFICIENCY
(Alerts on training aids during training and actual searches)

DAY OF MONTH	1	2	3	4	5	6	7	8	9	10	11	12	13	14	15	16	17	18	19	20	21	22	23	24	25	26	27	28	29	30	31	PROFICIENCY = a / (a + b) X 100
a. TOTAL CORRECT ALERTS	0	0	7	5	8	0	0	0	26	63	0	0	0	0	0	29	0	28	0	0	0	0	0	4	8	6	0	0	0	7	0	191
b. TOTAL FALSE/MISSED ALERTS	0	0	0	3	0	0	0	0	1	0	0	0	0	0	0	3	0	1	0	0	0	0	0	0	0	0	0	0	0	0	0	199 96%

DA Form 3992-R, Oct 84

Figure D-2. Sample DA Form 3992-R (Continued)

NARCOTICS OR EXPLOSIVES DETECTOR DOG TRAINING AND UTILIZATION RECORD

NATZ/PEDD TATTOO: E309 MONTH: SEP YEAR: 2003

DETONATING CORD — DETECTOR DOG PROFICIENCY (ALERTS ON TRAINING AIDS DURING TRAINING AND ACTUAL SEARCHES)

DAY OF MONTH	1	2	3	4	5	6	7	8	9	10	11	12	13	14	15	16	17	18	19	20	21	22	23	24	25	26	27	28	29	30	31	PROFICIENCY A / A + B X 100
a. TOTAL CORRECT ALERTS			2	1						20						1									3							27
b. TOTAL MISSED ALERTS				1																												28 96%

C-4 PLASTIC EXPLOSIVE — DETECTOR DOG PROFICIENCY (ALERTS ON TRAINING AIDS DURING TRAINING AND ACTUAL SEARCHES)

DAY OF MONTH	1	2	3	4	5	6	7	8	9	10	11	12	13	14	15	16	17	18	19	20	21	22	23	24	25	26	27	28	29	30	31	PROFICIENCY A / A + B X 100
a. TOTAL CORRECT ALERTS				1					22	1						3		2								1						30
b. TOTAL MISSED ALERTS				1					1																							32 94%

POTASSIUM CHLORATE — DETECTOR DOG PROFICIENCY (ALERTS ON TRAINING AIDS DURING TRAINING AND ACTUAL SEARCHES)

DAY OF MONTH	1	2	3	4	5	6	7	8	9	10	11	12	13	14	15	16	17	18	19	20	21	22	23	24	25	26	27	28	29	30	31	PROFICIENCY A / A + B X 100
a. TOTAL CORRECT ALERTS									1							20		2						2	1							26
b. TOTAL MISSED ALERTS																1																27 96%

SODIUM CHLORATE — DETECTOR DOG PROFICIENCY (ALERTS ON TRAINING AIDS DURING TRAINING AND ACTUAL SEARCHES)

DAY OF MONTH	1	2	3	4	5	6	7	8	9	10	11	12	13	14	15	16	17	18	19	20	21	22	23	24	25	26	27	28	29	30	31	PROFICIENCY A / A + B X 100
a. TOTAL CORRECT ALERTS					1				2									21						1	1							26
b. TOTAL MISSED ALERTS																		1														27 96%

WATER GEL (TOVEX) — DETECTOR DOG PROFICIENCY (ALERTS ON TRAINING AIDS DURING TRAINING AND ACTUAL SEARCHES)

DAY OF MONTH	1	2	3	4	5	6	7	8	9	10	11	12	13	14	15	16	17	18	19	20	21	22	23	24	25	26	27	28	29	30	31	PROFICIENCY A / A + B X 100
a. TOTAL CORRECT ALERTS				1						21						1		1								2				2		28
b. TOTAL MISSED ALERTS				1																												29 97%

DA FORM 3992-R, OCT 84 (CONTINUATION SHEET)

Figure D-2. Sample DA Form 3992-R (Continued)

NARCOTICS OR EXPLOSIVES DETECTOR DOG TRAINING AND UTILIZATION RECORD

NATZ/PEDD **TATTOO:** E309 **MONTH:** SEP **YEAR:** 2003

	1	2	3	4	5	6	7	8	9	10	11	12	13	14	15	16	17	18	19	20	21	22	23	24	25	26	27	28	29	30	31		
SEMTEX — DETECTOR DOG PROFICIENCY (ALERTS ON TRAINING AIDS DURING TRAINING AND ACTUAL SEARCHES)																																PROFICIENCY A / A + B X 100	
DAY OF MONTH	1	2	3	4	5	6	7	8	9	10	11	12	13	14	15	16	17	18	19	20	21	22	23	24	25	26	27	28	29	30	31		
a. TOTAL CORRECT ALERTS			1		2				1							2		2							1					2		11	
b. TOTAL MISSED ALERTS																																11	100%
AMMONIUM NITRATE — DETECTOR DOG PROFICIENCY (ALERTS ON TRAINING AIDS DURING TRAINING AND ACTUAL SEARCHES)																																PROFICIENCY A / A + B X 100	
DAY OF MONTH	1	2	3	4	5	6	7	8	9	10	11	12	13	14	15	16	17	18	19	20	21	22	23	24	25	26	27	28	29	30	31		
a. TOTAL CORRECT ALERTS																																0	
b. TOTAL MISSED ALERTS																																0	#DIV/0!
NITROGLYCERIN DYNAMITE — DETECTOR DOG PROFICIENCY (ALERTS ON TRAINING AIDS DURING TRAINING AND ACTUAL SEARCHES)																																PROFICIENCY A / A + B X 100	
DAY OF MONTH	1	2	3	4	5	6	7	8	9	10	11	12	13	14	15	16	17	18	19	20	21	22	23	24	25	26	27	28	29	30	31		
a. TOTAL CORRECT ALERTS				2						21															1							24	
b. TOTAL MISSED ALERTS																																24	100%
TNT (Trinitrotoluene) — DETECTOR DOG PROFICIENCY (ALERTS ON TRAINING AIDS DURING TRAINING AND ACTUAL SEARCHES)																																PROFICIENCY A / A + B X 100	
DAY OF MONTH	1	2	3	4	5	6	7	8	9	10	11	12	13	14	15	16	17	18	19	20	21	22	23	24	25	26	27	28	29	30	31		
a. TOTAL CORRECT ALERTS			3		3											2										1				2		11	
b. TOTAL MISSED ALERTS																																11	100%
SMOKELESS POWDER — DETECTOR DOG PROFICIENCY (ALERTS ON TRAINING AIDS DURING TRAINING AND ACTUAL SEARCHES)																																PROFICIENCY A / A + B X 100	
DAY OF MONTH	1	2	3	4	5	6	7	8	9	10	11	12	13	14	15	16	17	18	19	20	21	22	23	24	25	26	27	28	29	30	31		
a. TOTAL CORRECT ALERTS			1		2																			1	1	2				1		8	
b. TOTAL MISSED ALERTS																																8	100%
Totals from TNG Aids (Cont. Sheet)																																	
TOTAL PLANTS	0	0	7	8	8	0	0	0	27	63	0	0	0	0	0	30	0	29	0	0	0	0	0	4	8	6	0	0	0	7	0	197	
TOTAL FINDS	0	0	7	5	8	0	0	0	26	63	0	0	0	0	0	29	0	28	0	0	0	0	0	4	8	6	0	0	0	7	0	191	97%
a. TOTAL FALSE ALERTS																2																2	1%
Totals from TNG locations (3992 Front)																																	
TOTAL PLANTS	0	0	7	8	8	0	0	0	27	63	0	0	0	0	0	30	0	29	0	0	0	0	0	4	8	6	0	0	0	7	0	197	
TOTAL FINDS	0	0	7	5	8	0	0	0	27	63	0	0	0	0	0	29	0	28	0	0	0	0	0	4	8	6	0	0	0	7	0	192	97%

DA FORM 3992-R, OCT 84 (CONTINUATION SHEET)

Figure D-2. Sample DA Form 3992-R (Continued)

DA Form 3992-R
Continuation Sheet

DATE	REMARKS
3-Sep-03	TRAINING AT SCHWEINFURT BOWLING ALLEY. (TRAINING AIDS PLANTED FROM 6 FT H x 5 FT D)
4-Sep-03	TRAINING AT SCHWEINFURT FMO. (TRAINING AIDS PLANTED FROM 8 FT H x 3 FT D)
	DF1: MWD MISSED AID (1 STRAND DET CORD - 1 FT H x 1 FT D) DUE TO HANDLER ERROR.
	CA1: HANDLER TOLD TO WORK A REVERSE PATTERN UNTIL THE TEAM REACHED THE AID LOCATION.
	CR1: MWD FOUND AID AFTER HANDLER MADE PROPER PRESENTATION.
	DF2: MWD MISSED AID. (1/4 BLOCK C4 2 FT H x 1 FT D)
	CA2: MWD BROUGHT BACK FOR RE-PRESENTATION.
	CR2: MWD FOUND AID AFTER HANDLER MADE SECOND PRESENTATION.
	DF3: MWD MISSED AID. (1 STICK NITRO DYNAMITE- 7 FT H x 1 FT D)
	CA3: MWD BROUGHT BACK FOR RE-PRESENTATION.
	CR3: MWD FOUND AID AFTER HANDLER MADE HIGH PRESENTATION AND DRAGGED HAND UPWARD.
	DF4: MWD MISSED AID. (1 STICK WATER GEL - 5 FT H x 1 FT D)
	CA4: MWD BROUGHT BACK FOR RE-PRESENTATION.
	CR4: MWD FOUND AID AFTER HANDLER MADE SECOND PRESENTATION.
	MWD WILL BE PROTOCOLED ON ALL MISSED AIDS AT A LATER DATE DUE TO TIME CONSTRAINTS.
5-Sep-03	TRAINING AT 630TH MILITARY POLICE COMPANY MOTOR POOL. (TRAINING AIDS PLANTED FROM 3 FT H x 7 FT D)
9-Sep-03	TRAINING AT SCHWEINFURT OPEN AREA BY CONN CLUB & OUTDOOR REC. (TRAINING AIDS PLANTED FROM 7 FT H x 4 FT D)
	CA: (REFERENCE DF2, 4 SEP) MWD WAS PROTOCOLED ON 1/2 BLOCK C4.
	CR: COMPLETED 4 HV ASSISTED FOR 4 TRIALS AND 15 CONSECUTIVE UNASSISTED TRIALS.
10-Sep-03	TRAINING AT ANSBACH CMFO WAREHOUSES (TRAINING AIDS PLANTED FROM 4 FT H x 1 FT D).
	CA: (REFERENCE DF2-4, 4 SEP) MWD WAS PROTOCOLED ON 1 STRAND DC, 1 STICK ND, AND 1 STICK WG.
	CR: COMPLETED 4 HV ASSISTED FOR 4 TRIALS AND 15 CONSECUTIVE UNASSISTED TRIALS FOR EACH AID.
11-Sep-03	NO TRAINING DUE TO MWD GOING TO VET. MWD HAS BEEN DIAGNOSED WITH A POSSIBLE PULLED/STRAINED MUSCLE IN HIS RIGHT LEG WHICH HE IS FAVORING AS HE WALKS. X-RAYS SHOW NO BREAKS OR FRACTURES. MWD CANNOT DO ANYTHING STRENUOUS FOR THE NEXT WEEK, BUT IS CLEARED TO WORK DETECTION NEXT TUES (16 SEP).
16-Sep-03	TRAINING AT KITZIGEN TISA WAREHOUSE & LEIGHTON THEATER. (TRAINING AIDS PLANTED FROM 9 FT H x 6 FT D.)
	DF1: MWD MISSED AID. (PC - 1 FT H x 6 FT D)
	CA1: MWD BROUGHT BACK FOR REPRESENTATION. WILL CONDUCT PROTOCOL.
	CR1: MWD RESPONDED ON AID AFTER HANDLER MADE SECOND PRESENTATION. MWD WAS PROTOCOLED ON PC COMPLETING 4 HV ASSISTED FOR 4 TRIALS AND 15 CONSECUTIVE UNASSISTED TRIALS.
	DF2: MWD FALSE SAT IN THEATER LOBBY.
	CA2: MWD WAS ESCAPED FROM AREA AND CONTINUED TO WORK.
	CR2: MWD CONTINUED PATTERN WITHOUT FALSE SITTING.
	DF3: MWD FALSE SAT IN THEATER BACK WALL.
	CA3: MWD WAS ESCAPED FROM AREA AND CONTINUED TO WORK. MWD WILL BE WORKED ON BLANK PROBLEM.
	CR3: MWD CONTINUED PATTERN WITHOUT FALSE SITTING.
18-Sep-03	TRAINING AT SCHWEINFURT STRIP LOT & OUTDOOR REC. (TRAINING AIDS PLANTED FROM 8 FT H x 6 FT D)
	CA1: (REFERENCE DF2, 16 SEP) CONDUCTED BLANK PROBLEM.
	CR1: MWD DID NOT ATTEMPT TO FALSE SIT.
	DF2: MWD MISSED AID. (SC - 7 FT H x 3 FT D)
	CA2: MWD BROUGHT BACK FOR RE-PRESENTATION. MWD WAS PROTOCOLED ON SC.
	CR2: MWD FOUND AID AFTER HANDLER MADE SECOND PRESENTATION. COMPLETED 4 HV ASSISTED FOR 4 TRIALS AND 15 CONSECUTIVE UNASSISTED TRIALS.
24-Sep-03	TRAINING AT SCHWEINFURT FMO. (TRAINING AIDS PLANTED FROM 5 FT H x 5 FT D)
25-Sep-03	TRAINING AT THEATER. (TRAINING AIDS PLANTED FROM 2 FT H x 2 FT D)
26-Sep-03	TRAINING AT COMMISSARY. (TRAINING AIDS PLANTED FROM 9 FT H x 6 FT D)
30-Sep-03	TRAINING AT FURNITURE WAREHOUSE. (TRAINING AIDS PLANTED FROM 6 FT H x 7 FT D)

Figure D-2. Sample DA Form 3992-R (Continued)

Appendix E
Deployment Equipment List

Table E-1 is a list to be used as a general guide to aid leaders in the planning process prior to MWD deployments. Mission, enemy, terrain, troops, time available, and civilian considerations (METT-TC) should be considered when determining what equipment is necessary for mission success. Required and authorized equipment is determined by tables of organization and equipment (TOEs), modification tables of organization and equipment (MTOEs), and tables of distribution and allowances (TDAs).

Table E-1. Deployment Equipment List

Worn or Carried		
M9 with LCE (LBV)	Identification card	Body armor
Canteen cup	Identification tag	Black gloves with liners
First aid pouch with dressing	Protective mask with a carrier	Military police brassard (subdued)
Kevlar® helmet		
MWD Equipment		
Choke chain	Grooming brush/comb	Feed pan
Leather or nylon collar	Reward ball/Kong	Kennel crate
Stake-out chain	360-inch nylon leash	16-foot leash
Plastic or leather muzzle	Water bucket	Harness
Alice Pack		
BDU (2 sets)	Towels (2)	Sleeping bag with a cover
Brown T-shirt (2)	Gloves, black, cold weather	Shower shoes
Undergarments (2)	BDU Cap	PT uniform (complete set)
Green and black socks (2)	Boot polish kit	MOPP gear in an NBC bag
Poncho with a liner	Flashlight	Boots
Shaving kit	Lip balm	Personal hygiene gear
Waterproof bag	Foot powder	
A/B Bag		
Duffel bag with a padlock	Brown T-shirt (5)	White socks (3)
Cold weather boots	BDU (2)	Green and black socks (5)
Cold weather gear	Towels (3)	Personal hygiene gear
Waterproof bag	Overboots	Undergarments (5)
Military police brassard	Laundry bag (2)	
Additional Equipment		
30-90 day supply MWD food	50-gallon trash can (MWD food)	5-gallon water can
MWD training records	Medical supplies	Engineer tape
MWD medical records	Carpenter kit/tools	Cots
MWD health certificate	Safe for narcotics aides	Fire extinguishers

Table E-1. Deployment Equipment List (Continued)

Additional Equipment		
MWD first aid kit	Explosives and narcotics training aides	Axes, shovels, rakes, brooms
Arm protector, sleeve (hidden)	Parasite preventative mediation	Cooling vest (MWD)
Soldier first aid kit	Light set	Paw pad protectors (2 sets)
Bite suit (optional)	Generator	Foot lockers (2) or blood boxes
Dog shampoo	Transformer	Equipment duffel bag
Heaters and fans	Warning signs (MWD)	Camouflage netting with poles

Appendix F
Force Protection Conditions

The FPCONs outlined below describe the progressive level of a terrorist threat to all US military facilities and personnel under *DOD Directive 2000.12.* As approved by the Joint Chiefs of Staff, the terminology and definitions in this appendix are recommended security measures designed to ease interservice coordination and to support US military antiterrorism activities. Each installation must localize these measures for its specific circumstances, accounting for service and commander requirements, local laws, and SOFAs as a minimum. See *AR 525-13* for further details concerning FPCONs. This appendix details what each condition is and the suggested actions of the MWD teams for each level and measure. Commanders are encouraged to employ MWD teams wherever an MWD team can enhance security and force protection measures that are already established.

FORCE PROTECTION CONDITION NORMAL

F-1. FPCON normal applies when a general global threat of possible terrorist activity exists. It warrants a routine security posture.

FORCE PROTECTION CONDITION ALPHA

F-2. FPCON Alpha applies when there is an increased general threat of possible terrorist activity against personnel or facilities, the nature and extent of which are unpredictable. Alpha measures must be capable of being maintained indefinitely. The following bullets explain the FPCON Alpha measures:

- **Alpha measure 1.** Inform personnel and family members of the general situation at regular intervals. Also, remind them to be alert for and report suspicious activities such as the presence of unfamiliar personnel and vehicles, suspicious/unattended parcels, and possible surveillance attempts.
- **Alpha measure 2.** Ensure that the duty officer or personnel with access to building plans as well as the plans for area evacuations are available at all times. Plans to execute access control procedures should be in place. Key personnel required to implement security plans should be on call and readily available.
- **Alpha measure 3.** Secure and randomly inspect buildings, rooms, and storage areas not in regular use.

Note: MWD should randomly inspect buildings after normal duty hours. The KM should ensure that no inspection time or location patterns are ever set by MWD teams.

- **Alpha measure 4.** Increase security spot checks of vehicles and persons entering installations under the jurisdiction of the United States.

Note: MWD should perform vehicle checks randomly at entry points.

- **Alpha measure 5.** Limit access points for vehicles and personnel commensurate with a reasonable flow of traffic.

- **Alpha measure 6.** Apply Alpha measures 4, 5, 7, or 8 either individually or in combination with each other as a deterrent.
- **Alpha measure 7.** Review intelligence, counterintelligence, and operations dissemination procedures.
- **Alpha measure 8.** Review and, if necessary, implement security measures for HRP.
- **Alpha measure 9.** Consult local authorities on the threat and mutual antiterrorism measures as appropriate.
- **Alpha measure 10.** Review all plans and be prepared to implement a higher FPCON.
- **Alpha measure 11 (spare).** Add additional measures as deemed appropriate by the commanders.

FORCE PROTECTION CONDITION BRAVO

F-3. FPCON Bravo applies when an increased or more predictable threat of terrorist activity exists. Sustaining Bravo measures for a prolonged period may affect operational capabilities and impact relations with local authorities. The following bullets explain the FPCON Bravo measures:

- **Bravo measure 1.** Continue, or introduce, all measures in the previous FPCON level.
- **Bravo measure 2.** Check plans for the implementation of the next FPCON level.
- **Bravo measure 3.** Identify critical and high-occupancy buildings. Move cars and objects (such as crates and trash containers) away from buildings to reduce vulnerability to bomb attacks. Consider centralized parking.
- **Bravo measure 4.** Secure and inspect all buildings, rooms, and storage areas not in regular use.
- **Bravo measure 5.** Inspect the interior and exterior of buildings in regular use for suspicious packages at the beginning and end of each workday, as well as at random intervals. Patrol teams search the exterior of high-occupancy buildings at the beginning and end of each day.

Note: Use PEDD to search buildings and areas.

- **Bravo measure 6.** Implement mail-screening procedures to identify suspicious letters and parcels.

Note: Search the mail distribution center with both PEDD and PNDD teams randomly.

- **Bravo measure 7.** Inspect commercial deliveries randomly. Advise family members to check home deliveries.
- **Bravo measure 8.** Increase presence around and surveillance of domestic accommodations, schools, messes, clubs, and other soft targets to improve deterrence and defense and build confidence among staff and dependents.

Note: Randomly patrol and perform walking patrols of populated areas with MWDs.

- **Bravo measure 9.** Implement plans to enhance off-installation security of DOD facilities. In areas with threat levels of moderate, significant, or high, coverage includes facilities (such as DOD schools and daycare centers) and transportation services and routes (such as bus routes) used by DOD employees and family members.
- **Bravo measure 10.** Inform local security committees of actions being taken.
- **Bravo measure 11.** Verify the identity of visitors and randomly inspect their suitcases, parcels, and other containers.
- **Bravo measure 12.** Conduct random patrols to check vehicles, people, and buildings.
- **Bravo measure 13.** Implement additional security measures for HRP.
- **Bravo measure 14.** Identify and brief personnel who may augment guard forces. Review specific rules of engagement including the use of deadly force.

- **Bravo measure 15.** Test the mass notification system.
- **Bravo measure 16.** Verify the identity of all personnel entering buildings as deemed appropriate.
- **Bravo measure 17.** Identify off-installation safe havens for use at the next FPCON.
- **Bravo measure 18.** Review and implement proper OPSEC, communications security (COMSEC), and information security (INFOSEC) status.
- **Bravo measure 19.** Consider using tactical deception techniques.
- **Bravo measure 20 (spare).** Add additional measures as deemed appropriate by the commanders.

FORCE PROTECTION CONDITION CHARLIE

F-4. FPCON Charlie applies when an incident occurs or intelligence is received indicating that some form of terrorist action or targeting against personnel or facilities is likely. Implementing FPCON Charlie measures creates hardship and affects the activities of the unit and its personnel. The following bullets explain the FPCON Charlie measures:

- **Charlie measure 1.** Continue, or introduce, all measures in the previous FPCON.
- **Charlie measure 2.** Ensure that duty personnel are cognizant of actions required to implement antiterrorism and force protection plans. Recall required personnel.
- **Charlie measure 3.** Reduce access control to a minimum and strictly enforce entry. Randomly search vehicles.

Note: Increase the PEDD and PNDD teams' search times at entry points.

- **Charlie measure 4.** Increase standoff from sensitive buildings based on the threat.
- **Charlie measure 5.** Issue weapons and appropriate communication equipment to guards according to SOFAs. Include specific orders on the issue of ammunition.
- **Charlie measure 6.** Increase patrols of the installation to include waterside perimeters, if appropriate.

Note: Use MWD assets to expand the perimeter patrol area and detect intruders.

- **Charlie measure 7.** Implement a barrier plan to prevent a vehicle-borne attack.
- **Charlie measure 8.** Protect all designated vulnerable points. Give special attention to vulnerable points outside the military establishment.
- **Charlie measure 9.** Consult local authorities about closing public (and military) roads and facilities to reduce the vulnerability to attack.
- **Charlie measure 10.** Consider searching suitcases, briefcases, and packages being brought onto the installation through ACPs and consider randomly searching suitcases, briefcases, and packages leaving.

Note: Use PEDD and PNDD teams to search the items above.

- **Charlie measure 11.** Review personnel policy procedures to determine a course of action for family members.
- **Charlie measure 12.** Review access procedures for all non-US personnel and adjust as appropriate.
- **Charlie measure 13.** Consider escorting children to and from DOD schools.
- **Charlie measure 14 (spare).** Add additional measures as deemed appropriate by the commanders.

FORCE PROTECTION CONDITION DELTA

F-5. FPCON Delta applies in the immediate area where a terrorist attack has occurred or when intelligence has been received that terrorist action against a specific location or person is imminent. Normally, FPCON Delta is declared as a localized condition. The following bullets explain the FPCON Delta measures:

- **Delta measure 1.** Continue, or introduce, all measures in the previous FPCON.
- **Delta measure 2**. Augment guards as necessary.
- **Delta measure 3.** Identify all vehicles within operational or mission support areas.
- **Delta measure 4.** Search all vehicles and contents before allowing entrance to the installation.

Note: Use PEDD and PNDD as needed to search vehicles and contents.

- **Delta measure 5.** Control access and implement positive identification of all personnel with no exceptions.
- **Delta measure 6.** Search all suitcases, briefcases, and packages brought into the installation.

Note: Use PEDD and PNDD as needed to search the items above.

- **Delta measure 7.** Escort children to and from DOD schools or close the schools.
- **Delta measure 8.** Make frequent checks of the exterior of buildings and of parking areas. Use MWD teams to make frequent checks of the exterior of buildings and parking areas.
- **Delta measure 9.** Restrict all nonessential movement.
- **Delta measure 10.** Close public and military roads and facilities if permitted.
- **Delta measure (spare).** Add additional measures as deemed appropriate by the commanders.

Appendix G
After-Action Reviews

The AAR process is used to document and discuss MWD operations and training. AARs are focused on performance standards that enable other MWD handlers, training NCOs, KMs, program managers, and concerned agencies to discover what happened, why it happened, and how to sustain strengths and improve on weaknesses. Leaders, planners, and developers can use AARs to get the maximum benefit from every mission or task conducted during predeployment, deployment, and redeployment.

AFTER-ACTION REVIEW PURPOSE

G-1. AARs must provide—

- Candid insights into specific soldier, leader, and MWD strengths and weaknesses from various perspectives.
- Feedback and insight critical to future successes.
- Details that evaluation reports alone often lack.

G-2. The USAMPS MWD Career Management NCO collects historical data, end-of-tour reports, and other organizational specific issues. With the submission of AARs, USAMPS can build a historical database to benefit all in the improvement of Army MWD team tactics, techniques, and procedures. Ultimately, information will be submitted to the Center for Army Lessons Learned (CALL) for distribution to the Army.

AFTER-ACTION REVIEW SUBMISSION

G-3. After the completion of TDY missions, special operations, deployments or other significant events, MWD handlers submit AARs to their respective MACOMs. MACOMs forward the AARs to HQDA Office of the Provost Marshal General (OPMG) and USAMPS.

POINT OF CONTACT

G-4. The point of contact for the MWD Career Management NCO is commercial: (573) 563-8040 and defense switched network (DSN): 676-8040/7946. Send hard-copy correspondence to: US Army Military Police School, ATTN: ATSJ-P, 401 MANSCEN Loop, Suite 1058, Fort Leonard Wood, MO 65473-8926.

EXAMPLE AFTER-ACTION REVIEW

G-5. *Figure G-1,* page G-2, should be used as the format for AARs submitted to the USAMPS MWD Career Management NCO. Refer to *TC 25-20* for comprehensive guidance on AARs.

DEPARTMENT OF THE ARMY
42^{D} Military Police Detachment
Ft. Bragg, NC 27380

OFFICE SYMBOL DATE

MEMORANDUM FOR RECORD

SUBJECT: After-Action Review, Task Force Iron Dog, Sep 03 – Nov 03

1. State what was supposed to happen: This paragraph contains a brief introduction of the mission that was being performed by the MWD team or section. Include operation objectives, the overall mission, and the intent from the task force commander as well as any other relevant information.

2. State what happened: This paragraph contains pertinent information about what actually happened during the mission. Was the mission accomplished as stated above or were other objectives established inside of the area of operation?

3. Determine what went right and wrong during the mission:

   a. Topic: What was right and wrong with what happened during the mission.

   b. Discussion: Give the who, what, when, where, and how for the mission. Give a good summary of what the issue was.

   c. Recommendation: Give recommendations to improve the issues that were identified as needing improvement.

4. Point of contact for this memorandum is the undersigned at DSN: 123-4567 or jon.doe@us.army.mil.

Jon Doe
SSG, USA
Kennel Master / PEDD Handler

Note: If you have a positive topic just use Topic and Discussion sections.

Figure G-1. Sample AAR Format

Appendix H

Health and Welfare Inspections Briefing Guide

Commanders frequently request MWD support during health and welfare inspections. Commanders should be made aware that the conduct and responsibility of health and welfare inspections lies within their purview and that the MWD team is not the authority for the mission. This appendix identifies the actions that need to be addressed with the commander to ensure a successful and lawful health and welfare inspection. These actions are needed to establish and act on probable cause determinations. Questions concerning these matters should be directed to the local Staff Judge Advocate Office for clarification.

COORDINATION

H-1. The KM coordinates and approves each health and welfare inspection. After the KM receives the request for assistance, he assesses the needs of the mission and determines which detector dog team (if any) will assist the commander. The KM notifies the unit commander or 1SG and accepts or denies the mission. Once the mission is accepted, the KM briefs them on their roles and responsibilities and informs them that if his criteria are not met, the KM will terminate the search.

HANDLERS BRIEFING

H-2. Prior to the inspection, the KM, plans NCO, or handler ensures that the commander requesting support is aware that the building being inspected must be unoccupied and that all doors are unlocked (or keys available on site). On the arrival of the MWD team at the location of the health and welfare inspection, the handler performs a commander's briefing and demonstration. The handler—

- Explains the probable cause folder to the commander.
- Shows the commander the training aids used for detection.
- Demonstrates the dog's ability to detect.

H-3. A briefing document shows handlers how to record the details of each health and welfare inspection (*Figure H-1*, page H-3). The commander is provided with a copy of the briefing document containing the following information:

- The MWD teams used for the inspection.
- The canines' names and tattoo numbers.
- The scope of the proficiency training of the MWD teams. Discuss with the commander the standards and scope of the detector training required to maintain the certification of assigned MWD teams.
- The MWDs' accuracy rates. Explain to the commander that the accuracy percentage rate is based on quarterly training.
- The MWDs' detection experience. Inform the commander of the amount of time that the MWDs have actually spent conducting detection missions. Include any possible finds in prior inspections.
- The MWDs' capabilities. Explain to the commander that the MWDs are capable of detecting either drugs or explosives.

- The MWDs' alert response. Describe what the MWDs will do when they indicate the presence or odor of found drugs or explosives.
- The unit representatives. Notify the commander that one SSG or above is required to escort each MWD team through out the inspection. The commander may not be a unit representative.

UNIT REPRESENTATIVE DUTIES AND RESPONSIBILITIES

H-4. Ensure that the unit representatives are informed that they—

- Escort the MWD team at all times.
- Unlock doors or areas as needed.
- Carefully observe the MWD team.
- Follow the handler's instructions.
- Obey safety precautions.
- Spot areas that are missed by the MWD team.

DOCUMENTATION

H-5. It is important that the MWD handler documents the commander's briefing on every occasion. *Figure H-1*, page H-3, gives an example for documenting the briefing given to the commander before a health and welfare inspection. Advise the commander to maintain the handlers briefing and other documentation concerning each response of the MWD. Advise the commander that a positive response by a narcotic detector dog provides probable cause for a command-directed urinalysis. Ensure that the commander signs the briefing form and then provide a copy for filing in case of a judicial action.

INSPECTION

H-6. After the briefing has been conducted, the inspection may begin. Remember to advise the commander and unit representatives that this health and welfare inspection is the unit's mission and that the MWD team is assisting in the mission. The MWD team is not in charge of the health and welfare inspection but can be used as a tool by the commander.

ATZP-PM-K DATE: ____________

SUBJECT: Commander's Briefing for Health and Welfare Inspections Utilizing Detector Dogs

1. (commander's name) has requested narcotics detector dog support for a health and welfare inspection on (dd/mm/yyyy) .

2. Prior to canine(s) being utilized, the building must be unoccupied and all doors must be unlocked (or keys available).

3. The following provides the commander with information essential to a successful inspection:

_____ A. Team(s) to be utilized: (number of PEDD and PNDD to be used) .

_____ B. MWD name(s) and tattoo number(s): (name and tattoo number for each MWD)

_____ C. The scope of formal training of the canine team(s). Discuss detector training. Discuss the standards that must be successfully completed to maintain certification of detector dog team(s).

_____ D. The canine's accuracy rate(s) is/are (accuracy rates for each MWD) . This is based on the previous 3 months training records.

_____ E. The extent of actual MWD detection field experience. (experience for each MWD)

_____ F. Explain the types of drugs or explosives that the canine(s) is/are trained to detect.

_____ G. Describe the canine's specific behavior(s)/actions when it responds, indicating the presence or the odor of narcotics (detector response of each MWD) .

_____ H. Ensure that the commander(s) is/are aware that one representatives (SSG or above) is needed to act as an escort for each team. The commander may not be one of the individuals accompanying canine team(s).

_____ I. Ensure that the escorts are briefed on their duties and responsibilities. Discuss the safety precautions that they must follow, and ensure the safety of others.

_____ J. Ensure that the commander(s) is/are aware that the unit is responsible for maintaining proper documentation (date, time, room number, common or personal area and occupant[s] name[s]), concerning the response of the canine(s). This information should be kept on file with the commander's copy of this briefing sheet, in case of a judicial action.

_____K. Ensure that the commander is aware that a positive response by a narcotic detector dog is probable cause for a command-directed urinalysis on the occupant or occupants of that room.

I (commander's name) have received a personal briefing covering all the information stated above and now understand the capabilities of the canine team(s). I am satisfied that the canine team(s) have the training and experience necessary to provide reliable information upon which to authorize a search and/or perform a command-directed urinalysis. I authorize the canine team(s) to inspect the following area(s): (areas to be inspected)

Figure H-1. Health and Welfare Briefing Document Sample

HANDLER(S) SIGNATURE

1. ______________________ 2. ______________________

3. ______________________ 4. ______________________

COMMANDER'S INFORMATION

NAME: ______________________

RANK: ______________________

UNIT: ______________________

SIGNATURE: ______________________

1SG'S INFORMATION

NAME: ______________________

RANK: ______________________

UNIT: ______________________

SIGNATURE: ______________________

UNIT REPRESENTATIVE

NAME: ______________________

RANK: ______________________

UNIT: ______________________

SIGNATURE: ______________________

UNIT REPRESENTATIVE

NAME: ______________________

RANK: ______________________

UNIT: ______________________

SIGNATURE: ______________________

Figure H-1. Health and Welfare Briefing Document Sample (Continued)

Appendix I

Terrorist Bomb Threat Stand-off Distances

Table I-1 contains information from the DOD technical support working group. Commanders, PMs, and MWD planners can use this information as a guide to determine the minimum distances necessary to evacuate personnel from suspected explosive threats.

Table I-1. Minimum Distances for Personnel Evacuation

Threat Description	*Explosives Capacity* (TNT Equivalent)*	*Building Evacuation Distance***	*Outdoor Evacuation Distance****
Pipe bomb	5 lbs/2.3 kg	70 ft/21 m	850 ft/259 m
Briefcase/suitcase bomb	50 lbs/23 kg	150 ft/46 m	1,850 ft/564 m
Compact sedan	500 lbs/227 kg	320 ft/98 m	1,500 ft/457 m
Sedan	1,000 lbs/454 kg	400 ft/122 m	1,750 ft/534 m
Passenger/cargo van	4,000 lbs/1,814 kg	640 ft/195 m	2,750 ft/838 m
Small moving van/delivery truck	10,000 lbs/4,536 kg	860 ft/263 m	3,750 ft/1,143 m
Moving van/water truck	30,000 lbs/13,608 kg	1,240 ft/375 m	6,500 ft/1,982 m
Semitrailer	60,000 lbs/27,216 kg	1,570 ft/475 m	7,000 ft/2,134 m

* Based on maximum volume or weight of explosives (TNT equivalent).
** Governed by the ability of an unstrengthened building to withstand severe damage or collapse.
*** Governed by either the fragment throw distance or glass breakage/falling glass hazard distance (whichever throws fragments farther than vehicle bombs).

NOTES:
1. All personnel must either seek shelter inside a building away from windows and exterior walls (with some risk) or move beyond the outdoor evacuation distance.
2. The preferred area is beyond the outdoor evacuation distance for the evacuation of personnel in buildings. This is mandatory for personnel outdoors.

Glossary

Acronyms and Abbreviations

1SG	first sergeant
AAR	after-action review
ACP	access control point
AF	Air Force
agg	aggression
AR	Army regulation
ASG	area support group
ASI	additional skill identifier
ASP	ammunition supply point
ATTN	attention
BDU	battle dress uniform
C2	command and control
C4	composition 4
CA	corrective action
CALL	Center for Army Lessons Learned
CCTV	Closed-circuit television
cert	certification
CI	civilian internee
CID	Criminal Investigation Division
clg	cleaning
CMFO	centralized furnishings management office
comm	communications
COMSEC	communications security
CONEX	container express
cont	continuted
CP	command post
CPT	captain
CR	corrective response
DA	Department of the Army
d	deep
DARE	Drug Abuse Resistance Education
DCSPER	Dputy Chief of Stagg for Personnel
DC	detonating cord
DD	Department of Defense
DEA	Drug Enforcement Administration

Dec	December
det	detonating
DF	deficiency
DOD	Department of Defense
DODMWDVS	Department of Defense Military Working Dog Veterinary Services
DSN	defense switched network
DST	drug suppression team
EAC	echelon above corps
EOD	explosive ordnance disposal
EPW	enemy prisoner of war
eval	evaluaiton
FM	field manual
FMO	Furnishing managemetn office
FPCON	force protection condition
ft	foot; feet
GE	Germany
GOV	government owned vehicle
GSA	General Services Administration
h	high
HF	hole, variable
HRP	high-risk personnel
HRT	high-risk target
HVAC	heating, ventilation, and air conditioning
I/R	internment and resettlement
IDS	Intrusion Detection System
IED	improvised explosive device
INFOSEC	information security
IPB	intelligence preparation of the battlefield
iso	isolation
kg	kilogram
KM	kennel master
LAFB	Lackland Air Force base
LAN	local area network
LBS	pounds
LBV	load-bearing vest
LCE	load-carrying equipment
LED	light-emitting diode
LIN	line item number
LTC	lieutenent colonel
m	meter

MACOM	major Army command
MANSCEN	Maneuver Support Center
MCI	military customs inspector
MCIPM	military customs inspector program manager
MCM	Manual for Courts-Martial
mech	mechanical
METL	mission-essential task list
METT-TC	mission, enemy, terrain, troops, time available, and civilian considerations
MEVA	mission-essential vulnerable areas
MILSTRIP	military standard transportation and issue procedures
min	minimum
MMSO	maneuver mobility support operations
MOPP	mission-oriented protective posture
MPR	military police report
MSR	main supply route
MTOE	modification table of organization and equipment
MWD	military working dog
MWDMS	Military Working Dog Management System
NBC	nuclear, biological, and chemical
NCO	noncommissioned officer
NCOIC	noncommissioned officer in charge
no	number
Nov	November
NSN	national stock number
OB	obstacle
Oct	October
ODCSOPS	Office of the Deputy Chief of Staff for Operaitons and Plans
off/L	off leash
on/L	on leash
OIF	Operation Iraqi Freedom
OPMG	Office of the Provost Marshal General
OPSEC	operational security
pam	pamphlet
PC	potassium chlorate
PD	patrol dog
PEDD	patrol explosive detector dog
PIO	police intelligence operations
PM	provost marshal
PMO	provost marshal office

PNDD	patrol narcotic detector dog
POV	privately owned vehicle
PT	physical training
prep	preparation
RAMP	random antiterrorism measures program
rec	recreation
S2	intelligence section
SC	sodium cholorate
sec	seconds
Sep	September
SF	standard form
SFC	sergeant first class
SGT	sergeant
SOFA	Status of Forces Agreement
SOP	standard operating procedures
SP	smokeless powder
SRT	special-reaction team
SSG	staff sergeant
STX	Semtex
TDA	table of distribution and allowancees
TDY	temporary duty
TISA	troop issue subsistence activity
TMP	transportation motor pool
TNT	trinitrotoluene
TOE	table of organization and equipment
TRADOC	United States Army Training and Doctrine Command
Tues	Tuesday
US	United States
USA	United States of America
USC	United States Code
USCS	United States Customs Service
USSS	United States Secret Service
v	version
VCO	veterinary corps officer
vet	veterinarian
VIP	very important person
VTF	Veterinary Treatment Facility
VTR	Veterinary Treatment Record
WT	weight

References

DA Forms are available on the Army Publishing Directorate web site <http://www.apd.army.mil>. DD forms are available from the Office of the Secretary of Defense web site <http://www.dtic.mil/whs/directives/infomgt/forms/formsprogram.htm>.

SOURCES USED

These are the sources quoted or paraphrased in this publication.

AR 190-12. *Military Working Dogs.* 30 September 1993.

AR 190-14. *Carrying of Firearms and Use of Force for Law Enforcement and Security Duties.* 12 March 1993.

DA Pam 190-12. *Military Working Dog Program.* 30 September 1993.

DA Pam 611-21. *Military Occupational Classification and Structure.* 31 March 1999.

DOCUMENTS NEEDED

These documents must be available to the intended users of this publication.

AR 700-81. *DOD Dog Program.* 5 May 1971.

DA Form 2807-R. *Military Working Dog Training and Utilization Record.*

DA Form 3992–R. *Narcotics or Explosives Detector Dog Training and Utilization Record.*

FM 3-19.15. *Civil Disturbance Operations.* 18 April 2005.

READINGS RECOMMENDED

These sources contain relevant supplemental information.

AF Form 1256. *Certificate of Training.*

AR 190-47. *The Army Corrections System.* 5 April 2004.

AR 190-8. *Enemy Prisoners of War, Retained Personnel, Civilian Internees and Other Detainees.* 1 October 1997.

AR 350-1. *Army Training and Education.* 9 April 2003.

AR 40-1. *Composition, Mission, and Functions of the Army Medical Department.* 1 July 1983.

AR 40-3. *Medical, Dental, and Veterinary Care.* 12 November 2002.

AR 40-905. *Veterinary Health Services.* 16 August 1994.

AR 525-13. *Antiterrorism.* 4 January 2002.

AR 600-8-101. *Personnel Processing (In-, Out-, Soldier Readiness, Mobilization, and Deployment Processing).* 15 July 2003.

AR 600-9. *The Army Weight Control Program.* 10 June 1987.

AR 614-200. *Enlisted Assignments and Utilization Management.* 30 September 2004.

DA Form 2028. *Recommended Changes to Publications and Blank Forms.*

DA Form 3161. *Request for Issue or Turn-In.*

DA Form 5513-R. *Key Control Register and Inventory.*

DA Form 6. *Duty Roster*

DA Form 7281-R. *Command Oriented Arms, Ammunition, & Explosives Security Screening and Evaluation Record.*

DA Pam 710-2-1. *Using Unit Supply System (Manual Procedures).* 31 December 1997.

DD Form 1834. *Military Working Dog Service Record.*

DEA Form 223. *Controlled Substance Registration Certificate.*

DOD Directive 2000.12. *DoD Antiterrorism (AT) Program.* 18 August 2003.

DOD Directive 5030.49. *DoD Customs and Border Clearance Program.* 4 May 2004.

DOD Directive 6400.4. *DoD Veterinary Services Program.* 22 August 2003.

DOD Instruction 3025.12. *Military Assistance for Civil Disturbances (MACDIS).* 4 February 1994

FM 21-20. *Physical Fitness Training.* 30 September 1992.

FM 27-10. *The Law of Land Warfare.* 18 July 1956.

FM 3-19.40. *Military Policy Internment/Resettlement Operations.* 1 August 2001.

Lackland AFB Form 375. *Patrol Dog Certification.*

Lackland AFB Form 375a. *Detector Dog Certification.*

MCM, Part III Military Rules of Evidence; Rule 311, *Evidence obtained from unlawful searches and seizures.* 2000.

MCM, Part III Military Rules of Evidence; Rule 313, *Inspections and inventories in the armed forces.* 2000.

MCM, Part III Military Rules of Evidence; Rule 314, *Searches not requiring probable cause.* 2000.

MCM, Part III Military Rules of Evidence; Rule 315, *Probable cause searches.* 2000.

MCM, Part III Military Rules of Evidence; Rule 315(f)(2), *Probable cause determination.* 2000.

MCM, Part III Military Rules of Evidence; Rule 316, *Seizures.* 2000.

SF 702. *Security Container Check Sheet.*

TC 19-210. *Access Control Handbook.* 4 October 2004.

TC 25-20. *A Leader's Guide to After Action Reviews.* 30 September 1993.

USC Title 18, Crimes and Criminal Procedure; Part I, Crimes; Section 1385, *Use of Army and Air Force as posse comitatus.*

Index

***FM 3-19.15 (FM 19-15)**

Field Manual
No. 3-19.15

Headquarters
Department of the Army
Washington, DC, 18 April 2005

Contents

Distribution Restriction: Approved for public release; distribution is unlimited.

*This publication supersedes FM 19-15, 25 November 1985.

Page

Preface

Field Manual (FM) 3-19.15 addresses continental United States (CONUS) and outside continental United States (OCONUS) civil disturbance operations. Today, United States (US) forces are deployed on peacekeeping, peace enforcement, and humanitarian assistance operations worldwide. During these operations, US forces are often faced with unruly and violent crowds intent on disrupting peace and the ability of US forces to maintain peace. Worldwide instability coupled with increasing US military participation in peacekeeping and related operations requires that US forces have access to the most current doctrine and tactics, techniques, and procedures (TTP) necessary to quell riots and restore public order.

In addition to covering civil unrest doctrine for OCONUS operations, FM 3-19.15 addresses domestic unrest and the military role in providing assistance to civil authorities requesting it for civil disturbance operations. It provides the commander and his staff guidance for preparing and planning for such operations. The principles of civil disturbance operations, planning and training for such operations, and the TTP employed to control civil disturbances and neutralize special threats are discussed in this manual. It also addresses special planning and preparation that are needed to quell riots in confinement facilities are also discussed.

In the past, commanders were limited to the type of force they could apply to quell a riot. Riot batons, riot control agents, or lethal force were often used. Today, there is a wide array of nonlethal weapons (NLW) available to the commander that extends his use of force along the force continuum. This manual addresses the use of nonlethal (NL) and lethal forces when quelling a riot.

Appendix A complies with current Army directives, which state that the metric system will be incorporated into all new publications.

The proponent for this publication is HQ, TRADOC. Send comments and recommendations on Department of the Army (DA) Form 2028 (*Recommended Changes to Publications and Blank Forms*) directly to Commandant, US Army Military Police School (USAMPS), ATTN: ATSJ-DD, 401 MANSCEN Loop, Fort Leonard Wood, MO 65473-8929.

Unless this publication states otherwise, masculine nouns or pronouns do not refer exclusively to men.

Chapter 1

Operational Threats of the Civil Disturbance Environment

As the Cold War ended with the fall of the Berlin Wall and the collapse of Soviet communism in 1989, the Army was no longer concerned with containing the aggressive Soviet Union. The US Army began to transform because the battlefield on which it would fight had changed. Operations Just Cause, Joint Endeavor, Desert Shield, and Desert Storm have all been recognized as containing nontraditional battlefields. Following these operations, the Army was called upon to conduct peacekeeping, peace enforcement, and humanitarian assistance operations. These operations required US forces to protect refugees and humanitarian workers and keep hostile factions separate to enforce peace.

GENERAL CAUSES FOR CIVIL UNREST

1-1. In these modern times, demonstrations, civil unrest, public disorder, and riots happen for a number of reasons. Some of these reasons are economic hardships, social injustices, ethnic differences (leading to oppression), objections to world organizations or certain governments, political grievances, and terrorist acts. An event can be triggered by a single cause or a combination of causes. For example, operations in the Balkans involving civil unrest and riots were the result of ethnic hatred, a lack of civil authority, food shortages, a revolution, and religious-based fighting factions.

1-2. Demonstrations may range from simple, nonviolent protests that address specific issues, to events that turn into full-scale riots. Gatherings in protest are recognized rights of any person or group, regardless of where US forces may be operating. This fundamental right is protected under the *Constitution of the US*. During peacekeeping or peace enforcement operations, US forces should never violate basic civil or human rights. Most protesters are law-abiding citizens who intend that their protests be nonviolent, but some protest planners insist that the event involve some kind of violence. Often, in the media, protesters can gain sympathy for their cause by prompting authorities to take physical action against them. Violence is often the result of demonstrators beginning to conduct unlawful or criminal acts and authorities (who are responsible for the safety and welfare of all) enforcing the laws of the municipality, state, or nation. The depth of violence is determined by the willingness of demonstrators to display and voice their opinions in support of their cause.

1-3. Commanders must be aware of the possibility that some individuals or groups within an organized demonstration may have the intent to cause disruption, incite violence, destroy property, and provoke the authorities. The tactical situation and actions of the crowd should dictate control and enforcement options. Agitators and criminal infiltrators within the crowd can lead to the eruption of violence. Inciting a crowd to violence or a greater intensity of violence through the use of brutish enforcement tactics should be avoided. Publicity can be detrimental to authorities and beneficial to crowds because it can further their causes.

1-4. Community unrest results in urban conflicts that arise from highly emotional social and economic issues. Economically deprived inner city residents may feel that they are treated unjustly or ignored by people in power and authority. Tensions can build quickly in a community over a variety of issues, such as hunger, poor employment opportunities, inadequate community services, poor housing, and labor issues. Tension in these areas creates the potential for violence. When tensions are high, it takes a small (seemingly minor) incident, rumor, or act of injustice to ignite groups within a crowd to riot and act violently. This is particularly true if community relations with authorities are part of the problem.

1-5. Significant ethnic differences in a community can create an atmosphere of distrust, even hatred. Unrest among ethnic groups competing for jobs, living areas, and sparse essentials can cause an eruption of civil disorder and/or riots. As emotions run high, violence becomes likely.

1-6. Terrorist organizations may infiltrate groups within a demonstrating crowd. These terrorist groups may intend to embarrass their government or other governments. Terrorist infiltrators can be used to provoke crowds as a diversion, as part of a demonstration, or as cover for terrorist acts.

CROWD DEVELOPMENT

1-7. Crowds are a gathering of a multitude of individuals and small groups that have temporarily assembled in the same place. These small groups are usually comprised of friends, family members, or acquaintances that represent a group belief or cause. Individuals assume a sense of anonymity—they are viewed as just another face in the crowd. People in small groups are known only to companions in their group and to others in the gathering that have come from the same neighborhood or community. Commanders must consider how the individuals assembled and how they are interacting during the gathering process.

GATHERINGS

1-8. The assembly process of a gathering refers to the movement of people from different locations to a common location within a given period. This largely determines who participates. Creating a gathering from a crowd is a process with a beginning, middle, and end. As shown in *Figure 1-1* a crowd has an assembly process that leads to the gathering of the crowd, which is always followed by a dispersal process.

1-9. Gatherings are often assisted by the activities of individuals or groups with a specific agenda, such as yelling catchy slogans and cheers that everyone can easily pick up and join in on. Some groups are so well organized that they can prestage leaders to infiltrate a gathering. This creates unity, even inciting newcomers to join their cause. The first phase of the gathering is the assembly process. It can occur in one of two ways—impromptu or organized.

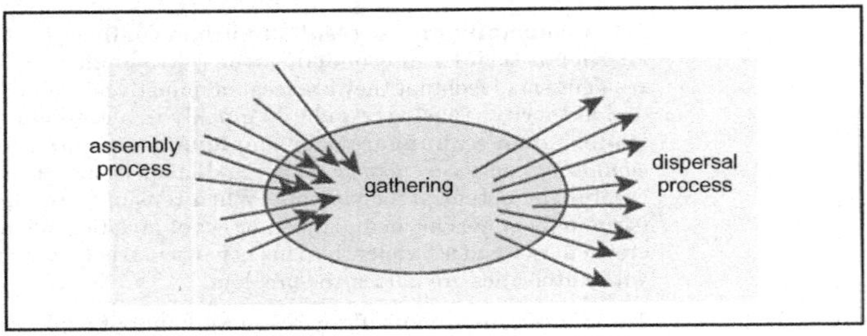

Figure 1-1. Crowd Building

IMPROMPTU GATHERINGS

1-10. Impromptu gatherings usually develop informally and are mostly done by word of mouth (one person to another or one group to another). Participants spread intelligence by telling one another when, where, and what is happening and inviting them to participate. An example of an impromptu gathering would be a gathering at a secured food distribution point after receiving information (by word of mouth) that a large truck carrying much-needed supplies and food is about to arrive. In this case, hunger would be the driving force causing the migration of people to the food distribution facility.

ORGANIZED GATHERINGS

1-11. Like an impromptu gathering, an organized gathering can also involve individuals and groups passing on information to one another. Passing intelligence of police activity and occurrences and on when and where events will take place helps organizers to prestage participants. Rarely is one represented group responsible for pulling together a gathering. Organized gatherings rely heavily on established groups that attract people to gather. Recent examples of these well-organized groups are anarchists, anti-globalization groups, and anti-free-enterprise groups. Groups representing extreme religious faiths and ethnic organizations have been common too. Some gatherings involve groups from rival, or even warring, factions.

1-12. Organized gathering processes rely more on centralized planning and organization. One or more of the groups offer organizers lists of individual names and groups, which they contact as potential participants. Modern technologies are available to these organizers through telephone banks, mailing lists, or e-mail addresses. In preparation for a long-term event, some group organizers have the means to provide logistical support of transportation, food, and water to participants.

> During the preelection period, Bosnian-Serbs (in a bid not to lose any of the territory they had won during the war) began to organize this fiercely contested city. Serbian mobs attacked US forces with bricks, rocks, metal bars, and firebombs. Organization for this altercation started days before, as many witnessed the prestaging of the items to be used in the attack. Wheelbarrows, 55-gallon metal drums, and other large containers were filled with the ammunition that would later be used to attack peacekeepers.
>
> Brcko, Bosnia
> September 1996

CROWD BUILDING

1-13. The second phase of the gathering process is the building of the crowd. Old concepts of crowds based on stereotypes are no longer true. Studies show that—

- Crowds are made of many entities, and all participants are not the same.
- Crowds are not made of isolated individuals but of a minority of individuals and a majority of small groups who may or may not be acquainted with each other.
- Groups and individuals in the crowd are not unanimous in their motivation.
- Groups and individuals in the crowd are usually not anonymous to each other, especially different groups gathered at the same place.
- Individuals within a crowd are not given to unique emotional displays; although known within a group, most individuals would not want to draw attention to themselves.
- Groups within a crowd will often act in unison in an attempt to achieve their agendas.
- Crowds are generally not unique or distinguished by violence or unlawful conduct.

DISPERSAL PROCESS OF A GATHERING

1-14. The last and final phase of any gathering of a crowd is the dispersal process. It involves the movement of people from the common location where they assembled to one or more alternate locations. The dispersal process ends the gathering of a crowd or at least begins its decline. The dispersal can occur on a routine, emergency, or coerced basis.

ROUTINE DISPERSAL

1-15. The routine dispersal may be specified in advance. It can also be included in the assembly instructions given by the organizers of an event.

EMERGENCY DISPERSAL

1-16. An emergency dispersal occurs when people evacuate an area in result of an unexpected crisis, such as a fire, explosion, bomb threat, or terrorist act. Individuals in such an emergency quickly recover from the initial shock. They keep their wits about them and improvise ways of extricating themselves and their companions from the dangerous situation. However, tragedies have occurred in some emergency dispersal situations. Therefore, it is important that forces on the scene are careful not to misinterpret what is occurring.

COERCION DISPERSAL

1-17. Coercion dispersal is caused by the use of force at some level. This is not necessarily the best way to force the dispersal of a crowd. The negotiated management of crowds is the preferred method and has proven to be highly successful in getting crowd organizers to police themselves, especially if the demonstration and/or protest leaders are available and willing to participate.

CROWD DYNAMICS

1-18. Understanding crowds and how individuals and groups form crowds through the gathering process is important because these issues apply to the dynamics of a crowd. Under most circumstances, gathered crowds are orderly and present little or no problems for authorities. Crowds consist of people who, although very motivated and passionate, are also subject to their own need for creature comforts. Rainy, cold, and nasty weather has a way of disheartening all but the few highly motivated and disciplined individuals. If problems exist, they usually fall into the following three categories:

- **Public disorder.** Public disorder is a basic breach of civic order. Individuals or small groups assembling have a tendency to disrupt the normal flow of things around them.
- **Public disturbance.** Public disturbance is designed to cause turmoil on top of the disruption. Individuals and groups assembling into a crowd begin chanting, yelling, singing, and voicing individual or collective opinions.
- **Riot.** A riot is a disturbance that turns violent. Assembled crowds become a mob that violently expresses itself by destroying property, assaulting others, and creating an extremely volatile environment.

1-19. Being part of a crowd of people has certain effects on different people. Each individual in a crowd is susceptible to behaving in a way that is contrary to their normal behavior. Some reasons for these behaviors are as follows:

- Crowds provide individuals with a sense of anonymity. With so many others, an individual realizes that he is just another face in the crowd, giving a sense of invulnerability.
- Crowd and individual behaviors are impersonal by nature. The "them-against-us" attitude affords those within the crowd the ability to freely (without hesitation or reservation) be verbally abusive, throw objects, or attack anyone who gets in the way.

- Crowds provide individuals with the idea that their moral responsibilities have shifted from themselves as a person to the crowd as a whole. Large numbers of people discourage individual behavior, and the urge to imitate others within the crowd is strong. Individuals look to others around them for cues of what to do next, disregarding their own background and training. Often, it is only the strong, well-disciplined person who can resist the prevailing behavior of a crowd.
- Crowd behavior influences the actions of both the disorderly individuals of the crowd and the authorities tasked to control them.

1-20. Individuals within a crowd are at times driven by deep-felt emotions. Emotional contagion is the most dramatic psychological factor of crowd dynamics. It provides the crowd with a temporary bond of psychological unity. Lasting long enough, this unity can push a simple organized crowd into a mob. Normal law and authority are rejected en masse under these conditions, increasing the potential for violence and panic to erupt.

1-21. Panic can erupt quickly, especially when crowds turn into mobs. Individuals within the mob can easily sense that their safety and well-being are at risk, putting them in a "fight-or-flight" mentality. Adding to the panic and confusion is the use of riot control agents (RCAs) by authorities in an attempt to gain control. Individuals in a mob, during the heat of confrontation, may attempt to leave and find that there are no escape routes and that roads are blocked. This can often lead to violent, physical attacks.

CROWD TYPES

1-22. Gaining an understanding of crowds and the dynamics that cause individuals or groups to join together in a crowd should be every leader's responsibility. Active, expressive, acquisitive, and hostile are not acceptable terms used to describe a crowd. They are the motives inferred from the actions the crowd takes. No gathering of a crowd is persistently or exclusively active, expressive, acquisitive, or hostile. For the sake of description, crowds can be identified as casual, sighting, agitated, or mob-like.

CASUAL CROWDS

1-23. Casual crowds are identified as individuals or small groups with nothing in common to bind them together. If they have an agenda, it is their own. They arrive separately and leave separately. Casual crowds are made up of individuals or small groups occupying the same common place, such as a shopping mall where these individuals or small groups meet.

SIGHTING CROWDS

1-24. Sighting crowds are similar to casual crowds with one additional element—an event. People migrate as a crowd to sporting events, are attracted to fires and accidents, and attend music concerts. Individuals or small groups gather at these events for the same purpose. It is the event and/or one's curiosity that compels a crowd to come together.

AGITATED CROWDS

1-25. Agitated crowds add responses that are based on the elements (people, space, and event). Individuals with strong emotional feelings within a crowd can quickly spread and infect the rest of the crowd. As more people within the crowd become emotionally involved, a sense of unity can develop, causing changes in the overall demeanor of the crowd. Yelling, screaming, crying, and profane name-calling are all associated with an agitated crowd.

MOB-LIKE CROWDS

1-26. Mobs have all the elements found in the first three types of crowds, with the addition of aggressive, physical, and sometimes violent actions. Under these conditions, individuals within a crowd will often say and do things they usually would not. Extreme acts of violence and property damage are often a part of mob activities. Mobs consist of (or involve) the elements of people and groups being mixed together and becoming fluid.

CROWD TACTICS

1-27. During a public disorder or disturbance, individuals and small groups within a crowd use any number of tactics to resist authority and disrupt and add turmoil in order to achieve their goals. These tactics can be unplanned or planned and violent or nonviolent. The more organized and purposeful a crowd becomes, the more likely a tactic will be used.

NONVIOLENT BEHAVIOR

1-28. As explained earlier in this chapter, most gatherings of individuals and small groups into a crowd do not involves violent behavior. A public disorder or disturbance usually involves some harmless name-calling, demonstrations to express views, corporate yelling and chanting, and even singing and dancing. In some instances, there may be property damage and the erection of makeshift barricades. All four types of crowds can be nonviolent, but mobs and agitated crowds have the greatest tendency to turn violent. Nonviolent actions of a crowd are disruptive because they are in direct conflict with what authorities want them to do, such as refusing to leave when directed, locking arms, and sitting in front of or around areas and buildings that the authorities are attempting to clear.

VIOLENT BEHAVIOR

1-29. A crowd that becomes a mob can be very violent and destructive. Although some nonviolent activities occur, violent crowds strike out physically at bystanders or others in the crowd, destroying both private and government property, setting fires, and employing bombs (in extreme cases). The only limitations for violent crowds are their own imaginations, the training of their leaders, and the materials readily available.

1-30. Violent crowds may erect barricades and physical barriers to impede the movement of authorities and prevent them from entering certain areas and/or buildings and to hide their activities. Common materials used for barricades include vehicles, felled trees, furniture, felled light poles, fencing materials, or anything else on hand and readily available.

1-31. Setting fires and preventing firefighters access to the structure by blocking it is another violent crowd tactic. Often, fires are set to create confusion in authorities or as a diversion to another activity.

RIOTS

1-32. A riot is one or more groups or individuals who are part of a larger crowd that involves threats of violence against persons or property. In some cases, a crowd will continue to gather until it evolves into a riot.

1-33. Riots vary considerably in both targets and players. A communal riot, for instance, deals with deep-seated ethnic, religious, and language differences. Commodity riots involve an attack on property by acts of vandalism, looting, and arson. Protest riots, such as the riot around the World Trade Organization (WTO) Assembly in Seattle, Washington (30 November 1999), illustrated individuals and groups aggressively and sometimes violently acting out or voicing their opposition to the assembly. The Democratic Convention in Chicago, Illinois, in 1968 reflects a riot that directly targeted police and authority in general. Celebration riots occur across the US as a result of home team victories in sporting events, among other reasons. Celebrating crowds look to make the moment more memorable through raucous acts that demonstrate their joy or happiness, for example, the riots that took place in Chicago in 1992 as a result of the Chicago Bulls winning the National Basketball Association (NBA) Championship.

1-34. Commanders and leaders must be aware that highly organized groups of protesters and/or demonstrators have developed tactics to disrupt the control force. These crowd tactics were published in handbooks for communist organizers during the Cold War, and today the Internet contains sites that are devoted to sharing these tactics with the general public and various protest groups. Although these tactics are somewhat outdated, commanders and leaders must be aware of them and their potential consequences. Many of these groups, such as anarchists, often mock the tactics employed against large crowds; for example, they may identify the police as "Darth Vader Cops" when the control force response is in full riot control gear. Commanders must be aware that well-organized crowds may attempt to engage the control force, surround it, and overpower it by sheer numbers. To avoid this, commanders must do a detailed terrain analysis (intelligence preparation of the battlefield [IPB]) of the area to include all approach and exit routes in the area. The control force flanks have to be protected to avoid being enveloped by the mob. To avoid being enveloped, the on-site commander should keep the crowd at a comfortable distance from the control force formation, which is usually the lethal zone of his NL munitions. Establishing predetermined rally points for the control force is critical in the event that this type of tactic is used by the crowd.

TACTICS USED TO DEFEAT AUTHORITIES

1-35. Organized mobs will try to defeat the control force by employing several different types of tactics. These tactics include the following:

- Constructing barricades.
- Using Molotov cocktails, smoke grenades, rocks, and slingshots.
- Feinting and flanking actions.

BARRICADES

1-36. By constructing barricades, the rioters are trying to protect themselves from assault by the control force. If the control force does advance toward the barricade, rioters hope that the barricade will split the control force and leave it vulnerable to actions by the mob.

MOLOTOV COCKTAILS, SMOKE GRENADES, ROCKS, AND SLINGSHOTS

1-37. Angry mobs will often attempt to disrupt the control force by throwing rocks and other projectiles. As the situation deteriorates, the mob may escalate the violence by using a battery of slingshots that will pellet the control force with a barrage of projectiles ranging from small rocks to marbles. They may also use smoke grenades (homemade or store-bought) to mask their movements. Finally, Molotov cocktails are used against personnel, employed vehicles, and portions of the control force. The mob will attempt to gain a position above an armored vehicle to enable them to drop a Molotov cocktail into an open hatch.

FEINTING AND FLANKING ACTIONS

1-38. Organized mobs may attempt to disrupt the movement of the control force by feinting an assault. When the control force moves to blunt the assault, the mob will assault the exposed flanks in an attempt to split the control force and envelop a portion of it.

WEAPONS

1-39. Mobs will often use various types of weapons against authorities to achieve their agendas and goals. It is extremely important that leaders train soldiers to recognize possible threats of the various weapons used by crowds or mobs, enabling the soldiers to react with the appropriate minimum force.

VERBAL ABUSE

1-40. In almost every instance of a civil disturbance or riot, verbal abuse will be an aggressive tool. Obscene language, racial remarks, taunts, ridicules, and jeers should always be expected. It is apparent that the purpose for using verbal abuse is to anger, demoralize, and provoke a physical response. Undisciplined, untrained soldiers who face such an attack could cause the situation to escalate. Just one provoked action of a soldier could be interpreted as an act of brutality by the media.

PERCEIVED INNOCENT PEOPLE OR THE WEAK

1-41. Women, children, and the elderly have often been used as pawns in civil disturbance operations. On many occasions, the innocent and the weak are moved to the front of the crowd and used as a barrier. With the innocent and the weak directly facing authorities, aggressive and violence-prone individuals or groups behind them attempt to provoke authorities to react. Often, this tactic is to exploit the situation in the eyes of the media for propaganda value. Despite the perception of sometimes being weaker, women, children, and the elderly can be just as committed to the insurgency as those behind them.

BARRICADES

1-42. Barricades are made of anything large or heavy enough to impede or prevent the movement of authorities. Trees, vehicles, fires, and furniture are all examples of items used to erect a barricade.

THROWN OBJECTS

1-43. In many cases, a crowd will throw anything in their reach that can be picked or picked up. Rotten fruit and/or vegetables, rocks, bricks, bottles, sticks, and pieces of lumber are most often used because they are readily available. More sophisticated types of thrown objects are improvised bombs, such as petrol bombs or tennis balls with nails sticking out of them. Never discount that these individuals and groups could acquire hand-thrown explosives, such as sticks of dynamite or grenades.

VEHICLES AND OTHER OBJECTS

1-44. Vehicles and other objects can be placed on higher ground (atop a hill or in a building) to disrupt civil disturbance operations. For example, a vehicle or tire can be set on fire and rolled down a hill, carts or barrels filled with hazardous or flammable materials can be pushed down a hill, or burning furniture and petrol bombs can be thrown from two-story or higher buildings or toward a formation of soldiers.

FIRE

1-45. Rioters have set buildings, vehicles, and other structures on fire to block the advances of authorities and to create confusion and diversions. If property damage is the goal of a crowd, fire is an effective tool. A sniper, for example, may set a building on fire to divert attention or provide maximum target possibilities. Other tactics used, particularly in the Balkans, were to flood the lower floors or basements of structures with flammable liquids or gas and then ignite it in an opportune moment.

FIREARMS

1-46. Sniping or massing fire from within the crowd, in buildings, or other covered positions can (at times) be effective tools for terrorists using a crowd to cover their acts. Firearms are anything that fires a projectile toward a target of opportunity.

EXPLOSIVES

1-47. A bomber's imagination is the only limiting factor in the use of explosives. Explosives can be used as simple diversions to block the advances of formations or provide an escape for rioters. The worst use of explosives is to purposely kill, injure, or demoralize authorities that are protecting and assuring the safety of everyone.

Chapter 2
Control Force Operations

Information is the key to developing civil disturbance plans. One must know who the demonstrators are; when, where, and why they are demonstrating; what their capabilities are; and what their possible course of action (COA) is. A commander's need for current, valid information cannot be overemphasized. The commander must learn as much as he can about the participants (their motivations, strategies, tactics, targets, and dedication). The more knowledge a commander has about the participants, the better equipped he is to counter their actions. He needs sound information to decide how to best use his available resources.

CIVIL DISTURBANCE PLANNING CONSIDERATIONS

2-1. The side that possesses better information and uses that information to gain understanding has a major advantage over its opponent. Forces that have this advantage can use it to bring about changes in attitudes, decisions, and actions in assembled demonstrators. In planning for crowd control or civil disturbance operations, planners must decide what data is needed to develop threat assessments. This decision is based on a multitude of information sources.

STANDARDIZED POSTURES IN COMPANY TACTICAL STANDING OPERATING PROCEDURES

2-2. Within squads, platoons, and companies, equipment may be increased or decreased, as needed. Some examples are as follows:

- Choose the M9 pistol for extraction and apprehension teams. The use of a long weapon (for example, an M16 with an M203, an M4 with an M203, or a 12-gauge shotgun) with NL munitions capability is also recommended, especially for overwatch personnel.
- Add nonstandard weapons such as shotguns for greater NL capabilities.

NOTE: The shotgun is used to protect the M203 gunner as he reloads.

- Add NL munitions to existing organic weapons systems, such as the M203.
- Arm soldiers in the front line of the formation with their standard weapon. If the weapon is a long weapon, it should be carried across the back from left to right or vice versa, with the butt up and muzzle down. Ensure that the weapon is cleared and that the magazine is in the appropriate ammunition pouch.

- Balance the mix of weapons and munitions according to the mission, enemy, terrain, troops, time available, and civilian considerations (METT-TC).
- Establish command relations and determine who has or gives the authority to fire NL munitions.
- Maximize the distance and barriers between the crowd and control formations. Use NL munitions to create a standoff distance.
- Maintain a lethal overwatch of the control force at all times.
- Maintain a reserve force at all times to reinforce the control force, as needed.
- Create NLW range cards for static positions.
- Consider environmental conditions and their effect on the performance of NL munitions. Proper storage, periodic inspections, and the rotation of NL munitions are critical to maintaining the effectiveness and viability of NL munitions.

WEAPON AND AMMUNITION CONFIGURATIONS

2-3. Individuals designated as NL shooters must have the means to transition to lethal rounds, if required. Lethal rounds are carried separately from NL rounds so that the shooter will not confuse them in the heat of confrontation. However, in the heat of confrontation where a lethal option is necessary, NL rounds can be used lethally by adjusting the point of aim to a vulnerable part of the body and closing in on the distance to the target.

2-4. Squad leaders should designate NL shooters within their squad. Ideally, the squad should not change its organization to accommodate NLW.

2-5. Commanders should not dispatch NL patrols. They should plan a combat or security patrol with NL capabilities. Soldiers are never sent in harm's way without lethal protection. NLW are only considered additional tools for the mission and not a mission in itself. There is no such thing as a NL mission.

2-6. Soldiers manning a static position should have NL capabilities depending on METT-TC. Shotguns and the M203 work well at static positions.

2-7. Recovery, apprehension, and/or extraction teams should be established before deployment. Team members should be equipped with personal protective equipment (PPE), an M9, NL munitions and weapons, and Flex-Cufs®. They should have some training in open-hand control, pain compliance, and handcuffs and/or Flex-Cufs.

2-8. Crowd control formations (see *Chapter 6*) should be well trained and well rehearsed. Rapid, coordinated movements of a well-trained and well-rehearsed control force can often be a strong enough deterrent. A lethal overwatch marksman always covers the control force.

2-9. During a NL engagement, the use of designated marksmen (DM) provides confidence and safety to those facing a riot. The DM in an overwatch position scans the crowd to identify threats and designate personnel for recovery and lethal rounds firing (if a lethal threat is presented). They are ideally suited for flank security and countersniper operations.

NONLEHTAL PLANNING

2-10. The fundamental principles of mission planning are not changed by the incorporation of NLW into a plan. NLW are intended to be operational-enhancement tools. The first concern of a commander is the successful completion of the mission. NL munitions provide a commander with a wider range of response options, but these munitions are not a replacement for lethal capabilities. Commanders must always be capable of answering a lethal attack with a lethal response.

2-11. When a commander commits his soldiers and equipment to a crowd control situation, he commits his forces with the additional tools of NL capabilities. This is an addition to the force continuum that the force commander now has available to him. Commanders that properly employ NL munitions and weapons have a tactical advantage over those who rely completely on lethal means.

2-12. When developing a tactical standing operating procedure (TSOP) (see *Chapter 6*), units should consider the following:

- Crowd control formations.
- Extraction teams.
- Apprehension teams.
- Lethal overwatch marksmen and/or observer teams.
- Reserve and/or security forces.

MILITARY WORKING DOG TEAMS

2-13. Military working dog (MWD) teams may be employed with a control force formation as a method of increasing crowd apprehension about approaching or engaging the formation. The teams should be in the rear of the formation in plain sight of the crowd, but in front of the command element and the M33A1 squad RCA disperser. The MWD teams work back and forth behind the formation as an intimidation measure. The presence of the MWD may produce a profound psychological effect on the crowd. These teams may also be used to help control individuals who have been captured by the recovery and apprehension teams.

CAUTION

Do not unleash an MWD on a crowd.

COMBAT CAMERAMEN

2-14. Video and still cameramen should make a photographic record of the individuals in the crowd who are leaders and instigators. Events must be documented to hold personnel, factions, gangs, or groups accountable for acts that violate law, destroy property, or cause physical harm. Electronically recording events aids in the prosecution of such cases and eliminates the sense of anonymity that people in large crowds often feel.

INFORMATION OPERATIONS

2-15. Information superiority helps forces to anticipate problems and requirements. It allows commanders to control situations earlier and with less force, creating the conditions necessary to achieve the optimal end state. Public affairs, psychological operations (PSYOP), and civil military operations are activities that will allow the commander to control situations earlier and with less force.

2-16. Information is available from a multitude of sources. A diversity of sources is the best approach because it prevents biased perspectives. Primary sources are as follows:

- Open sources.
  - Libraries.
  - Newspapers and news periodicals.
  - Radio and television.
  - Internet.
- Law enforcement sources.
  - Local law enforcement agencies.
  - National law enforcement agencies.
- Military sources.
  - Department of Defense (DOD) intelligence community (most restrictive source).
  - Local military intelligence (MI) field offices.

2-17. Police intelligence operations (PIO) is one of the five military police functions. This function is a process of actively and passively collecting information that is of a police, criminal, or combat nature. As military police perform the other four functions of maneuver and mobility support (MS), area security (AS), law and order (L&O), and internment and resettlement (I/R), they are gathering information that supports, enhances, and contributes to the commander's protection program, situational awareness, and battlefield visualization by portraying relevant threat information that may affect operational and tactical environments.

2-18. IPB is a continuous process for analyzing the threat and the environment of a specific geographic area. During the IPB process, the Intelligence Officer (US Army) (S2) or Assistant Chief of Staff, G2 (Intelligence) (G2) uses all available databases, intelligence sources and products, and related MI discipline to analyze the threat and the environment. The PIO function supports this process by providing the S2 with collected police, criminal, and combat information that can directly and significantly contribute to the success of the MI effort. In addition to combat information, the PIO function provides additional information on possible criminal threats and COAs. This is intended to support the S2 IPB process and can be used by the commander to upgrade force protection. (See *FM 19-10*.)

THREAT ANALYSIS

2-19. Crowd control requires its own thought process. Emphasis should be on prevention rather than confrontation. In combat, military forces are taught to fight and eliminate threats. In crowd control, military forces must deal with noncombatants that have internationally recognized rights. These rights must be respected while maintaining public order. This is an issue that law enforcement agencies (LEAs) have been struggling with for years.

2-20. Dealing with crowd control incidents is a matter of using some basic guidelines. It also includes asking the right questions in a logical manner so that key issues are not omitted. The commander must use these guidelines in context with METT-TC and the location of the incident (CONUS or OCONUS).

PREINCIDENT PLANNING

2-21. This phase of planning begins before the incident and is initiated at the operational level with guidance from the strategic level. The planning includes guidance on crowd control and addresses responsibilities, training, organization, operating procedures, use of force (CONUS), and/or rules of engagement (OCONUS). The most difficult and productive decisions are those made in the preincident planning process.

Avoid Confrontation

2-22. Crowd situations are highly unpredictable, but one thing seems certain—confrontation will likely cause crowd resistance. When pushed, people tend to resist opposition to the realization of their purposes.

Focus on Prevention

2-23. Planning should key in on the prevention of unfavorable outcomes. Experience has shown some LEAs attempting to help crowds accomplish their goals within the law that have been beneficial and even led to conceding some violations for the purpose of avoiding confrontation. However, LEAs maintain a law enforcement presence, which signifies social restraint. LEAs also decide when and where they will not compromise and the amount of force to use.

Define Goals

2-24. When defining a goal, deciding what must be accomplished is the first step. Defining the goal is fundamental. However, actually working toward and accomplishing the goal is easier said than done, as the process can drive the situation. Commanders and leaders must be aware of this. The military force must focus on what they are trying to accomplish.

BACKGROUND INFORMATION AND INTELLIGENCE

2-25. Seek to know as much as possible about social protest groups (within the limits of the law, see *Appendix B*) before an incident. This will provide insight into the organization and its functions and provide a warning as to what to expect. When you know as much about these organizations as possible, then it is possible to anticipate their next move. See *Appendix C*.

2-26. Continually assess what is to be accomplished against what others are trying to accomplish. Some groups may not have goals that conflict with those of US forces, but their actions may. Other groups may have goals and activities that do conflict. In either case, the assessment of group goals compared to the goals of US forces helps to understand and avoid potential conflicts.

CIVIL DISTURBANCE OPERATIONS PREPARATION

2-27. Counterdemonstration workgroups are a mixture of traditional and nontraditional staff proponents brought together to coordinate the resources to be employed by task force (TF) units before a potential civil disturbance. Meetings of the counterdemonstration workgroup are held on a regular basis (generally weekly) and are chaired by the Assistant Chief of Staff G3 (Operations and Plans) (G3). The following staff proponents may be involved, and others may be involved as needed:

- Public affairs office (PAO).
- Joint military commission (JMC), if organized.
- Provost marshal office (PMO).
- PSYOP.
- G2, Assistant Chief of Staff G5 (Civil Affairs) (G5), Assistant Chief of Staff G6 (Signal) (G6).
- Civil affairs (CA).
- Army airspace command and control (A2C2).
- Chaplain.
- Engineer.
- Surgeon.
- Fire support element (FSE).

RELATIONSHIPS DEVELOPMENT

2-28. Working relationships between commanders and protest group leaders are increasingly seen as the best means for preventing bad outcomes in crowd situations. This is called the negotiated management model of crowd control. LEAs in large metropolitan cities in the US and Europe practice it.

2-29. Open dialog helps develop working relationships between commanders and protest group leaders, providing an opportunity to communicate clearly. Talking allows group leaders to tell authorities and tactical commanders what they want to accomplish. It also allows authorities and commanders to tell group leaders what they are prepared to do and how they might respond to certain crowd behaviors. Such communications can do much to resolve issues and prevent violence.

2-30. Commanders may not be able to talk to all leaders before a crowd assembles. Some groups do not have recognized leaders and are ad hoc organizations. Other groups may have several leaders, but only some of those leaders will negotiate. In such cases, meetings with some leaders may tell you who the other leaders are and provide critical information. Negotiations may also encourage more moderate leaders to do things that will support the commander.

2-31. Initiate communication with group leaders to work out issues before an event. Commanders should make a concentrated effort to maintain a working relationship with the leaders of protest groups.

2-32. Communication also means persuasion. Commanders should make a concerted effort to win over demonstration leaders. Commanders may tell group leaders that they want to help them complete their mission in a safe manner. They should encourage demonstrators to protest in an acceptable way and, if possible, offer favors to get them to do just that.

NOTE: Commanders must only offer what they are willing and able to deliver.

2-33. Several rules apply in negotiations, and credibility is key. Only communicate necessary information and those actions that authorities and commanders intend to do. Negotiations are made from a position of strength, and by negotiating an agreement may be reached. However, contingencies should be developed in the event that the agreement is violated.

2-34. Working relationships between commanders and protest group leaders often result in protest groups policing themselves. This is one of the basic premises of negotiated management. Protesters are more likely to listen to their own leaders, as opposed to listening to US forces. Protest organizers are likely to buy into a cooperative effort and agree to ground rules set during preprotest negotiations. Many protest leaders seek to stay on the right side of the law.

ACCEPTABILITY

2-35. Deployed US forces will find themselves engaged in crowd control operations under difficult circumstances. The host nation (HN) may have groups that do not accept the presence of US forces, which will make imposing order and the protection of citizens difficult at best. US forces will be under intense media and political scrutiny (an environment much like what LEAs operate in on a day-to-day basis).

2-36. Winning in this environment is not like winning in combat. US forces may appear to be invincible and formidable, but they risk being portrayed as oppressors. Thus, US forces can lose by appearing to win. Groups that perceive themselves as oppressed will readily seek victimhood in an effort to gain the support of public opinion. Winning in this environment is about seizing and holding the moral high ground. US forces must maintain the authority and legitimacy of what they are doing.

2-37. Projecting a favorable image will require outreach to local leaders and citizens. It will also require developing a relationship with the media. For example, the Los Angeles Sheriffs Department often invites the media to accompany their tactical commander during crowd control situations. This shows that they have nothing to hide. It also provides an opportunity for individuals to see the commander's side of an event.

2-38. In this environment, commanders must consider how actions will play among several audiences—local, allied, US, and international. The media effect is inescapable. In some cases, it will be difficult to accommodate all of these audiences. However, the most important audience will be the US public, which is key to the continued support of US forces.

CROWD ASSESSMENT QUESTIONS

2-39. Assessing crowds requires its own set of questions. These questions should be answered before a crowd assembles or as quickly as possible if a gathering occurs without notification. In some cases, all the questions may not be answered. These questions represent a logical way to think about crowds. The questions are as follows:

- **Who are they? What is the overarching identity of the crowd? Are they strikers, ethnic factions, or social protesters? Do they identify themselves as strikers, ethnic groups, religious factions, or protesters against some perceived social injustice?** Understanding who they are will indicate what they may do. It may be possible to determine identities and goals from advance assembling instructions, leaflets distributed to bystanders, placards and banners, and chants and songs.
- **What are their goals?** What the group wants to accomplish by assembling could determine the extent to which they can be accommodated once they have assembled. They may only seek recognition for their cause (being seen and heard). If so, this goal is usually easy to accommodate. However, some groups may have more demanding goals, for example the demonstrators in Seattle who sought to stop the WTO. Goals that cannot be accommodated make confrontation very likely.
- **What is the composition of the crowd and are there any known factions?** Seattle demonstrations against the WTO were comprised of groups that were protesting environmental issues, wages, and child labor laws. Differing goals and the resulting friction were evident between such organizations as the Ruckus Society and the American Federation of Labor-Congress of Industrial Organizations (AFL-CIO). Factions within a crowd represent threats and opportunities.
- **What are they capable of doing?** Protest groups often claim that they will assemble large numbers of people to produce some disruptive action. However, there is often a difference between the claim and reality. An organization may claim that it can mass a 100,000 people, but in reality can only get 40,000. Organizers exaggerate for the following reasons: they want to boost the morale of their own people and they want the media to report that they have strength in numbers. Studying the past activities of a group may provide indications of what they are capable of doing in the future.
- **What are their traditional behaviors or cultural repertoires?** What people do during protests is not universal. It varies with the group and the culture. Social protest organizations and striking unions will carry placards and banners. Other groups will protest in a more quiet way, like the 5,000 women in Sarajevo whose standard Sunday behavior was to sit and block traffic. They were protesting the loss of male relatives in the Bosnian war. Understanding the goals of

the protesters can be helpful in deciding how to respond to their behaviors.

- **When and where will they assemble?** Every protest organizer has a time and place for assembling and perhaps a destination for the crowd to move toward. If the organizer attempts to mobilize large numbers of participants, the time and place for assembling and dispersing must be made known in the instructions. This information may be stated in the mobilization instructions or disseminated by an informal network, such as word of mouth. Such impromptu networks in densely populated areas can enable rapid assembling.
- **Where will they go?** Many crowds have destinations. Organizations may march a specified distance to ensure that their cause gets sufficient attention. Commanders need to know the route to minimize disruptions to the rest of the community. In Los Angeles, LEAs attempt to reroute traffic and prevent congestion caused by protest marches. It is also necessary to provide security along the route to prevent counterdemonstrators from confronting the marchers. This will prevent an even greater problem for the community.
- **What are the possible targets of violence?** Riots, in particular, may focus on target facilities. In the 1992 Los Angeles riots, gun stores were major targets. In developing countries, targets may be more basic. In Somalia, throngs of people stormed food supply facilities. In Haiti, riots broke out over trash dumps where US forces had discarded the remnants of meals, ready to eat (MREs).
- **What is the worst-case scenario?** The worst-case scenario must be recognized for the sake of avoidance. This may be when a peaceful crowd degenerates into a violent riot.
- **When and where will they disperse?** Crowds have a life cycle that includes how they disperse (see *Chapter 1*). Commanders must consider this. It is essential that there be one or more avenues that individuals can use to disperse. These dispersal routes should be clearly marked, visible, and open-ended. There may be a need for US forces to assist the crowd with dispersal. For example, in Los Angeles a 70,000-man march went from east to central Los Angeles. Once at their destination, they engaged in peaceful activities and then dispersed. The LEAs provided transportation back to the assembly area (AA); otherwise, the protesters would have returned to the AA in a disorganized and unsupervised manner.
- **Are there plans for subsequent gatherings?** A crowd may disperse for a short time so participants can take care of personal needs. This could be days or only a few hours.

LEGAL CONSIDERATIONS

2-40. The legal authority in which a peace operation is conducted defines the parameters of the operation. All commanders, leaders, and soldiers must know the legal authority that regulates their operations.

MANDATES

2-41. There are two types of mandates—United Nations (UN) and non-UN. Mandates are the authority under which an operation is conducted. UN mandates are created from resolutions by the UN Security Council (UNSC) or the UN General Assembly (UNGA) that authorize and define an operation. Non-UN mandates are created from treaties, accords, resolutions, or agreements evolving from other international or regional organizations.

STATUS OF FORCES AGREEMENT AND STATUS OF MISSION AGREEMENT

2-42. A status of forces agreement (SOFA) and a status of mission agreement (SOMA) are key documents that define the legal authority and responsibilities of a force and force personnel participating in an operation. A SOFA or SOMA may be a treaty or memorandum of understanding. It is an agreement negotiated between the UN and the host countries, which details the rights, privileges, immunities, and nature of services to be provided to the force and its personnel and their responsibilities and obligations.

INTERNATIONAL LAW

2-43. Legally, international law affects most stability operations and support operations (SOSO). These laws consist mainly of international agreements, treaties, and customary international law, which include the agreements and laws known as the law of war. International agreements prescribe the rights, duties, powers, and privileges of nations relative to particular undertakings. International agreements will affect US participation in SOSO in the—

- Right of US forces to enter a foreign country.
- Status of US forces while in the foreign country.
- Construction and operation of US bases of operations.
- Aircraft overflights and landing rights.
- Processing of claims for any damage to persons and property of the foreign country.

2-44. These agreements or customary international laws govern all aspects of operations carried out. In the absence of a viable HN government or proper international agreement, many aspects of the operations are controlled by domestic (US) or customary international law.

UNITED STATES LAW

2-45. Operations in peace and conflict must comply with US law, whether as a statute, executive order, regulation, or other directive from a branch or agency of the federal government. The *Uniform Code of Military Justice (UCMJ)* applies to questions of military justice. Various statutes, *Executive Order 12333; Section 3, Code of Federal Regulations (CFR);* and DOD and service regulations govern all intelligence-gathering activities. The staff judge advocate (SJA) must actively advise and participate in all activities from initial planning to redeployment.

HOST NATION LAW

2-46. If a viable HN government exists, whether at the national or the local level, all laws of the HN will apply to US forces in that country unless an international agreement provides otherwise. The types of laws that may inhibit US operations are in the fields of immigration, labor, currency exchange, the procurement of goods and services, customs and taxes, and criminal and civil liabilities. A SOFA usually covers these areas before an operation begins. Therefore, planners must understand the law to assess it and determine if it will adversely affect the operation. Assistance may be available from the local US embassy or the command judge advocate; if not, the command may have to rely on other sources for guidance. If local law hinders the operation, the commander must inform the US embassy and request that it negotiate a solution.

CIVIL AUTHORITIES

2-47. There will be times that US forces will deploy to nations where the infrastructure of civil authorities has survived the conflict and is functional. Local authorities and police are responsible for ensuring the safety of their citizens, including the UN High Commissioner for Refugees (UNHCR). US forces assist in this by providing general military security and by facilitating negotiations. In countries where the civil authority infrastructure has collapsed, the UN mandate may require the coalition of forces to help ensure the safety of citizens until their own civil authorities can be rebuilt.

RULES OF ENGAGEMENT

2-48. The rules of engagement (ROE) are the commander's rules for the use of force (RUF). Commanders are responsible for interpreting, drafting, disseminating, and training on the ROE. Ensure that the SJA assists throughout the ROE development process. This will ensure that the ROE will not improperly constrain actions, but are still consistent with domestic and international law and policies and orders of the chain of command.

2-49. Clearly stated ROE are published before Army forces are committed. The highest military authority, with input from subordinate commanders, will continually evaluate the ROE and modify them as appropriate. When the use of NLW has been authorized in the ROE, the decision to use NLW should be delegated to the lowest possible level, preferably the platoon or squad level. This requires that all personnel, not just leaders, have a clear understanding of the ROE and the commander's intent (see *FM 3-22.40*).

2-50. In peace operations, the use of force is restrained while the diplomatic solution is sought through negotiation or mediation. Accordingly, ROE are more restrictive in peace operations, making them more politically sensitive. In today's world, peace operations are usually conducted by a coalition of forces operating under the purview of the UN charter and customary international law. Therefore, the UN may mandate certain restraints. Using overwhelming force could compromise diplomatic efforts to reach a peaceful settlement. Even a single, relatively small engagement could jeopardize the legitimacy of the peacekeeping force and upset negotiations or mediations. Commanders must beware that in any confrontation ROE decisions made by soldiers can have strategic political implications on current and future operations. Therefore, it is vital that leaders and soldiers hold a common understanding of when, how, and to what degree force is to be used. This requires training.

MULTINATIONAL OPERATIONS

2-51. Multinational operations include military forces from different nations. These operations are usually interagency operations. Conducting operations with foreign military partners, like operations with civilian partners, is uncommon to many soldiers, so a clear understanding of this different environment is necessary.

2-52. The majority of US peace operations will be part of a UN peace operation. Their multinational character merits particular attention because national interests and organizational influence may compete with doctrine and efficiency. Commanders can expect contributing nations to adhere to national policies and priorities, which at times complicate the multinational effort.

2-53. When working within a multinational force and with allies, do the following:

- Train coalition partners and allies on tactics and techniques.
- Remember that teamwork and trust are essential.
- Achieve unity of effort.
- Consider equipment and communications capability and interface.
- Strive to eliminate sources of confusion and misunderstanding.
- Remember that military doctrine varies from country to country.
- Remember that national culture influences the way units operate.
- Remember that language barriers represent a significant challenge.

SCALABLE EFFECTS

2-54. Scalable effects concepts are a measured approach in response to a crowd gathering. By recognizing a use-of-force policy, soldiers must be taught and understand that they use the minimum force necessary. Without the appearance of a graduated response, the gathering crowd may consider actions as excessive, causing a possible escalation of hostilities or violence. Do the following to aid in the scalable effects process:

- Try to persuade the crowd to quietly disperse by talking with leaders.
- Use translators as necessary.
- Let the first approach be by local authorities (the mayor or police).
- Pass out handbills requesting that the crowd return home.
- Use video and still cameras to photograph individuals and events for later use at trials.
- Give warnings before moving to the next level of force.

2-55. Ideally, the force should be positioned out of sight of the crowd. Have the troops move into position with few shouted commands. For maximum effect on the crowd, have them form their formation decisively and professionally. Do the following to aid in the formation process:

- Display force in a graduated manner (for example, a UH 60, hardened vehicles, and soldiers with charged weapons).
- Exploit the psychological effects of show of force.
- Demonstrate force (do not use on unarmed civilians).
- Escalate the PSYOP message.
- Highlight the target pointer.
- Demonstrate sniper precision strike capability.
- Employ riot control measures.
- Use NL munitions.
- Use RCAs (if approved).
- Move through the crowd using riot control formations and movement techniques.
- Use the graduated response matrix (GRM)

2-56. With the increased participation of US forces in contingency operations around the world, it accentuates the need to establish procedures for applying graduated military responses to situations that threaten these missions. Numerous GRMs and similar products exist throughout the military. These products graphically portray available responses in a graduated manner. The intent of these products is to give the on-scene commanders a list of options with which to control or defuse a situation before it gets out of hand. Most threats can be eliminated without loss of life or collateral damage by effectively applying the resources available.

PLAN A GRADUATED RESPONSE MATRIX

2-57. The planning for and development of a GRM begins with the mission analysis portion of the military decision-making process (MDMP). Some missions require US forces to enforce treaties or accords to protect the lives of civilians in uncertain or hostile environments (such as noncombatant evacuation operations [NEO]) or to provide large-scale humanitarian assistance. These missions require some sort of graduated response criteria to maintain order and prevent uncertain environments from becoming hostile. The development of a GRM requires the following seven steps:

- Identify the need for a GRM.
- Establish a team to develop the GRM.
- Develop targets.
- Coordinate staff functions.
- War-game.
- Receive command approval.
- Operate deliberately.

Identify the Need for a Graduated Response Matrix

2-58. Not all missions require a GRM. The decision to use a GRM requires careful consideration. Once it is agreed that a GRM is necessary, it requires guidance from the commander regarding the response options available.

Determining the appropriate responses is based on the facts, assumptions, and limitations identified during mission analysis.

2-59. Planners (staff) must agree on the intent of the GRM. The GRM can be used as a training and rehearsal tool. It provides leaders with the most likely vignettes that can be incorporated into a COA analysis, predeployment training, and rehearsals. The GRM can also be used as a handy reference during situations that require graduated responses.

Establish a Team to Develop the Graduated Response Matrix

2-60. Establish a GRM development team with a broad range of skills, knowledge, and professionalism. The following list of individuals makes up a typical GRM team:

- Provost marshal or senior military police advisor.
- FSE.
- Brigade legal officer.
- PSYOP representative.
- Land information warfare or information operations officer.

2-61. Since the GRM is designed to give commanders graduated options for dealing with both hostile and nonhostile threats to the mission, this team composition allows for target selection, application of the ROE, and attack methods using both NL and lethal means.

Develop Targets

2-62. The FSE, in conjunction with the S2 section, develops targets for both lethal and NL attacks. In the case of stability operations, these targets are usually not the conventional specific point or piece of equipment on the ground. They are more situational than specific. The GRM identifies situations or acts that subordinate elements could face during the mission. The sample GRM in *Figure 2-1* shows three possible situations or acts that on-scene commanders could expect to encounter. From the targeting standpoint, these are groups of more specific targets.

2-63. During mission analysis, the fire support officer (FSO) identifies both lethal and NL assets available to the unit. A tactical PSYOP team (TPT) attached to the unit is an example of a NL attack asset that should not be overlooked. The following are examples of what the FSO should look for:

- RCAs.
- TPTs.
- Electronic warfare assets.
- CA teams.
- Information operations (IO) teams.
- Artillery smoke projectiles.
- Aircraft (AH-64s, OH-58Ds, and AC-130s).
- Mortars.

2-64. The lethal assets described could be used in a NL show of force or demonstration to defuse a situation before it requires lethal force. The critical element of this mission analysis is not to focus solely on lethal assets. In

<table>
<tr><th>Crowd Action</th><th colspan="4">Nonlethal Response</th></tr>
<tr><td rowspan="2">Unarmed Civilians</td><th>Command Presence/PSYOP</th><th>Show of Force</th><th>Demonstration of Force</th><th>Riot Control Means (if approved)</th></tr>
<tr><td rowspan="3">- Ensure that the on-site commander of the TPT directs the broadcast of the dispersal proclamation and/ or passes handbills.
- Escalate the tone of the dispersal proclamation from information to a warning of force.</td><td rowspan="3">- Display force along with escalating the dispersal proclamation.
- Display force in a graduated manner, such as a helicopter hovering over a crowd or soldiers with charged weapons.
- Exploit the psychological effect of a show of force.</td><td>- Do not demonstrate force toward unarmed civilians.</td><td rowspan="2">- Employ RCAs at the point of penetration.
- Use PSYOP to exploit the psychological effect.
- Move through the crowd using riot control formations and movement techniques.</td></tr>
<tr><td>Armed Civilians (knives, clubs)</td><td rowspan="2">- Display force along with escalating the dispersal proclamation.
- Highlight the target pointer.
- Demonstrate sniper precision strike capabilities.</td></tr>
<tr><td>Armed Crowds/ Military (firearms)</td><td>- Do not use RCAs—they may escalate the situation.</td></tr>
<tr><th colspan="5">Hostile intent/hostile act occurs by armed threat.</th></tr>
<tr><th>Crowd Action</th><th colspan="4">Lethal Response</th></tr>
<tr><td rowspan="2">Unarmed Civilians</td><th>Sniper Response</th><th>Small Arms Direct Fire</th><th>Aerial Fires</th><th>Indirect Fires</th></tr>
<tr><td rowspan="3">- Ensure that target leaders or troublemakers are targeted.
- Use the minimum response necessary.
- Exploit the psychological effect of an attack.</td><td rowspan="3">- Determine that sniper attack is insufficient.
- Consider demonstrating capabilities.
- Exploit the psychological effect of a lethal response.
- Escalate gradually, starting with a small caliber, single round and work up to a large caliber, automatic.</td><td>- Determine that small arms direct fire is ineffective.
- Use the minimum response necessary.
- Use a minimal precision strike initially; use subsequent fires based on the situation.
- Exploit the psychological impact of each strike.</td><td>- Determine that air assets are unavailable or ineffective.
- Use the minimum response necessary.
- Ensure that the response is directed by the on-site commander.</td></tr>
<tr><td>Armed Civilians (knives, clubs)</td><td colspan="2" rowspan="2">- CAS/indirect fires must be authorized by the MACOM commander.
- Consider requesting permission for use when—
• All lesser means have been ineffective.
• There are physical eyes on target.
• Proximity to civilians has been considered.
• Risk to friendly forces/evacuees outweighs the risk of collateral damage.</td></tr>
<tr><td>Armed Civilians/ Military (firearms)</td></tr>
</table>

Figure 2-1. Sample GRM Card

stability operations, the Army wants to prevent acts of hostility first and then be prepared, if necessary, to apply lethal force.

2-65. Graduated responses can range from command presence through the show of force, a demonstration, the use of RCAs, and other techniques (such as the application of lethal force using snipers, small arms, AC-130s, and indirect fires).

Coordinate Staff Functions

2-66. To coordinate staff functions the rest of the GRM team assembles to complete the escalation sequence for each response. PYSOP and legal representatives are critical attendees during the escalation sequencing process. In the area of psychological operations, the TPT must exploit the effects of all responses.

2-67. The legal officer evaluates each escalation of force option and graduated response to ensure that it is consistent with the ROE. The GRM is designed to recommend applications of force consistent with the ROE, yet not limit the leader or individual soldier's right of self-defense. A GRM should clearly show that if a hostile act occurs, lethal options would be first and foremost.

2-68. In the case of lethal responses, the commander's guidance must again be applied. For example, lethal responses are allowed only in self-defense.

War-Game

2-69. Once the types of escalations for each potential graduated response are determined and annotated, the GRM must then be war-gamed. The staff must walk through each act or situation from the on-scene commander's standpoint.

Receive Command Approval

2-70. Once the GRM has been war-gamed, it must be submitted to the commander for approval. This is the final check to ensure that the GRM team has applied the commander's guidance correctly and met his intent.

Operate Deliberately

2-71. Operating deliberately affords commanders ample opportunity to plan and prepare for possible civil disorder situations. Through the effective gathering of information and a working cooperation with local government and police officials, commanders can often be made aware of dates, times, locations, and what groups may assemble before the operation. The purpose is to gather as much information as possible who is involved, where they are assembling, what incident promoted the activity, and what seems to be the prevailing attitude of the assembling crowd.

2-72. Civil disturbance operations are dynamic, ever-changing environments requiring effective communications both up and down the chains of command. They require advance preparation and planning using the established troop-leading procedures (TLP) outlined in *FM 3-19.4*. The eight-step process of TLP aids commanders and their subordinate leaders in planning and preparing for an operation. Most steps can be accomplished concurrently, but rarely is there enough time to go through each step in detail. Despite this, leaders must follow these steps to ensure that nothing is left out of the planning process. As outlined in *FM 3-19.4*, the following are the eight steps of TLP:

- Receive and analyze the mission.
- Issue a warning order.
- Make a tentative plan.
- Initiate movement.
- Conduct a reconnaissance.
- Complete the plan.
- Issue the order.
- Supervise, refine, and rehearse.

CONDUCT HASTY OPERATIONS

2-73. Conducting hasty operations seems to be the norm rather than the exception. Unlike deliberate operations, hasty civil disorder operations are reactionary in nature with little or no time for planning. In these situations, the event of a crowd gathering is already underway. There is very little, if any, advanced warning of the unfolding situation, and commanders are usually put in the position of sending their soldiers into an already volatile, and perhaps hostile, environment.

2-74. Commanders should be in immediate and constant communication with local civil and police authorities. The purpose is to gather as much information as possible about who is involved, where they are assembling, what incident promoted the activity, and what seems to be the prevailing attitude of the assembling crowd.

2-75. It is essential that commanders operating in environments where civil disturbance operations are likely ensure that their soldiers receive training at all levels. Commanders and subordinate leaders need to instinctively rely on the same TLP that they would in a deliberate operation. See *FM 3-19.4*.

CONDUCT TACTICAL OPTIONS

2-76. In the peacekeeping and peace enforcement environment, the use of conventional firearms or the threat of their use may not be the solution to a situation where US forces must separate two belligerent, hostile ethnic groups or prevent a similar group from entering an area that is off limits to them.

2-77. Crowd control options are often combined. Commanders choose their options based on an evaluation of the particular crowd. Commanders select any combination of control techniques and force options they think will influence the particular situation (METT-TC). Commanders must always try to choose the response that can be expected to reduce the intensity of the situation.

Monitoring

2-78. Monitoring an assembled crowd consists of gathering necessary intelligence and watching them to determine progress and development. It is this gathered information that helps commanders select the appropriate actions. Gathered information can often be used to help defuse a situation through persuasion.

2-79. Monitoring is continuous. Without timely information, responses could cause the situation to escalate beyond what it should have or cause a response to be inadequate based on the situation.

2-80. Commanders may task teams with specific missions to monitor crowd activity and note any new developments. Monitoring done by these observation teams gives the commander up-to-the-minute information so he can gauge crowd activity and intent in relation to the overall situation.

2-81. Observation teams monitor crowd activities to gather information. They observe and report on crowd size, location, and mood and on the developing situation. An observation team may consist of a marksman, a radio operator, and an observer equipped with binoculars. They may be posted strategically on rooftops and other high terrain that overlooks the crowd. Sometimes observers use helicopter-mounted observation devices. This also affords security for the operational forces. Commanders must know where their observation teams are located so they are not mistaken for teams of snipers.

2-82. The timely flow of information may allow the commander to influence the outcome of the situation with simple negotiations. Monitoring is appropriate when more decisive action is not feasible because of crowd size or when the intensity of a situation might escalate. It is particularly useful in large, nonviolent demonstrations. Monitoring can serve as an interim measure until more control forces arrive. It includes establishing communication with crowd leaders to convey official interest and intent to the crowd. Monitoring also includes efforts to gain the cooperation of crowd leaders.

2-83. Communication with crowd leaders and participants can help a commander control a situation without more severe measures. If communications exist with crowd leaders, the authorities may be able to divert the leaders or the crowd from their stated or apparent goal. Pressure can be put on leaders to channel the crowd into an area that minimizes disruption to the community and aids crowd control operations. March routes and demonstration areas can be limited to those that will help contain the crowd and reduce their potential for disrupting the community. Pressure can be positive (offering concessions) or negative (with deterrents).

2-84. If the commander can gain the cooperation of crowd leaders, it can decrease the potential for disorder, and if crowd leaders seek cooperation from authorities, officials should try to be accommodating. Crowd leaders can be placed in liaison positions between the crowd and the control force. Crowd leaders can be made responsible for managing the crowd by policing their own activities.

2-85. Taking still pictures or videotaping the faces of individuals within a crowd may prevent or reduce unlawful and violent acts. This could lessen their sense of anonymity. If needed, photographs or videotapes can be used as evidence for prosecution. To be effective, crowd members must see their presence being recorded. The photographer or cameraman should be in uniform to let the crowd know who he or she is. Recorders must be close enough to the crowd to be seen easily, but not close enough to be endangered.

Blocking

2-86. It is not uncommon for protests and demonstrations against the US or its forces to occur outside the main entrances to military installations or US Embassies. Because of the sensitive nature of equipment and information that can be found at these locations, blocking unauthorized entry is critical.

2-87. Blocking is the physical denial of advance upon a facility or an area by a crowd. Commanders may have to task their forces to block a crowd. Blocking physically denies crowd advance. Crowd control formations, particularly line formations, along with barricades can be used to block advancing crowds. Barricades of vehicles, concertina wire, and water-filled barrels are used to block or channel the movement of the crowd.

Dispersing

2-88. Dispersing is taking deliberate actions to fragment an assembled crowd in order to prevent the destruction of property or prevent injury. It is extremely effective against smaller crowds in congested urban environments. However, this may increase and spread lawless activity rather than reduce it. Thus, the commander's forces must control dispersal routes and the areas in which the dispersal will occur. Forces must protect the facilities that could be considered likely targets of opportunity for dispersing individuals or groups. Dispersal of the crowd may require apprehension of small groups still active in the area. Proclamations, shows of force, crowd control formations, and RCAs can disperse crowds.

2-89. Avoid (if possible) the dispersion of crowds into wide-open areas because it gives the crowd the opportunity to grow rapidly in size. The direction of crowd movement is difficult to anticipate and channel.

2-90. Once the crowd has started to disperse, it may occasionally be necessary for the forces to advance, keeping the crowd on the move and in the right direction. A crowd on the move should not be hurried to avoid panic. At no time should the crowd be cornered in a position where there is the perception of no escape. This invokes the "fight-or-flight" syndrome, possibly escalating violent activity.

2-91. Issuing a proclamation (see *Figure 2-2,* page 2-20) can help disperse a crowd. Proclamations officially establish the illegal nature of crowd actions and put the populace on notice that the situation demands special measures. Proclamations prepare the people for the presence of military authority and it tends to inspire respect. Proclamations support law-abiding elements and psychologically bolster military forces trying to restore order. They also demonstrate the gravity of the situation to all concerned and are excellent ways to make a commander's intentions known to a crowd. It is also a good way to reduce crowd size before direct action is taken. Commanders can make a verbal proclamation similar to the following statement: "*In the name of the President of the United States, I command that you disperse and retire peaceably to your homes.*" Such a proclamation may even make direct action unnecessary.

I. DISPERSAL

Unlawful Gathering

"Attention! Attention! This area must be cleared at once! Further unlawful behavior will not be tolerated. Clear this area at once or the necessary force to do so will be used."*

Demonstration

"Disperse and retire peaceably! Disperse and retire peaceably! Attention all demonstrators!"

"The demonstration in which you are participating ends at _______. The permit that was agreed to by the leaders of the demonstration expires at that time. All demonstrators must depart from the _______ by _____. All persons who wish to leave voluntarily may board the buses. These buses will go to the _____. Those who wish to take buses should move to ______. Those demonstrators who do not leave voluntarily by ______ will be arrested and taken to a federal detention center. All demonstrators are urged to abide by the permit."*

Warning of Looting

"Return to your homes! Someone may be looting them at this moment! During a disturbance, criminal activity is at its peak. Your family or your property may be in danger."*

II. EMPLOYMENT OF TROOPS

Employment

"Attention! Attention! Troops are present in this area. They are preparing to advance. Order must and will be maintained. Disperse peaceably and leave this area. To avoid possible injury, leave at once."

"Disperse now and avoid possible injury! Disperse now and avoid possible injury!"*

(Repeat until troops are committed.)

When Troops Are Committed

"Troops are advancing now. They will not stop until this crowd is dispersed and order is restored. To avoid injuries, leave the area at once. Return to your homes as peaceful citizens. Troops have their orders, and they will not stop until the crowd is dispersed. Do not get hurt. Leave this area."*

III. PRESENCE OF CHILDREN

(Used in conjunction with other announcements.)

"Attention! Attention! Do not attempt to cause further disorder. Disperse now in an orderly manner and avoid possible injury to children. Return at once to your homes."*

IV. ADDITIONAL INSTRUCTIONS

*Indicate the method, streets, and direction that the crowd should use when dispersing.

Figure 2-2. Sample Proclamation

2-92. In making a proclamation, a commander may consider imposing a time limit. However, the situation may change, and not imposing a time limit leaves the commander free to choose other actions when he wants. A proclamation must be specific in its instruction.

2-93. If a time limit is stated, it must be a reasonable length of time for the crowd to comply with the instructions. When drafting a proclamation, the commander must consult closely with the SJA. He must use the simplest language possible to maximize the effects of the proclamation. If proclamations must be translated to a local language, the translation must be made with great care.

2-94. The commander at the scene may direct that a proclamation be issued over public address (PA) systems. The force of the words used in the proclamation must be gauged to the composition of the crowd. If the crowd consists of normally law-abiding citizens who are presently assembled to show disagreement with an existing situation, the proclamation requires less force. On the other hand, if the crowd consists of militant rioters, the proclamation requires more force. The text may take a number of forms, depending on the situation.

2-95. A show of force is often a useful measure for dispersing a crowd. When troops arrive, the psychological impact of their arrival can be used. Soldiers can dismount from the helicopters, buses, or trucks in plain sight of the crowd, but they must be far enough away to prevent a provoked attack of thrown objects. The first echelon to dismount from the vehicle secures the surrounding area.

2-96. When small groups are scattered throughout a large disturbance area, a show of force can be made by marching troops conducting motor marches, conducting saturation patrolling, and setting up static posts. Sometimes marching well-equipped, highly-disciplined soldiers in view of a crowd may be all that is needed to convince them to disperse and retire peaceably. On the other hand, a show of force may attract people to an event. It may also provoke a nonviolent crowd into a violent confrontation. Intelligence can help best in preparing a show-of-force response.

Containing

2-97. Containment is the process of limiting a crowd to the area they are presently occupying. It is a suitable option when the crowd must be prevented from spreading to surrounding areas and communities. Additionally, it prevents those outside the already assembled crowd from joining the gathering. Containment is useful when apprehensions become necessary for preventing those in the crowd from escaping. Crowd control formations, perimeter patrols, and barriers are effective methods to accomplish containment. In all instances, caution must be used to avoid the "fight-or-flight" syndrome common to people feeling trapped with no escape.

2-98. Armored vehicles, armored security vehicles (ASVs), and up-armored high-mobility multipurpose wheeled vehicles (HMMWVs) are adaptable to roadblock operations as they can serve as barriers. They also provide added protection for the soldiers inside and those outside manning the roadblocks because they provide an easily accessible barrier for them to crouch behind.

2-99. Military vehicles traveling at close intervals in a column formation next to a crowd are largely a psychological barrier (see *Chapter 6*). They can be used to contain a large, fast-moving crowd. The moving cordon creates a temporary obstacle between the crowd and the line beyond which they will not be allowed to cross. A well-trained driver in a mobile cordon can do a better job than dismounted soldiers. In order to execute safe, quick reversals of direction for mobile cordons on narrow roads, the soldiers executing this formation must be highly-trained vehicle operators.

2-100. By-the-number commands issued over vehicle radios are the most common method for coordinating a cordon movement. Vehicles equipped with PA systems can also prove effective to issue both commands to troops and directives to the crowd. The cordon speed is no more than 5 miles per hour with an interval of at least 20 feet. Blue and red lights, sirens, blinkers, and horns may or may not be used. Armored vehicles can also serve as mobile command posts. When used properly, vehicles provide security, communications, and mobility.

DECIDE TO CONTAIN OR DISPERSE

2-101. Current crowd control doctrine places emphasis on crowd dispersal. Forced dispersal may result in a crowd breaking up into multiple groups that scatter over a large area. This may pose even greater public order problems and may pose a continued threat to control forces. A crowd is often better controlled by means of containment (confining its activities to a given area). A crowd has a limited duration, and its numbers are likely to diminish as individual needs take precedence over those of the crowd.

2-102. During the G8 summit protest in Ottawa, Canada, in June 2002, several thousand people showed up to demonstrate against the evils of world corporate leaders. Canadian Police determined that their best response to this event was a "soft response." Throughout the two-day event, hundreds of officers in standard patrol uniforms operated in four-officer teams, prestaged throughout the city. As the demonstrators began to assemble and eventually took to marching, officers provided escorts both in the front and rear of the enormous crowd along with officers marching to the flanks of the demonstration. This "soft approach" frustrated demonstrators, especially those who had it within their agenda to cause property damage and personal injury. Even more frustrating to the demonstrators was the heavy downpour of rain they endured on both days. Their organizations quickly disbanded out of concern for their own creature comforts. In the end, 10,000 people participated in the protest, with only three arrests made in two days.

USE PATROLS

2-103. Alert, aggressive patrolling of the disturbance area deters the gathering of crowds. The use of saturation patrolling is most effective for this purpose. It allows information to be collected, and it creates the psychological impression of the control force being everywhere at once.

2-104. Standard military transport vehicles provide mobility and communications for area coverage. Soldiers must be deployed with enough vehicles to provide the flexibility to handle all situations in the disturbance area. The vehicles can be modified with sandbags, wire screens, or similar materials to protect against sniper fire and thrown objects. If sniper fire is expected or has been encountered, then up-armored HMMWVs or light armored vehicles should be used for patrolling.

2-105. Armored vehicles patrolling an area of violence provide an added psychological effect and allow troops to maneuver close to snipers to make an apprehension. They may also be used as rescue vehicles to extract hostages or people surrounded by a hostile crowd.

2-106. Varying patrol routes and times for mounted and dismounted patrols keep lawbreakers from being able to pick a safe place or time to act. Patrols are particularly useful in preventing overt arson and violence. Patrol members can also spot and promptly report fires. Whenever possible, military patrols are integrated with civil police patrols. Joint patrols conserve military forces and aid civilian military communications. They also help troops become familiar with an area quickly. In addition, the civil police are available to make apprehensions, if needed. Patrol members must practice proper standards of conduct and fair treatment of civilians at all times. They are performing an important community relation and control function.

2-107. Motorized patrols, because of their speed and mobility, provide timely reconnaissance and broad area coverage. Motorized patrols are in radio contact with TF headquarters (HQ). They can make periodic contact with foot patrols and stationary posts, while patrolling their own areas. Moreover, motorized patrols can respond quickly to calls for help from other patrols and guard posts. Motorized patrols should have at least three vehicles with three men in each. Strength in numbers is necessary for protection. When planning patrol routes, avoid areas where the patrols can be isolated or placed in jeopardy. In addition, motorized patrols are equipped with fire extinguishers to put out small fires, thus reducing the burden on the fire department.

2-108. Besides foot and motorized patrols, air patrols provide a third dimension in directing the overall control effort in the disturbance area. They can perform reconnaissance and surveillance and provide near real time formation over the disturbance area. They are an excellent means of providing timely information to the commander concerning the demonstration area and its surrounding perimeter. They can monitor the actions of rioters, the extent of damage, the status of access routes, locations and conditions of road barriers, and other important conditions.

USE DELAYING TACTICS

2-109. AOs that units are assigned to patrol are often larger than the unit has resources to manage effectively. In addition, units must manage taskings, contingency plans, and other commitments, which will quickly drain available manpower. This problem will manifest itself when factions threaten to riot or gather in large numbers to demonstrate. In order for the commander to stall for time while he shifts unit assets, he may want to delay the arrival of buses and vehicles full of demonstrators. This technique is based on the assumption that demonstrators are moving to the demonstration site by vehicle. Key and secondary routes into the demonstration site must be analyzed by the S2 when he does the IPB.

NOTE: There should be an overwatch when this technique is attempted.

2-110. As demonstrator buses and/or vehicles move toward the demonstration site, soldiers in two HMMWVs block the road just ahead of the approaching vehicles. Once the demonstrators dismount their vehicles and begin to close in on the soldiers, the soldiers leapfrog (see *Figure 2-3*) 1 kilometer farther down the road to repeat the process again. This is repeated as often as necessary. This process, if repeated enough times, should discourage the demonstrators enough that they give up and go home. It should also provide enough time to shift assets to the demonstration site.

2-111. Demonstrators are just as capable of blocking secondary and key routes leading into a demonstration or riot site to prevent relief forces from assisting with riot control as US forces are capable of preventing demonstrators access to the site. One must remember that demonstrators are often well organized and have handheld radios and other items. A great example of this was the 1999 demonstration against the WTO in Seattle, Washington. Police had erected temporary portable fences in the downtown areas to keep demonstrators from disrupting the delegates. Determined to be disruptive, the demonstrators relocated the temporary fences by not allowing police to enter the area. Police were forced to back patrol cars over their own fences to gain access.

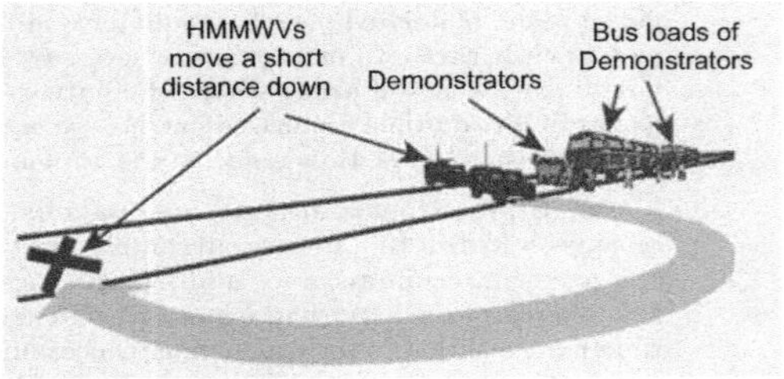

Figure 2-3. Delaying Tactics

Chapter 3

Apprehension, Search, and Detention

During the course of a civil disturbance, some members of the crowd may take part in unlawful activities, such as looting and assault. This will require the civil authorities or attached elements to apprehend, search, and detain people who are participating in the demonstration. Circumstances of the unfolding situation may require US military forces to be called upon to search, take custody of, and detain people who are participating in some form of protest and violating the law. Search, apprehension, and detention operations are conducted to halt these violations and to deter future violations.

LEGAL CONSIDERATIONS

3-1. All apprehensions are made by the civil police force, unless it is not possible for them to do so. Individuals must be known to have committed an illegal act, or there must be probable cause to believe that an individual committed such an act to be apprehended. If it becomes necessary for a military control force to apprehend or temporarily detain such violators, control force members often do so with the approval of the civil authorities. This ensures that all searches, apprehensions, and detentions (if necessary) are conducted within the parameters of the law.

3-2. Commanders operating overseas, similar to Bosnia and Kosovo, may find themselves without an established government or LEAs to work with. Commanders on the ground are then thrust into positions of authority, making them responsible for the actions, conduct, searches, apprehensions, and detentions performed by their troops.

3-3. If US military forces are called upon by the local authorities, certain policies have to be observed during search, apprehension, and detention operations. First, treat all people firmly but with reasonable courtesy and dignity. Remember that participating in a legal demonstration to express views is a right of all people, not just Americans in the US. The action, attitude, and behavior of US soldiers performing these operations are very important.

3-4. Our forces should never be seen as a military policing force on our own soil or as an occupying force in another country. Treating people with contempt, hostility, or excessive force increases the likelihood of resistance and violence. Searching people, placing them under apprehension, and detaining them without probable and just cause or without concern for their constitutional rights create problems and hinder due process. Numerous cases exist where authorities violated an individual's rights, resulting in the prosecution of police authorities or civil suits being waged against the officers and their leaders. This includes the military personnel assisting those officials.

3-5. It is very important that civil law enforcement (if available) be present and supervise all activities. Military personnel conducting a search or making an apprehension must carry out all the procedures carefully within the parameters of their training and the authority given to them. For each search and apprehension performed, the apprehension element must document the specific conduct of the person that violated the law.

3-6. Apprehending officers and supporting US servicemen must provide careful descriptive data for each subject. This description must be sufficient so that, at a later time, the suspected person can be clearly identified as the subject in question. Names and addresses of witnesses to the violation must be obtained. This information can be recorded on *Department of Defense (DD) Form 2708 (Receipt for Inmate or Detained Person).* Apprehension teams must retain and tag each item of physical evidence, such as weapons or stolen goods, that supports the apprehension on *DA Form 4137 (Evidence/Property Custody Document).* The team must give the owner a receipt for all property collected as evidence and retained.

TYPES OF SUBJECTS

3-7. There are two types of subjects (cooperative and uncooperative); they are as follows:

- **Cooperative.** Cooperative subjects actively and willingly comply with instructions. Cooperative subjects are usually compliant and offer no resistance to verbal directions or physical actions. However, if they provide any resistance to authority, it is usually passive. Some examples of this resistance are as follows:
  - Yelling.
  - Screaming.
  - Tightening the body.
  - Pulling away.
  - Letting the body go limp.
- **Uncooperative.** Uncooperative subjects will display preattack cues. Uncooperative subjects will actively resist being apprehended. Often, this resistance comes in the form of physical actions, such as—
  - Assaulting verbally.
  - Refusing to be handcuffed.
  - Refusing to be searched.
  - Attempting to push away, forcing a pursuit.
  - Wrestling with the apprehension team.
  - Striking out at the apprehension team with fists, hands and elbows, and feet.
  - Striking out at the apprehension team with weapons, such as knives, clubs, bats, bricks, rocks, or firearms.

APPREHENSION

3-8. US military forces will be in a supporting role to the local civilian authorities or acting as the control force in civil disturbance operations. When making apprehensions, it is best to use apprehension teams. These teams provide quick, organized responses to developing situations. Teams can be organized at the team, squad, or platoon level depending on the number of apprehensions expected. The intelligence information gathered helps to prepare for events where apprehensions may be necessary.

3-9. Each team (if possible) consists of a civilian police or apprehension officer, security element, and recorder. Team apprehension officers make the actual apprehension with the help of the security elements. If there is not a civilian police officer available to act as the apprehension officer, then it may be an officer or noncommissioned officer (NCO).

3-10. Ensuring that the apprehension process is properly conducted and documented, determine who is to be apprehended. Then inform the subject of why he is under apprehension and supervise his handling.

3-11. Security elements help the apprehension officer handle subjects. Security elements move, restrain, and search subjects under the supervision of the apprehension officer. The primary responsibility of the security elements is to watch the crowd and act as a blocking force, keeping the crowd from interfering with the apprehension. Recorders document the event, preferably by video, and complete *DD Form 2708*. If possible, recorders also photograph the apprehension officer with each subject. This aids in the identification process because it links the apprehension officer to the subject and provides the apprehension officer with documentation of the apprehension for use in court.

3-12. Electronically capturing the events as they unfold is essential in all civil disturbance operations. If personnel and equipment are available, videotape the events as they happen or, at a minimum, take still photographs to provide pictorial documentation that may be used later in court. Recording the scene before, during, and after the apprehension provides necessary evidence and can also help eliminate additional hostile and illegal activities. Having control force members take videos or pictures can help control crowd reaction. It reduces the feelings of anonymity that a crowd often enjoys and makes crowd members less prone to unruly or illegal acts.

3-13. Verbal commands given to the subject should be in a voice and manner appropriate for the purpose and the situation. Commands should be simple, concise, and in an understood language. Verbal commands should be convincing and convey an expectation that the instructions will be obeyed. With desperate, frightened people, sympathy combined with an attitude of firm assurance on the soldier's part may avoid additional violent behavior or trouble. An attitude of forceful authority may be needed with an uncooperative person. Commands given by soldiers must be done in a manner and context so that the subject has no confusion as to what the soldier is ordering them to do.

3-14. A subject will need to be transported from the area. In CONUS operations, civil authorities will usually be responsible for providing the transportation of a subject. Exceptions to this may be if the demonstration occurs on an installation. OCONUS operations may require US forces to provide vehicles, such as trucks, buses, or sedans, that will be used to transport a subject. Whenever possible, vehicles used to transport a subject should be modified with barriers to separate them from the drivers. If a barrier cannot be provided, a guard is to be placed in the vehicle. Passenger compartments are checked for items that could be used as weapons. When guards are required, the guard always sits with his weapon away from the prisoner. Prisoners must be placed where they can be best controlled, but they are not fastened to the vehicle with hand irons or Flex-Cufs. When seat belts are available they must be fastened. See *Soldier Training Publications (STPs) 19-95C1-SM and 19-95C24-SM-TG.*

3-15. Female escorts should always be used to transport a female subject. If female escorts are not available, measures must be taken to avoid false charges of molestation or sexual misconduct. Net control stations should be notified of the departure and arrival times and vehicle mileage readings (before and after). At least two male escorts must be present at all times when a female escort is not available. These procedures also apply when the male and female roles are reversed.

FLEX-CUFS

3-16. The two main positions used to place Flex-Cufs on a subject are the standing-supported and the prone.

3-17. When using the standing-supported position, the search man—

- Escorts the subject to a wall or other sturdy vertical surface (see *Figure 3-1*).

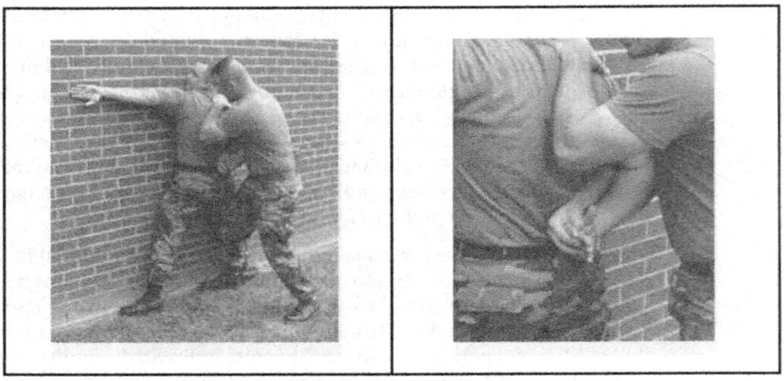

Figure 3-1. Standing-Supported Position

- Instructs the subject (verbally) to do the following:
  - Turn his head away.
  - Place his free arm (palm out, thumb down, and elbow locked) against the surface of the wall.

- Spread his feet until instructed to stop.
- Place his heels against the wall.

- Uses the leg closest to the controlled arm to pin the subject's leg (above the knee) to the wall.
- Reinforces verbal commands with pain compliance techniques if the subject fails to follow instructions.
- Maintains a double 90° hold on the controlled arm with his inside hand. With his free hand, the search man grasps the middle portion of the Flex-Cuf, keeping one cuff up and one cuff down.
- Places the first cuff (bottom cuff) on the subject's controlled arm, maintaining wrist manipulation (see *Figure 3-2*).
- Reaches through the free cuff with his outside hand and instructs the subject to place his free hand in the center of his back. With his outside hand, the search man takes the subject's free hand and completes the restraint by securing the cuff. He ensures that there is at least one-finger distance between the cuff and the subject's wrist (see *Figure 3-3,* page 3-6).

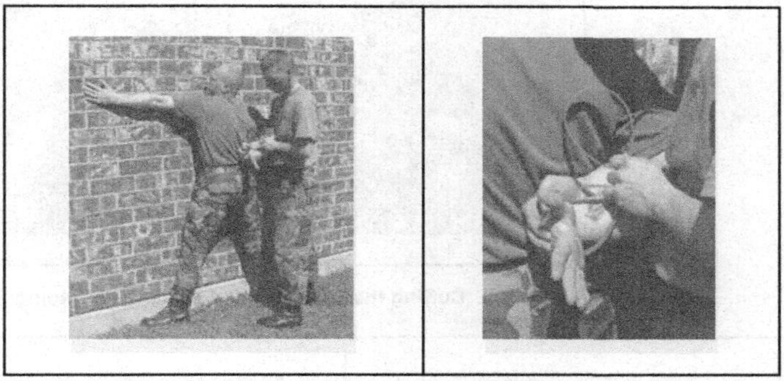

Figure 3-2. Cuffing the First Hand From the Standing Supported Position

3-18. When using the prone position, the search man—

- Grasps the Flex-Cuf with his outside hand, maintains tension on the subject's arm, and places the first cuff on the subject's controlled wrist (see *Figure 3-4,* page 3-6). Next, the search man secures the cuff, keeping at least a one-finger distance between the cuff and the wrist.
- Reaches through the free cuff with his outside hand and instructs the subject to place his free hand in the center of his back (see *Figure 3-5,* page 3-7).
- Reinforces verbal commands with pain compliance techniques if the subject fails to follow instructions. The search man bends the subject's controlled arm (using the natural break of the elbow as a hinge) toward the hand in the middle of the subject's back. With the outside hand, the search man takes the subject's free hand and completes the restraint by securing the cuff.

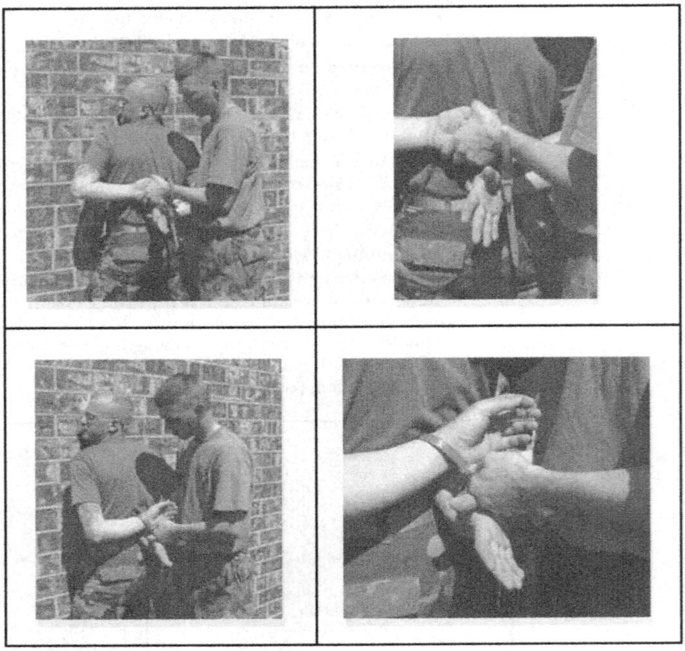

Figure 3-3. Cuffing the Second Hand From the Standing Supported Position

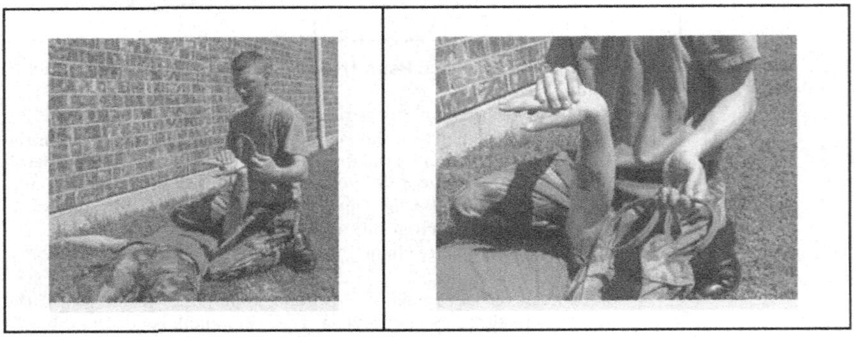

Figure 3-4. Cuffing the First Hand From the Prone Position

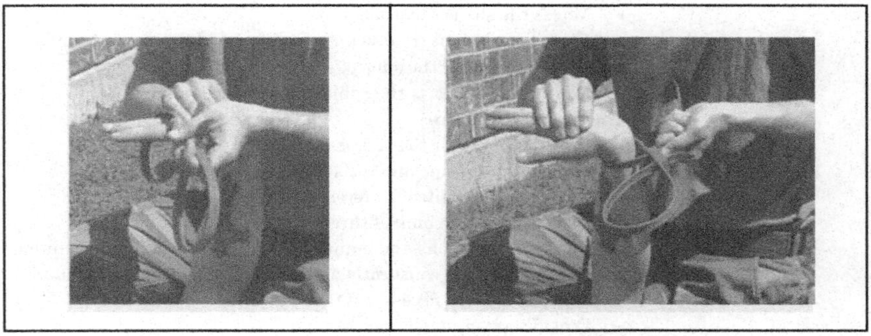

Figure 3-4. Cuffing the First Hand From the Prone Position (continued)

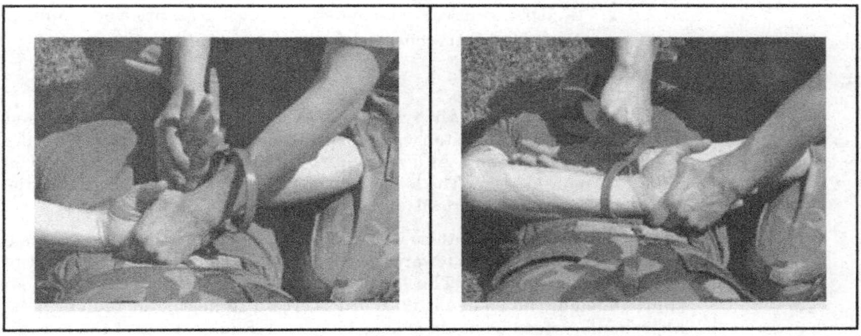

Figure 3-5. Cuffing the Second Hand From the Prone Position

SEARCHES

3-19. Initial contact with the subject is the most dangerous time during any apprehension; therefore, initial contact must be a swift and decisive team effort. Do not give the subject time to think or react. Distract his attention when possible. Apprehension teams are comprised of a search man, threat cover man, and security man (if necessary). Their responsibilities are as follows:

- The search man—
  - Issues verbal commands.
  - Conducts searches.
  - Handles radio communication.
  - Holsters his weapon before approaching a subject and informs the threat cover man that he is holstering his weapon.
  - Performs control procedures and handcuffs the subject.

    - Moves the subject to a safe area.
    - Removes weapons (if any) and sensitive evidence from the subject.
    - Disengages after the subject is restrained.
    - Controls and escorts the subject to a detention area.
- The threat cover man—
    - Provides cover for the search man.
    - Discourages escape and resistance.
    - Discourages hostile interference.
    - Alerts the search man of threats.
    - Resists distractions and maintains target acquisition or scanning.
    - Assesses threats constantly and prepares for the unexpected.
- The security man (if necessary)—
    - Provides cover for the search man.
    - Discourages hostile interference.
    - Alerts the search man of threats.
    - Resists distractions and maintains target acquisition or scanning.
    - Assesses threats constantly and prepares for the unexpected.

Search Principles

3-20. A subject should always be put at a disadvantage and searched immediately, but circumstances may dictate postponing the search until a safer place can be found. At a minimum, a quick pat down can be done en route to a safer location. The longer the apprehension takes, the greater the risk of drawing a hostile crowd and violent acts toward the control force.

3-21. Crowds often sympathize with a subject and try to help him or escalate the activities of the crowd toward violence. A subject's actions may also incite the crowd to violence. The longer a subject is the focal point of an apprehension, the easier it is to stir a crowd to hostile or violent acts. Apprehension teams should choose to move the subject out of the crowd's view. For the safety of the team, always apply Flex-Cufs or handcuffs and escort the subject to an area where a search can be done safely.

3-22. Search team members never operate alone; searches are conducted by a minimum of two team members. Initial steps must be taken to put the subject at a physical disadvantage. Before conducting a search (regardless of rank, sex, position, age, or physical condition), a team member applies the handcuffs or Flex-Cufs. Verbal commands must be short, clear, and distinct.

3-23. A search is conducted on everyone that has been apprehended to ensure that the team and those being apprehended are safe. The search man must abide by the following procedures when conducting a search:

- Search in a clear, secure area (when possible).
- Be ready to counter any resistance or an assault by the subject. Consider the subject armed until proven (by search) otherwise.
- Employ the appropriate level of force immediately, based on the actions of the subject.

- Do not step over the subject when he is at a disadvantage on the ground; go around him and remain out of sight.
- Do a 360° search of the subject's waistline for weapons. Search obvious and inconspicuous areas. Inconspicuous areas such as the groin, armpits, chest, and small of the back must be searched on both men and women; however, do not linger in these areas.

NOTE: Conduct same-gender searches when possible. If mixed gender searches are necessary for speed or security, conduct them in a respectful manner and avoid any action that could be interpreted as sexual in nature. To prevent allegations of sexual misconduct, the apprehension officer shows careful control of the soldiers who perform mixed-gender searches.

- Maintain security.
- Brief subjects quickly on what to do and who to follow.
- Remember that when conducting a search, a subject's body can be divided for a more efficient and thorough search. For example, search from head to toe or from side to side.

NOTE: The search man announces the items found and secures weapons and contraband.

- Military forces (when supporting civil authorities) may perform searches of people and property that are not usually subject to military law during a civil disturbance.
- Remember that it is always better for civilian police to search civilians and civilian property (when possible).
- Remember that if military forces are called upon to support police it is usually a last resort and there are not enough civilian officers available to accomplish the mission.
- Military personnel may search people incident to an apprehension.

3-24. The prescribed methods of searching someone are the frisk, the standing, and the prone. The method of search used largely depends on the situation and the mission. Searches are made not only of the subject and his personal effects such as purses, bags, and wallets, but also of the immediate surrounding area. This prevents the subject being searched from grabbing a weapon or destroying evidence. It may also be necessary for military forces to search private property, including vehicles, if there is—

- Reasonable belief that a person has committed or is committing a violent crime and is hiding in a building or a vehicle.
- Reasonable belief that a vehicle contains weapons or instruments of violence.
- Probable cause for searching a building or a vehicle.
- Probable cause to believe that (unless immediate action is taken) evidence of a crime will be destroyed before a warrant can be obtained.

NOTE: Search teams should always consist of two personnel—one to conduct the search, and the other to provide security and overwatch for the soldier conducting the search. Before being searched, a subject should always be restrained and put at a disadvantage by

securing his hands with handcuffs or Flex-Cufs. Ask the subject if there is anything in his pockets that could hurt the searcher, such as knives or needles.

Frisk Search

3-25. The frisk search (also known as the stop and frisk, pat down, and terry stop) is a search incident to a lawful stop. Subjects are not under apprehension at this point, but if it were an apprehension situation, handcuffs or Flex-Cufs would be applied first. Frisk searches are used when there is reasonable belief that a subject may be armed with anything that could be considered a safety threat.

3-26. Legal and regulatory circumstances for conducting a frisk search are covered in the *Manual for Courts-Martial (MCM)*. When a lawful stop is performed, the person stopped may be frisked for weapons when that person is reasonably believed to be armed and dangerous.

3-27. The primary purpose of a frisk search is safety. Frisk-search techniques are to be used only on a compliant subject. Frisk searches only cover the outer garments. Do not search the pockets of the individual unless a weapon is suspected. A frisk search begins when the search man informs the subject of the intent to search.

- During the preparation phase, the search man—
  - Tells the subject that he must perform a search for weapons for the subject's safety and his own.
  - Tells the subject to turn and face away from him and instructs the subject to keep his hands out to his sides where they are visible.

NOTE: The cover man providing security moves around the subject and positions himself out of reach and in front of the subject.

  - Instructs the subject to spread his feet until instructed to stop (more than shoulder width) and point his toes outward compromising his balance (see *Figure 3-6*).

NOTE: Do not kick at the subject's feet to spread them apart; it is unnecessary and unprofessional.

  - Tells the subject to bring both of his hands back to the small of his back with palms out, thumbs up, and the back of his hands together.
  - Instructs the subject not to move (see *Figure 3-7*).
- During the performance phase the search man—
  - Approaches the subject from behind (if possible) at a 45° angle, off center. The search man approaches with caution to ensure that his feet are to the outside of the subject's stance. See *Figure 3-8,* page 3-12.
  - Stays just inside arms reach and grips the subject's fingers with his support hand. The search man firmly grips at least one finger on each of the subject's hands (both thumbs, both index fingers, or an index and middle finger). An overhand or an underhand grip can be used effectively.

Figure 3-6. Approaching the Subject to Perform a Frisk Search

Figure 3-7. Giving Instructions to Subject Before Frisk Search

NOTE: The primary purpose of gripping the fingers is to control the subject's hands while performing the frisk search (see *Figure 3-9*, page 3-12). If the subject is going to resist or become aggressive, he will attempt to free his hands first. If the subject attempts to pull his fingers from the search man's grip, the search man must respond instantly by increasing his grip. If the grip on the fingers is lost, the search man should immediately push away from the subject and create a safe distance. This will allow the search team to respond with the necessary force to regain control of the situation.

Figure 3-8. Performing a Frisk Search

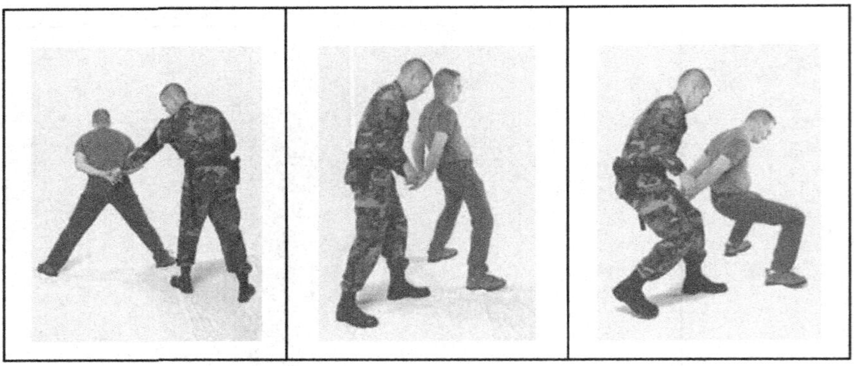

Figure 3-9. Gripping the Subject's Fingers During a Frisk Search

- Ensures that the subject always looks away from him.
- Places his strong hand on the outside of the subject's elbow (similar to the handcuffed escort position).

NOTE: The search man searches one side (the front, back, or side) at a time from head to foot.

- Divides the subject (visually) straight down his middle and does not cross the line while conducting the search.

NOTE: The search man is just feeling or searching for possible weapons, so an open hand should be used.

- Works at arm's reach. It is the sense of touch not the sense of sight that is being used. While performing the search, search men should stay behind and to the side of the subject. This provides the search man with a tactical advantage and prevention from being struck by the subject's elbows or knees if he should begin to resist.

3-28. When searching below the subject's knees, the search man takes a half step back and to the outside. He squats down, but does not drop his knees to the ground.

3-29. If no weapons are found and the subject is to be detained, the hands are secured with Flex-Cufs and the subject is escorted to the holding area. If the subject is to be released, collect information (name and address) and escort him away from the crowd.

Standing Search

3-30. Safety is the primary concern when dealing with a subject under apprehension. Therefore, the search man must protect himself by restraining subjects under apprehension before conducting the standing search. All subjects under apprehension have the propensity to turn violent or hostile. Putting a subject at a physical disadvantage provides safety for the subject and the search man. It limits the subject's ability to hide, destroy, or remove any evidence he may have on him.

3-31. Apply Flex-Cufs to the subject quickly during the apprehension. It is impossible to predict what a subject will do when put under apprehension. A subject may initially be cooperative and compliant, but as the subject becomes aware of his situation he may be overcome with emotion and begin to struggle or pull away.

3-32. There is no legal or regulatory prohibition against opposite-gender searches. Common sense prevails and whenever possible, searches should be done by soldiers of the same gender to avoid obvious questions and complaints. Once a subject is put at a physical disadvantage by cuffing and is more easily controlled, it affords the necessary time to get someone of the same gender there to conduct the search.

- During the preparation phase, the search man—

NOTE: Standing searches always begin with the subject in the handcuffed escort position (see *Figure 3-10,* page 3-14).

  - Takes his outside hand (farthest from subject) and places it on the outside of the subject's elbow while his inside hand (closest to subject) has a reverse-handshake grip of the subject's hand. At this point, both hands are on the same arm (see *Figure 3-10,* page 3-14).
  - Ensures that his position is on the same side of the subject's body being controlled (behind and to the side). When changing sides, change hand position.
  - Instructs the subject to spread his feet until instructed to stop (usually more than shoulder width) and point his toes outward, compromising the subject's balance.

Figure 3-10. Handcuffed Escort Position

NOTE: Do not kick at a subject's feet to spread them; it is unnecessary and unprofessional. If the subject becomes noncompliant or aggressive, pain compliance is usually the best tool to use, but only use enough force to gain compliance.

- Uses his outside hand to conduct the search. His inside hand is always used to maintain control during the search.

- During the performance phase, the search man—

NOTE: Only one side (the front, back, and side) of the subject is searched at a time from head to toe (see *Figure 3-11*).

- Divides the subject (visually) (straight down the middle using the spine as the centerline). He does not cross this line while conducting the search. Searches must be systematic, from head to foot. If there is an interruption, it must be started all over again.
- Searches with an open hand, using a rub down or squeezing method. Do not only pat down like in the frisk search.
- Works at arms reach. It is the sense of touch, not the sense of sight that is being used. To prevent the possibility of being struck by an elbow or knee, the search man stays behind and to the side of the subject.
- Takes a half step back and to the outside when searching below the subject's knees. The search man squats down, but does not drop his knees to the ground. Maintaining positive control is essential throughout the search, so never the reverse handshake is never released.
- Change sides when satisfied that all items have been removed from the side in which the search started. Without losing positive control, the search man conducts the search on the other side of the subject.

Figure 3-11. Performing the Standing Search

NOTE: Changing sides is accomplished by moving the outside hand (the searching hand) to the subject's uncontrolled handcuffed hand located on the outside. The outside hand and forearm will cross over the top of the reverse-handshake grip of the inside hand (see *Figure 3-12,* page 16).

- Crosses over the reverse-handshake grip with the outside hand and grips the subject's uncontrolled handcuffed hand in a reverse-handshake grip.
- Releases (quickly) the first reverse-handshake grip moving that hand to the subject's elbow as he sidesteps to the side yet to be searched. The search man repeats the standing search technique used on the first side of the subject.
- Gives strong verbal commands and uses only the force necessary to control the subject if he resists or becomes aggressive at any time during the search.
- Removes all items found during the search to be evaluated as potential evidence or a weapon. Items not considered a weapon or of evidentiary value will be returned to the subject at a later time.

Prone Search

3-33. Safety is the primary concern when dealing with a person under apprehension. Therefore, the search man must protect himself by restraining subjects under apprehension before conducting this search. Remember that all subjects under apprehension have the propensity to turn violent or hostile. That is the reason a prone search would be necessary. After performing an arm bar takedown on a noncompliant subject, it is especially important for the search man to put him at a physical disadvantage to provide for the subject's safety and his own. By doing so, the subject's ability to hide, destroy, or remove any evidence is limited.

Figure 3-12. Performing the Standing Search (Changing Sides)

3-34. The safest way to handcuff or Flex-Cuf a noncompliant subject is in the prone position, when the search man has greater control of the subject. It is usually the safest position in which to conduct a search

3-35. Handcuff the subject quickly during the apprehension. It is impossible to predict what a subject will do when put under apprehension. Often a subject may initially be cooperative and compliant; however, as the he becomes aware of his situation, he may become overcome with emotion and begin to struggle or pull away.

3-36. There is no legal or regulatory prohibition against opposite-gender searches. Common sense prevails and whenever possible, searches should be done by the same gender to avoid obvious questions and complaints. Once a subject is at a physical disadvantage by cuffing and is more easily controlled, it affords the necessary time to get someone of the same gender there to conduct the search.

NOTE: The subject is put into a physical disadvantage when he is taken facedown to the ground using an arm bar takedown and cuffed.

- During the preparation phase, the search man—
  - Ensures that the Flex-Cufs are not too tight.
  - Rotates his entire body around the subject's shoulder and center at the outside of the subject's hip with his shin over the subject's shoulder and one knee on the ground. At least one knee should remain over the subject's back (near side), with the other on the ground.
  - Pins the subject down by pressing his knee on the side of the subject's back or down on his closest elbow if resistance or aggression is encountered. He simultaneously grips the subject's closest handcuffed hand, grabs it with a reverse handshake, and applies pain for compliance.

- Uses strong, forceful verbal commands, such as "Stop resisting, stop moving around."
- Ensures that the subject's feet and ankles are crossed.

CAUTION

Stay alert. If the subject uncrosses his feet and ankles, it is usually a sign of preattack.

- Removes all items found during the search so they can be evaluated as potential evidence. Items not considered a weapon or of evidentiary value will be returned to the subject at a later time.

- During the performance phase, the search man—

NOTE: The search man must be systematic from head to foot while conducting the prone search. If interrupted during the search, it is important to start from the beginning to ensure that nothing is missed. It is better to go over the same area two or more times than to miss something that can be used as a weapon.

NOTE: The search man searches only one side (the front, back, or side) of the subject at a time until that side is complete.

- Divides the subject (visually) straight down the middle, using the spine as the centerline.

NOTE: He does not cross this line while conducting the search.

- Searches with an open hand, using a rub down or squeezing method for best results. He does not just pat down the subject like in the frisk search.
- Works at an arm's reach, searching only the area of the subject's body that is not in contact with the ground. The sense of touch, not the sense of sight, is being used.
- Stops the search as the subject's knees are reached. He taps one of the subject's legs and firmly commands him to raise his foot up and back, allowing for an inspection of the bottom of his foot. He then searches from the knee to the ankle area. This is performed on both the subject's legs and feet.

NOTE: At this point, the entire back side of has been searched. What remains to be searched is the front side of the subject (the chest and front of the hips, waist, groin, and thighs).

- Pivots back and faces the subject. Using the hand closest to the subject's head, the search man reaches across the subject's back and firmly grips the shoulder, rolling the subject up on the his side toward him to expose the areas not yet searched.
- Rolls the subject back down on his chest once all areas have been searched.

3-37. The search man must be completely satisfied that he has found everything. He must be especially sensitive to areas where weapons are often concealed like the waistband, the small of the back, ankles, pockets, shoes, the tongue of the shoes, and the groin area.

NOTE: Before removing subjects from the area, a brief summary must accompany them. The summary must include the subject's name, what he is suspected of doing, the time and date of the incident, and who saw him do the action.

DETENTION

3-38. Civil authorities must provide adequate detention facilities for all subjects. Authorities must be prepared to detain large numbers of people. They may choose to expand existing detention facilities or set up temporary facilities to accommodate the extra load. If possible, large-scale arrests are delayed until sufficient detention facilities have been set up.

3-39. If US military forces are committed to support local authorities, commanders should coordinate with them to ensure that adequate detention facilities are available and to learn their locations and capacities. If there are more detainees than civil detention facilities can handle, civil authorities may ask the military to provide support by setting up and operating temporary facilities. Army correctional facilities cannot be used to detain civilians. A temporary military detention facility can be set up if—

- Federal troops have been employed according to provisions.
- The TF commander has verified that available civilian detention facilities can no longer accommodate the number of prisoners awaiting arraignment and trial by civil courts.
- The Army Chief of Staff has granted prior approval.

3-40. Use of the temporary facility ends as soon as civil authorities can take custody of the detainees. Military forces are responsible for the custody, health, comfort, and sustenance of all detainees in their facilities until custody is transferred to civil authorities. Temporary facilities cannot be used to confine people arraigned or convicted in civil courts.

3-41. Women are detained in their own facility. Temporary facilities are supervised and controlled by officers and NCOs trained and experienced in military correctional operations. Females guard females. If two female guards are unavailable, use two male guards or one male and one female guard (see *FM 3-19.40*).

3-42. The same operational procedures that apply to the management of an installation confinement facility (see *FM 3-19.40*) apply to the management of temporary detention facilities, except for training, employment, and administrative discipline. Guards and support personnel under direct supervision and control of military officers and NCOs need not be trained or experienced in military correction operations. They must be specifically instructed and closely supervised in the proper use of force, custodial procedures, and the completion of military and civilian forms and reports that may be used. Soldiers that may be tasked to operate a temporary detention facility should be familiar with the forms and reports used for civilian apprehensions too.

3-43. Temporary facilities are set up on the nearest military installation or on suitable property under federal control. Ideally, the facility should be close enough to the disturbance area to minimize transportation and escort needs. However, it should be far enough away not to endanger those being detained. Whenever possible, existing structures are adapted for this use, but construction may be needed to provide segregation for ensuring effective control and administration.

3-44. The basic structure must include the following:

- Search areas.
- Holding areas for incoming men and women.
- A processing area.
- Holding areas for men and women who have committed misdemeanors.
- Holding areas for men and women who have committed felonies or are violent.
- A holding area for detainees' property and evidence obtained during a search.
- A holding area for administrative support and records.
- A medical station.
- Latrines.

For more information on these structures see *FM 3-19.40.*

3-45. Facility personnel must ensure that proper sanitation is maintained. When large numbers of people are detained or processed through a facility, sanitation becomes a problem. Medical personnel must conduct regular health inspections to detect unsanitary practices and conditions.

3-46. Facilities must be organized for a smooth flow of traffic. Processing stations must be set up so there is a linear or circular sequence of movement. These stations are set up out of sight of the holding areas. If possible, they are separated by a door to reduce noise. To reduce distractions, each station may be partitioned. Detainees may be more cooperative if they are out of sight of each other.

3-47. Injured people are given prompt medical treatment and transportation to medical facilities when necessary. A medical aid station for screening detainees and treating minor injuries is set up inside or next to the detention facility. Treatment areas, however, must be out of sight of the processing and holding areas. If possible, access to the medical facility bypasses the holding and processing areas. Facility personnel may also consider setting up a separate holding area for injured detainees.

3-48. Detention facility operation plans must contain emergency procedures. As a minimum, the following areas must be addressed:

- Fire evacuation.
- Disturbance control.
- Facility security.

NOTE: Depending on the situation, other emergency procedures may be needed.

3-49. Upon arrival at the detention facility, the detainee is logged in and searched, regardless of whether a complete search has already been conducted by the apprehending team in the field. Separate search areas are set up for men and women. Weapons, contraband, flame-producing devices, suspected evidence, money, and high-value items are confiscated. Medications are also confiscated. Medical personnel screen detainees that have had medications confiscated. Receipts are provided for any property or evidence that is confiscated. Confiscated items are tagged, and the items are stored in a controlled-property area.

3-50. As a detainee is brought to the facility, a file is initiated. Assigned detainee case numbers are used on all paperwork, such as logs, evidence tags, reports, and visual documentation. All paperwork (including photographs) that began outside the detention facility is marked with the case number. Facility personnel may also use hospital identification tags. Using indelible ink, the case number is written and attached to the tag on the detainee's wrist. Different colors may be used to identify different subject classifications, such as misdemeanors, felonies, or violent offenders. If opposing factions are involved, procedures are established to ensure that members of opposing factions are not processed together or detained in the same holding area.

3-51. After a detainee has been searched and classified, he is taken to a processing station where his paperwork is processed. If subjects passively resist by going limp, they may be moved by wheelchair. This reduces the number of escorts needed and the fatigue to facility personnel. Paperwork is reviewed to ensure that all information is complete, including charges, witnesses, and reasons for the apprehension.

3-52. Equipment such as height charts, scales, fingerprinting equipment, and cameras must be available for completing the police report and identifying the subject. Detainees may use aliases and not carry identification, so pay special attention to obtaining as much information as possible about the detainee's physical characteristics.

3-53. After processing is complete, the detainee is placed in the appropriate holding area. Paperwork for the detainee is forwarded to the administrative section. Files are reviewed for completeness and to determine the disposition of the detainee. Information from the file may be placed in a computer to find repeat subjects. Computers may be used to file criminal information only.

3-54. Custody transfers and release procedures must be coordinated with civil authorities and the appropriate legal counsel. Every effort must be made to arraign subjects quickly. The purpose of a detention facility is not to keep people off the streets, but to aid in processing subjects through the legal system. To speed up the release process, planners should consider issuing citations or subpoenas for minor offenses.

3-55. Military force members may have to respond to writs of habeas corpus. Writs are court orders addressed to a prisoner's custodian. Such a writ directs the custodian to bring the prisoner to court to determine the legality of the prisoner's apprehension and detention. Custodians must ensure that all documentation concerning the case is brought to court with the detainee.

3-56. Military personnel must obey writs issued by federal courts. For writs issued by a state court, the custodian or his legal advisor should respectfully reply that US authority is holding the prisoner. The Office of the SJA can answer any questions and explain the correct procedures.

Chapter 4

Riot Shield and Riot Baton Techniques

A soldier may be required to be part of a riot control formation where his primary weapon is a riot baton and his only protection a riot shield. A riot baton in the hands of a well-trained soldier is an invaluable weapon in a crowd control situation and may also be the most appropriate weapon. A riot baton is a very versatile weapon that can be used as an offensive and/or defensive weapon. The riot shield offers protection from the top of the head to just below the knees.

OVERVIEW

4-1. Use of the riot baton and riot shield is based on METT-TC and the commander's choice-of-force option. Formations committed to a riot with riot batons must have lethal overwatch, lethal weapons, and NLW employed within the formation and reserve forces with additional NL and lethal capabilities. These forces are positioned to rapidly reinforce the control force formation.

4-2. During the termination phase of the disturbance (when the violence has subsided and is expected to remain that way), soldiers are placed in a defensive posture. They may be equipped with a riot shield and a riot baton to perform their tasks. The riot shield is primarily a defensive weapon, which may be used in an offensive mode. It is primarily used as an offensive weapon when the formation is in close contact with rioters. The riot baton can also be used when formations are on the offensive and in contact with the crowd. However, crowd control elements employed with riot shields and riot batons must have lethal weapons and NLW available.

WARNING

Riot control formations are never employed with riot shields and riot batons only. There must always be lethal overwatch elements in position and lethal and NL capabilities within the riot control formation with a ready reserve force armed with lethal and NL capabilities.

RIOT SHIELD

4-3. In the hands of a properly trained soldier, the riot shield has proven to be the first line of defense for a formation. The riot shield may be used as an offensive weapon when in contact with an aggressive crowd.

4-4. The riot shield may be held with one arm. It is secured to the weak side arm with a self-sticking strap system and a handle. This system is designed to make the riot shield an extension of the soldier's arm. The riot shield is held at a slightly inward angle. This allows for debris thrown at the soldier to fall harmlessly to the ground (see *Figure 4-1*).

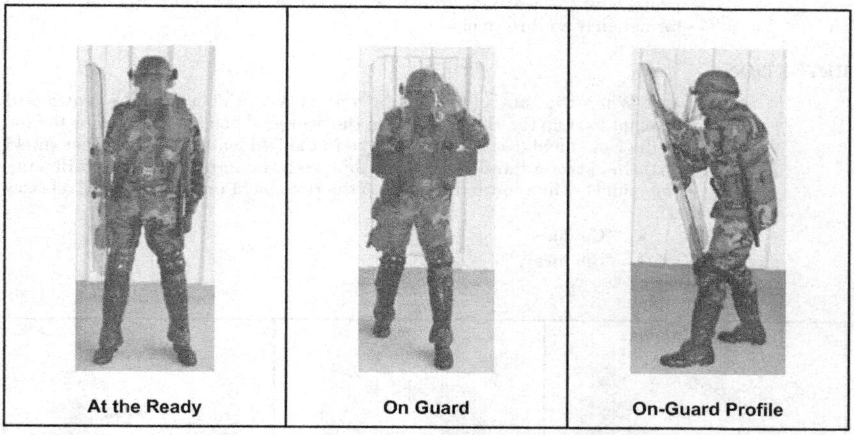

Figure 4-1. Riot Shield Positions

RIOT SHIELD CARRIES

4-5. The two basic riot shield carries are as follows:

- **At the ready.** The at-the-ready position is used while in a column formation or on the move toward hostilities. This position is designed for rapid movement. It should not be used when in contact with a crowd.
- **On guard.** The on-guard position is used when the riot control formation is in contact with the crowd or contact is eminent. It is used primarily in the following formations:
  - Line.
  - Wedge.
  - Echelon.
  - Diamond.
  - Circle.

DEFENSIVE AND RETENTION TECHNIQUES

4-6. The primary purpose of the riot shield is for defense. However, each riot shield holder must be proficient in riot shield retention techniques.

DEFENSE

4-7. The first line of defense for the formation is the riot shield. It is designed to withstand strikes from objects thrown by the crowd. In addition to protecting the riot shield bearer, it also protects other members of the formation from thrown objects.

RETENTION

4-8. When the control force is in contact with the crowd, the crowd will attempt to strip the riot shield from the bearer. When a subject grabs the top of the riot shield (see *Figure 4-2*), the riot shield holder slaps the riot shield with his strong hand near the subject's hands and gives the following commands (while continuing to slap the riot shield until the subject releases it):

- "Get back."
- "Get away."
- "Stop."

Figure 4-2. Riot Shield Retention Technique (Top Attempt)

4-9. If the riot shield is grabbed from the bottom (see *Figure 4-3,* page 4-4), then the riot shield holder forcefully drops to one knee, pinning the subject's fingers and/or hands to the ground. This must be done quickly and with force.

RIOT BATON

4-10. A riot baton in the hands of a properly trained soldier is a formidable weapon. However, it must be used in conjunction with other measures to be most effective. Soldiers must be trained with the riot baton to the point that its various techniques are automatic to them. This training must also include

Figure 4-3. Riot Shield Retention Technique (Bottom Attempt)

learning the vulnerable points of the human body so they can avoid areas that may cause permanent injury or death when struck.

> **WARNING**
> **The riot baton is never raised above the head to strike a subject in a club fashion. Not only is it likely to cause permanent injury; but it also gives an unfavorable image of the control force. The soldier is also vulnerable to an attack on his rib cage when his arm is raised.**

TARGET AREAS OF THE HUMAN BODY

4-11. The target areas of the human body are divided into three areas that represent the primary-, secondary-, and final-target areas. These areas are color coded as green (shown as gray in *Figure 4-4*), yellow (shown as white in *Figure 4-4*), and red (shown as patterned gray in *Figure 4-4*).

Primary-Target Areas

4-12. The primary-target areas are those areas which, when struck or restrained, cause the least amount of trauma to the body. These include the meaty or muscular areas of the body, such as the—

- Foot, shin, and instep.
- Inside and outside of the thigh.
- Lower abdominal region.
- Fore- and upper-arms.

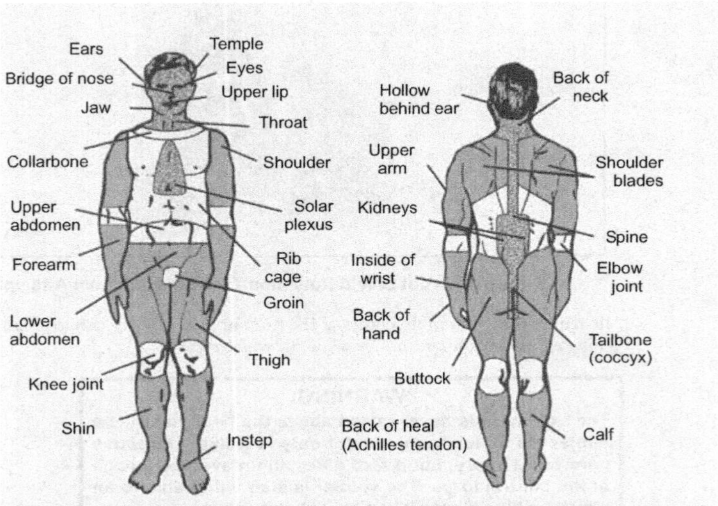

Green (lowest level of trauma) Primary-target areas:

Reasoning: Minimum level of resultant trauma. Injury tends to be temporary rather than permanent; however, exceptions can occur when applying as a restraint technique.

NOTE: When performing a block with a riot baton, the whole body is a green area, except for the head, neck, and spine.

Yellow (moderate level of trauma) Secondary-target areas:

Reasoning: Moderate level of resultant trauma. Injury tends to be more permanent, but may also be temporary when applying as a striking technique.

Red (deadly force) Final-target areas:

Reasoning: Highest level of resultant trauma. Injury tends to be permanent rather than temporary and may include unconsciousness, serious bodily injury, shock, or death.

Figure 4-4. Escalation of Trauma Chart

Secondary-Target Areas

4-13. The secondary-target areas are those areas of the body which, when struck or restrained, cause a moderate level of trauma to the body. This type of trauma tends to be longer lasting because the time to heal is longer than with other parts of the body. If these areas are struck with a riot baton, serious injury may occur. However, some of these areas are used for control or restraining holds that tend to lessen the threat of injury. Some secondary targets are the—

- Collar bone and shoulder blades.
- Elbows and knees.
- Upper abdominal region.

Final-Target Areas

4-14. The final-target areas are those areas which, when struck with a riot baton, cause serious trauma that tends to be fatal or critical. However, there are some areas that (when used for restraining only), will not cause death or serious injury. Some final-target areas are the—

- Front and back of the head (eyes, ears, nose, upper lip, and hollow behind the ear).
- Neck and throat.
- Upper chest.
- Xiphoid process.
- Spinal column.
- Groin area.

JOINTS OF THE HUMAN BODY

4-15. The joints of the human body are often described as hinges and, when manipulated in a certain fashion, allow for better control of the subject being restrained, especially if he is aggressive. When applied correctly, minimal pressure is needed to control the subject. When a subject attempts to resist, he creates additional pain for himself. By anatomically locking out these hinges, the subject will be forced to comply. When restraining a subject, emphasis must be placed on locking out two of the three joints of the upper body or two of the three joints of the lower body. The joints are as follows:

- **Shoulder.** Apply pressure to the shoulder to prevent the subject from freely moving the upper portion of his body.
- **Elbow.** Apply pressure on the elbow or just above the elbow on the ulner nerve.
- **Wrist.** Apply pressure by pancaking the hand to twist or lock out the wrist. Apply pressure at the break of the wrist and push the subject's thumb toward a stationary object (the wall or the ground) to lock the hand out.
- **Hip.** Apply pressure to the hip to knock the subject off balance or alter his center of gravity.

- **Knee.** Apply pressure to the knee by pushing outward on it to cause the subject to be placed off balance, or use the knee as a hinge for pain compliance.
- **Ankle.** Apply pressure to the ankle by clasping and turning it in a counterclockwise manner to lock the hinge out.

TYPES OF RIOT BATONS

4-16. The Army is currently fielding two types of riot batons, wooden and expandable. The most common riot baton is the 36-inch hickory riot baton with thong. There is also the 24- to 36-inch expandable riot baton, which has been added to the nonlethal capabilities set (NLCS). Regardless of which riot baton the soldiers are equipped with, all techniques for blocking and striking are identical. Similar to the riot shield, the riot baton offers soldiers both offensive and defensive capabilities.

4-17. Soldiers must be properly trained in all blocking and striking techniques. Improper use of the riot baton by an untrained soldier has the potential for creating a greater problem than what already exists.

NOMENCLATURE OF THE RIOT BATON

4-18. The riot baton is divided into four basic parts: the long end, long portion, grip portion, and grip end (see *Figure 4-5*). Although the expandable and wooden riot batons differ, their nomenclature is quite similar.

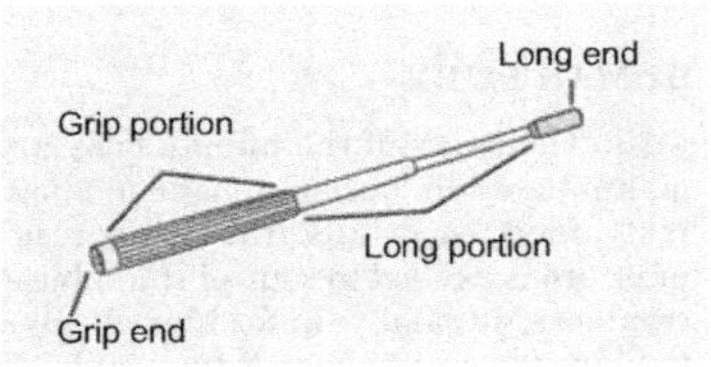

Figure 4-5. The Riot Baton

TYPES OF RIOT BATON CARRIES AND POSITIONS

4-19. As with any other weapon that the soldier will use, there are certain carries and positions from which the riot baton may be used offensively or defensively.

Two-Hand Carry

4-20. The two-hand carry is the most common carry for the riot baton. It is intended for executing all blocks and returning with strikes. When the expandable riot baton is in the collapsed position (at 24 inches), the long end of the riot baton is angled downward approximately 45°. When it is extended to 36 inches the long end is angled upward approximately 45°.

4-21. The wooden riot baton is secured with the soldier's hand by means of a leather thong. He hooks the thumb of his strong hand (palm down) in the thong, pulls the thong across the back of his hand, and grips the riot baton. This provides him with a very secure grip (see *Figure 4-6*).

4-22. The two-hand carry is intended for executing all blocks and returning with strikes. The soldier grips the riot baton with both hands. His strong hand grips the short end of the riot baton, and his weak hand grips the long portion of the riot baton, 2 to 4 inches from the end. His palm is down with his strong hand pulled in close against his strong side hip.

4-23. Wherever the long portion of the riot baton is pointed is where the soldier will strike or jab. The riot baton should be kept at a 45° angle to the ground. All blocks and two-hand jabs and strikes are executed from the two-hand carry.

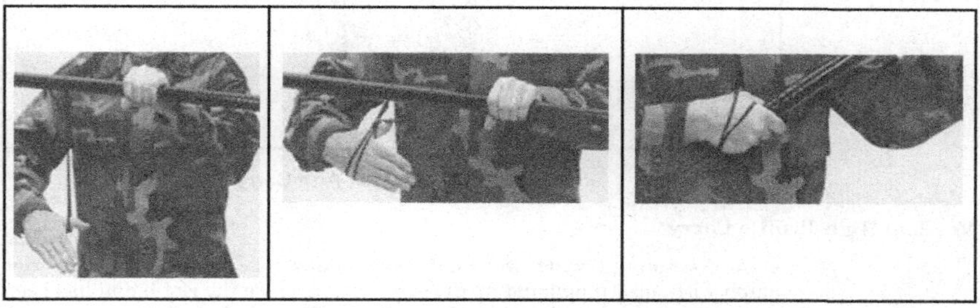

Figure 4-6. Two-Hand Carry

Outside-Arm Carry (Expandable Riot Baton Only)

4-24. The outside-arm carry (see *Figure 4-7*) is used with the expandable riot baton in the collapsed (24-inch) position only. Assuming a wide-based defensive stance (knees slightly bent and feet shoulder width apart), a soldier draws the grip portion of the riot baton with his strong hand. Without striking out or widely swinging the riot baton, he brings the long portion up against the outer part of his strong arm (the tricep area). The soldier's weak hand should be raised in a position to protect and block strikes from an aggressor. It is from this carrying position that a soldier can effectively use both the front and rear strikes. Front and rear strikes should never be performed with an expandable riot baton in the expanded position or with a 36-inch wooden riot baton.

CAUTION
The expandable and 36-inch wooden riot batons provide a greater risk of injury to the soldier's wrist.

Figure 4-7. Outside-Arm Carry

Vertical High-Profile Carry

4-25. Assuming a wide-based defensive stance, the soldier draws the riot baton with his strong hand by the grip portion. Once the riot baton has been drawn from the carrier, the soldier swings the riot baton in a downward motion, fully extending the arm and holding the riot baton down and parallel to the strong side leg. The soldier's weak hand is raised in a position to protect and block strikes from an aggressor (see *Figure 4-8*, page 4-10). Although this carry can be used with both riot baton systems, it is best suited for the expandable riot baton collapsed to 24 inches. The vertical high-profile carry clearly shows the soldier with a drawn riot baton in an effective defensive posture and ready for attack. A drawn riot baton in the hands of a trained soldier may offer just the kind of deterrence needed to stop the activity of an aggressor.

Vertical Low-Profile Carry

4-26. Assuming a wide-based defensive stance, the soldier draws the riot baton with his strong hand by the grip portion. Once the riot baton has been drawn from the carrier, the soldier swings the riot baton in a downward motion, fully extending his arm and holding the riot baton down and parallel to his strong side leg. His weak side hand is raised in a position to protect and block strikes from an aggressor (see *Figure 4-9*, page 4-10). Although this carry can be used with both types of riot batons, it is best suited for the expandable riot baton collapsed to 24 inches. This position portrays the soldier in an effective defensive position ready for an attack, but with a possible hidden weapon. The major difference between the vertical low-profile and high-profile carry is that in the low-profile carry the soldier hides the riot baton behind his strong side leg, allowing for some element of surprise to a potential aggressor. A drawn riot baton in the hands of a trained soldier may offer the deterrence needed to stop the activity of an aggressor.

Figure 4-8. Vertical High-Profile Carry

Figure 4-9. Vertical Low-Profile Carry

RIOT BATON BLOCKING TECHNIQUES

4-27. There are five basic types of riot baton blocking techniques that are executed from the two-hand carry position.

High Block

4-28. The high block is very effective in blocking a downward, vertical strike directed at the top of the head and shoulders. This type of downward, vertical strike may be from a club, pipe, or similar object. The high block is a two-step movement. To perform a high block the soldier performs the following steps:

Step 1. Raises the riot baton straight across his body until it is parallel to the ground (see *Figure 4-10*).

Step 2. Brings the riot baton straight-up in front of his body until the long portion of the riot baton is in front of his forehead (approximately 2 inches from his head). The riot baton should be in a horizontal position above his head and slightly angled down toward his weak hand, as it absorbs the shock from the strike (see *Figure 4-10*). The fingers of his weak hand should be open and behind the long portion of the riot baton when blocking the object to protect his fingers from being smashed by the strike.

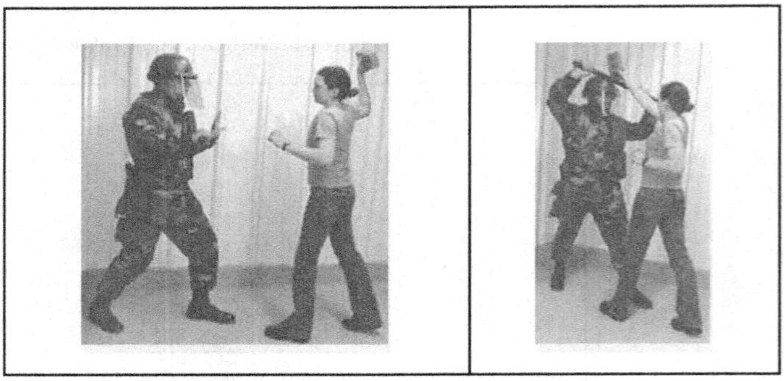

Figure 4-10. High Block (Steps 1 and 2)

Low Block

4-29. A low block is very effective in blocking an upward vertical strike directed at the groin, lower abdomen, chest, or chin. This upward vertical strike may be from an individual's foot, knee, or fist (for example, an upper cut to the chin). To perform a low block, the soldier performs the following steps:

Step 1. Raises the riot baton straight across his body until it is parallel to the ground (see *Figure 4-11*, page 4-12).

Step 2. Brings the riot baton straight down in front of his body, bending at his knees and keeping his body upright until the long portion of the riot baton is just below his knees (see *Figure 4-11*). The riot baton should be horizontal and parallel to the ground as it absorbs the shock from an upward strike. His weak hand fingers should be open and behind the long portion of the riot baton when blocking the object to protect his fingers from being smashed by the strike.

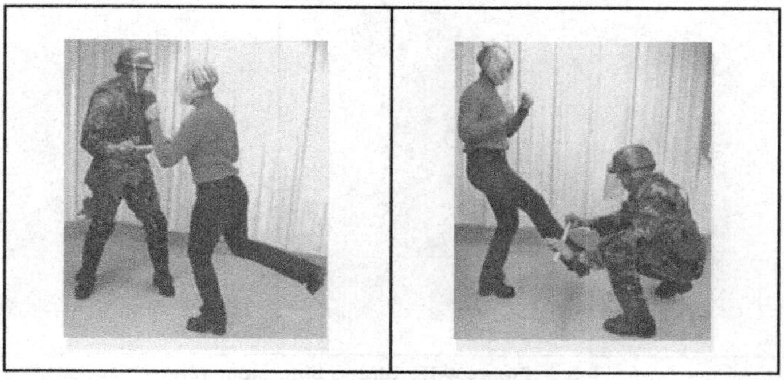

Figure 4-11. Low Block (Steps 1 and 2)

Strong Side Block

4-30. A strong side block is very effective in blocking a horizontal strike directed at the strong side of the head, neck, chest (or flank), or hip area. The horizontal strike may be from an individual's foot, fist, elbow, knee, or weapon. To perform a strong side block, the soldier performs the following steps:

Step 1. Raises the riot baton straight across his body in a vertical position where the grip portion is in his strong hand and the long portion is in his weak hand with the fingers straight up (see *Figure 4-12*).

Step 2. Turns his body laterally by rotating his hips toward his strong side and moving the riot baton across his body, keeping the riot baton vertical (see *Figure 4-12*).

4-31. The strong side block provides protection for a soldier from the front to the rear of his body. A riot baton held in the vertical position absorbs the shock of a strike coming from the strong side. The fingers of his weak hand should be open and behind the long portion of the riot baton, when blocking the object to protect his fingers from being smashed by the strike.

Weak Side Block

4-32. A weak side block is very effective in blocking a horizontal strike directed at the weak side of the head, neck, chest, or hip area. The horizontal strike may be from an individual's foot, fist, elbow, or knee. To perform a weak side block, the soldier performs the following steps:

Step 1. Moves the riot baton in a vertical position toward his weak side, ensuring that the grip portion is pointing straight down and the long portion is pointing straight up.

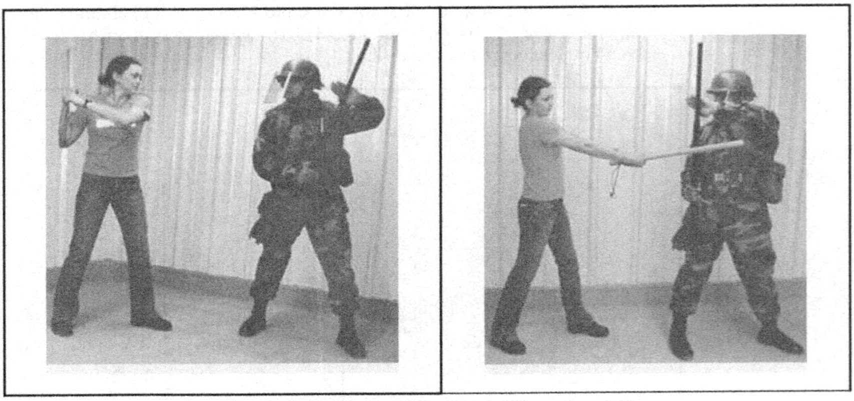

Figure 4-12. Strong Side Block (Steps 1 and 2)

Step 2. Keeps the fingers of his weak hand open and behind the long portion of the riot baton when blocking the object to protect his fingers from being smashed by the strike. At the moment of contact with the opposing strike, the blocking surface of the long portion should be at a right angle to the opposing object.

Step 3. Assess the situation following a weak side block, and take appropriate follow-up action, as needed.

Middle Block

4-33. A middle block is very effective in blocking a front horizontal strike directed at the face, throat, chest, abdomen, or hip. The horizontal strike may be from an individual trying to tackle near the waist or shove or punch at the face. To perform a middle block, the soldier performs the following steps:

Step 1. Moves the riot baton toward the middle of his body at a 45° angle, keeping the long portion pointed up and slightly forward of the grip portion (see *Figure 4-13*, page 4-14).

Step 2. Keeps the fingers of his weak hand open and behind the long portion of the riot baton when blocking the object to protect his fingers from being smashed by the strike. However, his weak hand may have to grip the long portion when pushing an individual away or forcing an individual down on the ground if the individual is trying to tackle.

RIOT BATON STRIKING TECHNIQUES

4-34. Seven strikes can be executed using the riot baton. Two of these strikes come from the outside-arm or vertical carry. The other five strikes come from the two-hand carry.

Figure 4-13. Middle Block

One-Hand Forward Strike

4-35. The one-hand forward strike is a very quick and effective offensive strike. It is usually employed as a countermeasure to an attack from the front that has been blocked effectively.

4-36. Target areas can be as high as the outside upper arm, down to the outer thigh region. An effective one-hand forward strike must be aggressively quick, with the purpose of distracting, disabling, or altering behavior. Therefore, those employing this technique must be sensitive to what part of the riot baton they use to strike the individual. Strike the target with the last 2 to 4 inches of the long end for optimal results.

4-37. A one-hand forward strike may be employed in one of the following three ways. When selecting available targets, use the principles in *Figure 4-4,* page 4-5.

- A horizontal manner.
- A downward diagonal manner to destabilize an individual.
- A vertical manner in front of the body to clear an aggressor's hands.

NOTE: When the one-hand forward strike is used in a vertical manner, keep the long portion straight up, not angled forward. By keeping the long portion up rather than angled forward, the chance of striking the individual in the face or head is significantly reduced. For the counterstrike to be effective, follow through with the technique when the baton makes contact with the aggressor and/or his object to destabilize the him. A pattern of movement such as a

forward shuffle, forward pivot, strong side step, or rear pivot may enhance this technique.

> **WARNING**
> **Do not execute the one-hand forward strike with the wooden or expandable riot baton in the extended position. It has been shown to cause injury to the wrist of the user.**

4-38. When using the vertical or outside-arm carry position (see *Figures 4-7,* page 4-9 and *4-8,* page 4-10), the soldier quickly moves the riot baton across his body using the strength and power of his hips, moving from his strong side to his weak side (see *Figure 4-14,* page 4-16). Ending this strike leaves the riot baton positioned under the weak side armpit in preparation for the one-hand reverse strike (see *Figure 4-14*). When delivering this strong side strike technique (in a horizontal or downward diagonal manner), ensure that the palm is facing up. In a vertical delivery, the palm will be toward the chest rather than up. Pause following the one-hand forward strike and assess the situation. Take appropriate follow-up action, as needed. If the situation does not call for additional strikes, avoid striking again.

One-Hand Reverse Strike

4-39. The starting position for the one-hand reverse strike is under the weak side armpit (see *Figure 4-14*). The one-hand reverse strike is used as a follow-up strike and is quick and effective.

4-40. Target areas can be as high as the outside upper arm, down to the outer thigh region. An effective one-hand reverse strike must be aggressively quick, with the purpose of distracting, disabling, or altering behavior. Therefore, those employing this technique must be sensitive to what part of the riot baton they use to strike the individual. Strike the target with the last 2 to 4 inches of the riot baton for optimal results.

4-41. A soldier may employee a reverse strike in one of the following three ways:

- A horizontal manner.
- A downward diagonal manner to destabilize an individual.
- A vertical manner.

4-42. In front of the body to clear an individual's hands. When the one-hand reverse strike is used in a vertical manner, keep the long portion straight up, not angled forward. By keeping the long portion up rather than angled forward, the chance of striking the individual in the face or head is significantly reduced. When selecting available targets, use the principles in *Figure 4-4,* page 4-5.

4-43. The one-hand reverse strike is considered a follow-up strike to the one-hand forward strike. As a follow-up strike, it is important that soldiers are taught the discipline of assessing the target. An aggressor may not require another strike, based on his condition and actions after enduring the one-hand forward strike.

Forward Strike	Target Assessment	Reverse Strike

Figure 4-14. One-Hand Forward and Reverse Strikes

4-44. When using the one-hand reverse strike, the solder quickly moves the straight baton across his body, using the strength and power of his hips (from weak side to strong side). When the strike is completed, the soldier's arm will be back in the outside-arm carry position (see *Figure 4-7,* page 4-9). The soldier ensures that his palm is facing down when delivering the counterstrike in a horizontal or diagonal manner. This will prevent injury to the user's wrist. In a vertical delivery, the palm will be toward the chest rather than up. Following the one-hand reverse strike, assess the situation and take appropriate follow-up action, as needed. If the situation does not call for additional strikes, the soldier should avoid striking again.

4-45. When the one-hand reverse strike is used in a vertical manner, keep the long portion straight up, not angled forward. By keeping the long portion up rather than angled forward, the chance of striking an individual in the face or head is significantly reduced. For the counterstrike to be effective, follow through with the technique when the baton makes contact with the aggressor and/or his object to destabilize him. A pattern of movement such as the forward shuffle, forward pivot, strong side step, or rear pivot may enhance this technique.

WARNING
This strike will not be executed with a 36-inch wooden or expandable riot baton in the extended position. It has been shown to cause injury to the wrist of the user.

Two-Hand Strong Side Horizontal Strike

4-46. Starting from the two-hand carry position, the two-hand strong side horizontal strike can be aggressively quick. The purpose of this strike is to create distance, distract, disable, or alter behavior. Strong side horizontal strikes are quick and effective offensive strikes, usually employed as a countermeasure to close frontal attacks that have been effectively blocked. When selecting available targets, use the principles in *Figure 4-4,* page 4-5.

4-47. The two-hand carry position enables soldiers to use riot batons at the 24- or 36-inch length. To effectively employ this strike, use the strength and power of the hips to thrust the riot baton in a horizontal manner. Simultaneously, pull back with the weak hand as the strong hand drives the grip end toward the target, striking a rib or the abdominal region of the aggressor (see *Figure 4-15,* page 4-18). Following the two-hand strong side horizontal strike, assess the target before following up with another strike. An aggressor may not require another strike, based on his condition and actions after enduring the one-hand forward strike.

NOTE: A pattern of movement such as a forward shuffle or forward pivot may enhance this technique.

Two-Hand Weak Side Horizontal Strike

4-48. An effective two-hand weak side horizontal strike must be aggressively quick to create distance, distract, disable, or alter behavior. The weak side horizontal strike is a quick and effective offensive strike usually employed as a follow-up strike to the strong side horizontal strike. It is a countermeasure designed to close frontal attacks that have been effectively blocked. When selecting available targets, use the principles in *Figure 4-4.* Maintaining a strong defensive stance after employing a strong side horizontal strike puts the soldier in position to execute a two-hand weak side horizontal strike.

4-49. Use of the two-hand carry position enables soldiers to use riot batons at 24 or 36 inches. To effectively employ this strike, use the strength and power of the hips to thrust the riot baton in a horizontal manner. Simultaneously, pull back with the strong hand as the weak hand drives the grip forward toward the aggressor, striking a rib or the abdominal region of the target (see *Figure 4-15*). Following the two-hand weak side horizontal strike, it is important to assess the aggressor. An aggressor may not require another strike, based on his condition and actions after enduring the one-hand forward strike.

NOTE: A pattern of movement such as a forward shuffle or forward pivot may enhance this technique.

Two-Hand Front Jab

4-50. The two-hand front jab starts from the two-hand carry position (see *Figure 4-16,* page 4-18). It is aggressively quick and creates distance, distracts, disables, and/or alters behavior. Front jabs are effective offensive strikes, usually employed as a countermeasure to charging and overpowering frontal attacks. When selecting available targets, use the principles in *Figure 4-4.*

Figure 4-15. Two-Hand, Strong Side, and Weak Side Horizontal Strikes

4-51. Use of the two-hand carry position enables soldiers to use riot batons at 24 or 36 inches. Use the strength and power of the hips to effectively employ this strike in a slightly downward or horizontal manner. Simultaneously, use both arms to shoot the long end of the riot baton straight out from the body to the target area. Immediately, pull back the riot baton to the two-hand carry position (see *Figure 4-16*).

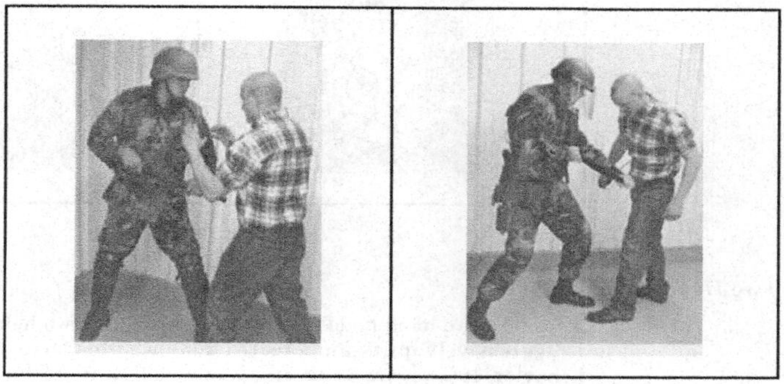

Figure 4-16. Front Jab

4-52. Forward jabs drive the long end of the riot baton toward the aggressor, striking a rib, a hip flexor, or the abdominal region (see *Figure 4-4*, page 4-5). Following the two-hand front jab, it is important that soldiers assess the aggressor before following through with another strike, as the aggressor may not require another strike.

NOTE: A pattern of movement such as a forward shuffle will enhance the power of this technique.

Two-Hand Rear Jab

4-53. Starting from the two-hand carry position, the two-hand rear jab is aggressively quick and creates distance, distracts, disables, or alters behavior. Rear jabs are quick and effective offensive strikes usually employed as a countermeasure to charging and overpowering attacks from the rear. When selecting available targets, use the principles in *Figure 4-4,* page 4-5.

4-54. Use of the two-hand carry position enables soldiers to employ riot batons at 24 or 36 inches. To effectively employ this strike, use the strength and power of the hips to thrust the riot baton in a slightly downward or horizontal manner. Simultaneously, turn the head to the rear to see and assess the target. Use both arms to shoot the riot baton straight back from the body to the targeted area and immediately pull back the riot baton to the two-hand carry position (see *Figure 4-17*).

NOTE: A pattern of movement such as a rear shuffle will enhance the power of this technique.

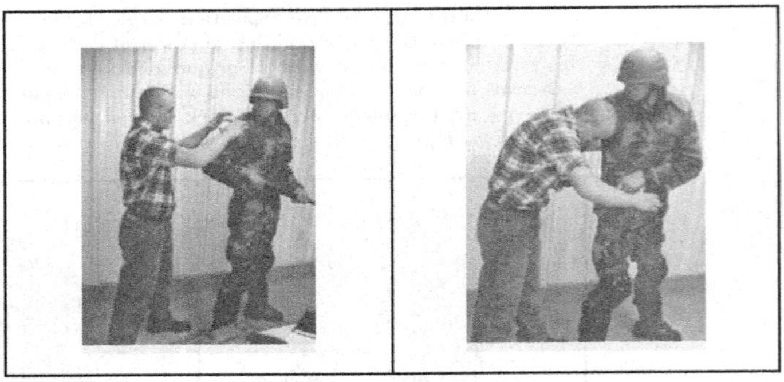

Figure 4-17. Rear Jab

Two-Hand Middle Strike

4-55. The two-hand middle strike starts from the two-hand carry position. It is aggressively quick and creates distance, distracts, disables, or alters behavior. It is proven to be an effective follow-up technique to a two-hand weak or strong side block or to destabilize an individual by pushing him back and away. The two-hand middle strike is used as a two-count movement. When selecting available targets, use the principles in *Figure 4-4.*

4-56. To perform a two-hand middle strike (see *Figure 4-18,* page 4-20) the soldier does the following:

Step 1. Brings the riot baton up to chest level and slightly off parallel to the ground.

Step 2. Steps forward aggressively, simultaneously thrusting the long portion of the riot baton forward in a horizontal manner. Both arms are fully extended at the end of the movement with a quick snappy return to the step 1 position. Use the entire body to generate the power.

NOTE: It is important that soldiers assess the aggressor before applying a follow-up strike. If the soldier steps forward as he employs the middle strike, it will afford greater impact and power.

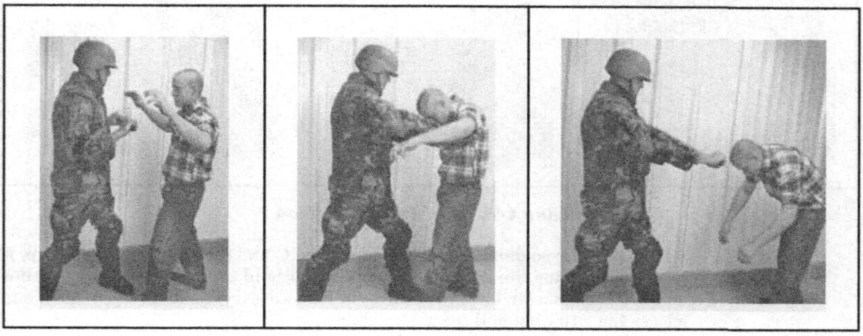

Figure 4-18. Middle Strike

Riot Baton Retention

4-57. A common technique an aggressor may use is grabbing for a soldier's riot baton or, in some cases, even taking the riot baton and using it against the soldier. This causes a disruption or distraction.

4-58. Maintaining physical control of the riot baton is essential for the safety of the soldier and the control force as a whole. Aggressors will generally attempt to grab the riot baton where they can get the strongest hold. Soldiers must resist the instinct of getting into a tug-of-war match over the riot baton, which often proves to be ineffective.

4-59. Using the momentum of the aggressor to pull the riot baton away, the soldier simply steps into or closer to the aggressor with the weak side foot. With the momentum created, the aggressor's hold on the riot baton is lessened. The soldier then rapidly uses the tracing-C technique to recover the riot baton.

4-60. A soldier using the tracing-C technique does the following:

Step 1. Pushes up on the riot baton.

Step 2. Pulls the riot baton around in an upward swing as if tracing a C in the air from bottom to top with the long end.

NOTE: This movement seems to wrap the aggressor's arms around themselves (see *Figure 4-19*).

Step 3. Drives (immediately with a quick and forceful movement) the long end of the riot baton straight down, as if slicing the C in half (see *Figure 4-19*).

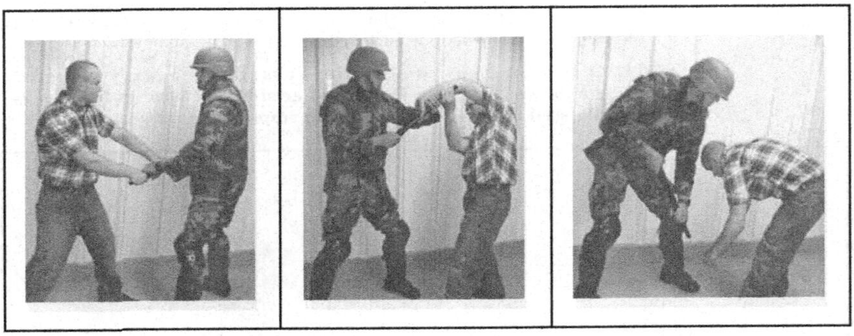

Figure 4-19. Riot Baton Retention

4-61. Once the soldier reaches the top of the C and the aggressor is tied up, it is impossible for the aggressor to maintain hold of the riot baton. Further actions such as a strike may or may not be necessary. Soldiers must be able to assess the situation and react accordingly.

WORKING AS A TEAM

4-62. Soldiers never have to rely solely on their own skills or abilities in riot control operations. Although individual skills were discussed in this chapter, it is important to remember that well-trained soldiers employing offensive, defensive, riot shield, and riot baton techniques will perform the collective task of crowd control well.

4-63. Riot control operations are dynamic and often chaotic. Soldiers armed with riot shields provide a first line of defense while those armed with riot batons and firearms provide overwatch as the formation moves toward engaging a hostile crowd. To work effectively and in concert, soldiers must be properly trained, drilled, and exercised. If an aggressor approaching a line of skirmishers violently grabs a riot shield, help is required to maintain control (see *Figure 4-20*, page 4-22). A soldier armed with a 36-inch wooden riot baton or an expandable riot baton (extended to 36 inches) can be of assistance. The soldier simply moves forward into a gap between the riot shields where he can quickly strike the top or side of the riot shield, slicing the long end across or straight down onto the hands and fingers of the aggressor (see *Figure 4-18*).

4-64. Riot batons are not the only way to effectively knock an aggressor's hands from a riot shield. Soldiers armed with 12-gauge shotguns or M16/M203s can just as effectively smash the hands or fingers of a persistent aggressor with the barrel of their weapons using the technique described above.

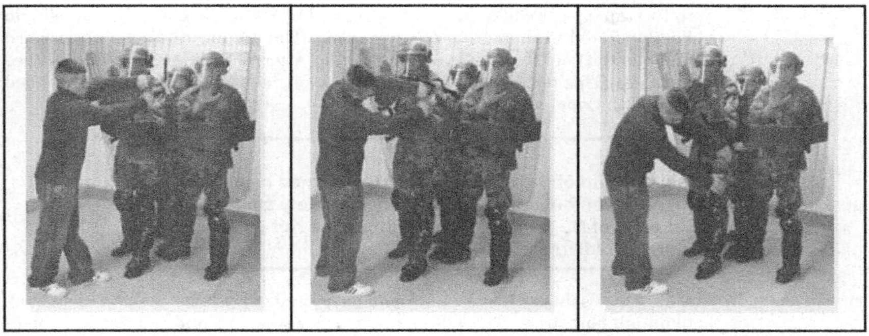

Figure 4-20. Clearing a Riot Shield

NONLETHAL MUNITIONS

4-65. With the inclusion of NL munitions, commanders now have additional tools that afford excellent standoff capabilities. This is critical because commanders now have intermediate options for dealing with a hostile crowd. No longer will a commander issue a proclamation, have it ignored by the crowd, and then be forced to use lethal munitions to backup the proclamation. Now he can use NL munitions to gain the same standoff distance but without the catastrophic results of lethal munitions.

4-66. NL munitions include (but are not limited to) a 12-gauge shotgun and 40-millimeter rounds. These blunt munitions have the capability of effectively engaging targets as close as 5 meters and as far as 50 meters. Range is dependent upon each round. These munitions are used from a distance to dissuade, discourage, and disperse the crowd. If the crowd disperses from 50 meters and soldiers are in close contact with the crowd, then the mission has been successful.

TAP-DOWN TECHNIQUE

4-67. Commanders must be familiar with the characteristics and capabilities of all rounds to use them effectively. When 12-gauge and M203 weapons are used within a line of skirmishers (usually located behind the riot shields), it is imperative that a method is developed, trained, and practiced to effectively operate as a team. Operators of these weapons may encounter problems in riot control situations when trying to identify and effectively engage legitimate targets. The tap-down technique is used to overcome target identification and engagement with NL munitions. If ordered to engage targets or the situation authorizes an engagement, soldiers armed with NL munitions may step forward, level their weapons, and fire upon legitimate targets (see *Figure 4-21*).

4-68. A soldier that is armed with a NLW and is using the tap-down technique must first tap on the shield holder's shoulder. Tapping on the shield holder's shoulder alerts him that the gunner is preparing to fire. In response to the tap, the shield holder drops to one knee while keeping the riot shield firmly affixed to their front for protection. The gunner takes a step forward and fills the gap. He then leans into the riot shield bearer's back with his knee and fires. This technique ensures that the long end of the weapon is extended beyond the riot shield.

WARNING

The long end of the weapon must be projected beyond the riot shield before firing. This is a critical safety step, which will prevent NL projectiles from hitting the riot shield and injuring soldiers.

4-69. Once soldiers firing munitions have fired, they raise their weapon back to a high port arms. The soldiers clear their weapons and tap the riot shield bearer a second time, signaling them to rise to a standing position.

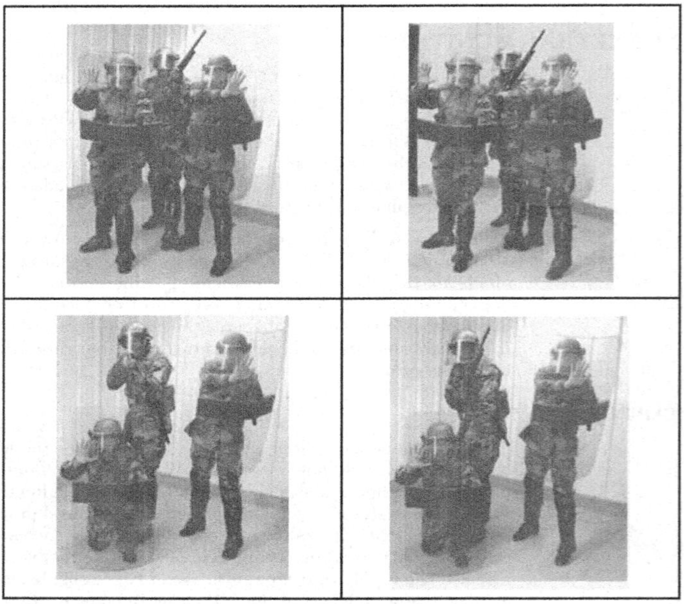

Figure 4-21. Tap-Down Technique

Chapter 5

Nonlethal Capabilities Set and Employment Considerations

During the past several decades, US forces have regularly performed peacekeeping and peace enforcement missions at a greater tempo than in the past. In the peacekeeping and peace enforcement environment, the use of conventional firearms or the threat of their use may not be the solution to a situation where US forces must separate two belligerent, hostile ethnic groups or prevent a similar group from entering an area that is off-limits to them. However, a show of force using NLW and NL munitions will cause crowds to disperse, separate, or leave the area with minimal casualities. The NLCS is a well-rounded, versatile package of both commercial off-the-shelf (COTS) and government off-the-shelf (GOTS) NL equipment and munitions. NLCS contents are divided into four distinct categories: personnel protectors, personnel effectors, mission enhancers, and training devices.

PERSONNEL PROTECTORS

5-1. Personnel protectors provide the individual soldier with added protection to the sensitive and vital areas of his body. They provide excellent protection for the individual soldier from trauma often inflicted by thrown objects (such as rocks, bricks, sticks, and bottles).

NONBALLISTIC RIOT FACE SHIELD

5-2. A nonballistic riot face shield (see *Figure 5-1,* page 5-2) has two adjustable positions and is constructed of hard durable plastic that securely attaches to the Kevlar® helmet. It may be worn with an M17 or M40 protective mask and is designed to protect the soldier's head, face, and neck from thrown objects. Nonballistic riot face shields will not protect from attacks with ballistic weapons, such as firearms. All soldiers operating in a civil disturbance environment should be outfitted with riot face shields.

NONBALLISTIC FULL-LENGTH SHIELD

5-3. A nonballistic full-length riot shield (see *Figure 5-2,* page 5-2) is a body shield constructed of hard durable plastic. Left- and right-handed personnel can easily use the nonballistic full-length riot shield. It is designed to protect the soldier's face, torso, and upper legs from thrown objects. Nonballistic full-length riot shields will not provide protection from ballistic weapons. Only the front line of skirmishers and those providing lateral protection to both flanks of the formation will be outfitted with nonballistic full-length riot shields.

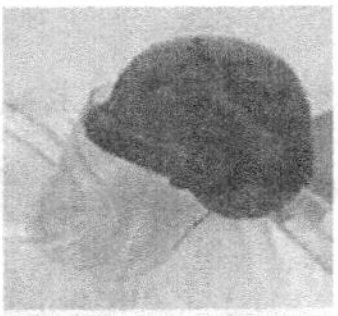

Figure 5-1. Nonballistic Riot Face Shield

Figure 5-2. Nonballistic Full-Length Riot Shield

NONBALLISTIC SHIN GUARD

5-4. The nonballistic shin guards (see *Figure 5-3*) are constructed of hard durable plastic and are easily fastened to a soldier's legs. They are designed to protect the soldiers, feet, shins, and knees from thrown objects. They will not provide protection from ballistic weapons, such as firearms. All soldiers operating in a civil disturbance environment should be outfitted with shin guards.

BALLISTIC RIOT FACE SHIELD

5-5. The ballistic riot face shield (see *Figure 5-4*) has two adjustable positions constructed of acrylic and bullet-resistant materials that provide the wearer with Level IIIA protection. The ballistic riot face shield easily attaches to the Kevlar helmet and is capable of being worn with either an M17 or M40 protective mask. It will protect the soldier from hand-thrown objects. Because of its excessive weight (3.4 pounds), the ballistic riot face shield should not be worn for standard riot control operations. Ballistic riot face shields are best suited for use within specialized teams associated with riot control operations; for example, teams responsible for the search and apprehension of identified criminals or building- and alley-clearing operations.

Figure 5-3. Nonballistic Shin Guards

Figure 5-4. Ballistic Riot Face Shield

BALLISTIC FULL-LENGTH RIOT BODY SHIELD WITH LIGHT KIT

5-6. The ballistic full-length riot body shield (see *Figure 5-5,* page 5-4) is constructed from Spectra Shield® bullet-resistant material that provides a soldier with Level IIIA protection. It will protect the soldier from thrown objects. Because of its excessive size (18 pounds and 20 by 36 inches), the ballistic full-length riot body shield should not be used as part of a standard riot control operation. Ballistic full length riot body shields are best suited for use within specialized teams associated with riot control operations; for example, teams responsible for the search and apprehension of identified criminals or building- and alley-clearing operations.

Figure 5-5. Ballistic Full-Length Riot Body Shield

BALLISTIC SHIN GUARDS

5-7. Ballistic shin guards (see *Figure 5-6*) are constructed of Kevlar KM2 materials and provide a soldier with Level IIIA protection to the feet, shins, and upper legs. They protect the soldier from hand-thrown objects. There are three different sizes, and because of their excessive weight (7 to 10 pounds), ballistic shin guards should not be deployed as part of a standard riot control operation. Ballistic shin guards are best suited for use within specialized teams associated with riot control operations; for example, teams responsible for the search and apprehension of identified criminals or building- and alley-clearing operations.

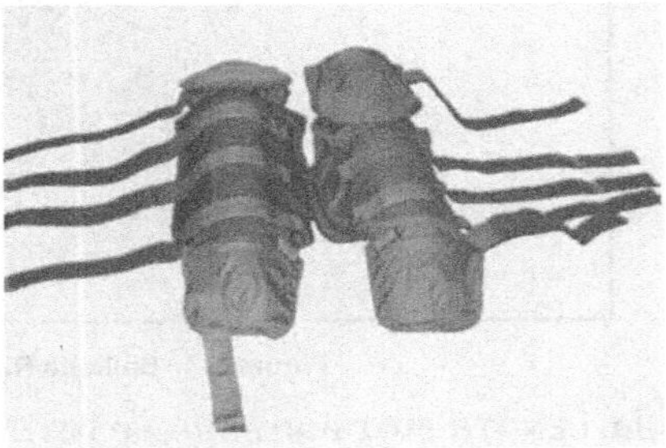

Figure 5-6. Ballistic Shin Guards

PERSONNEL EFFECTORS

5-8. Personnel effectors are those items that provide soldiers with an NL response without the necessity of always coming into direct physical contact with aggressors. Personnel effectors are 12-gauge and 40-millimeter ammunition that inflict blunt trauma. These munitions are compatible with the shotgun and the M203 grenade launcher already in the Army inventory. There are also diversionary and rubber ball grenades, which distract and inflict blunt trauma.

WOOD RIOT BATON WITH BELT RING (36 INCHES)

5-9. The 36-inch wood riot baton (see *Figure 5-7*) is made from solid oak. When held properly, the riot baton thong provides greater retention capabilities. Intended primarily for self-defense, the 36-inch wood riot baton can be used effectively as an offensive tool if it becomes necessary to keep rioters out of arm's reach of soldiers conducting crowd control operations.

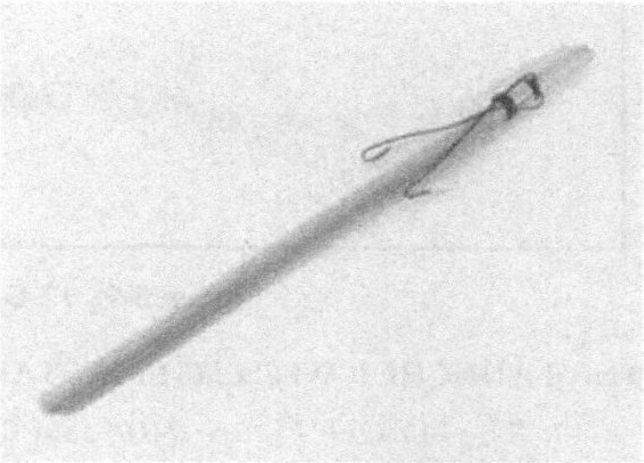

Figure 5-7. Wood Riot Baton

EXPANDABLE RIOT BATON WITH CARRIER (24 TO 36 INCHES)

5-10. The expandable riot baton (see *Figure 5-8*) provides the soldier with an NL means of crowd control and self-defense. The riot baton comes with a mounting device that attaches the riot baton to a belt. The riot baton is 24 to 36 inches in length and is primarily intended for self-defense. The expandable riot baton can be used effectively as an offensive tool if it becomes necessary to keep rioters out of arm's reach of soldiers conducting crowd control operations.

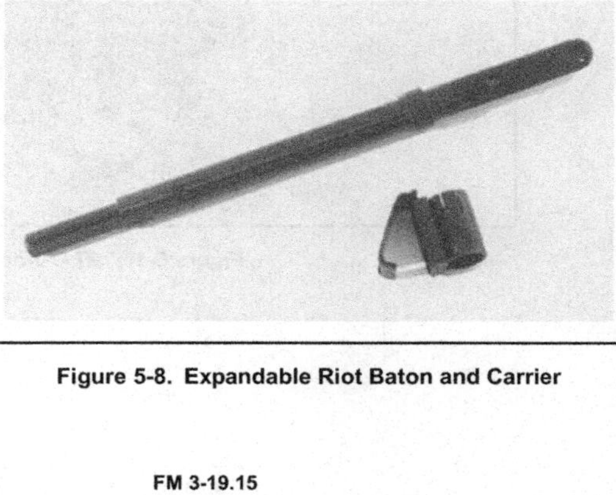

Figure 5-8. Expandable Riot Baton and Carrier

500 (3 INCH) 12-GAUGE SHOTGUN

5-11. The 12-gauge shotgun (see *Figure 5-9*) is a pump action shotgun currently in NLCS inventory. The pump action shotgun is chambered to take up to 3-inch shells. The 3-inch chamber allows for the use of M1012 and M1013 NL munitions. This shotgun also provides a visually distinct alternative to standard military weapons that may be desired based on mission considerations (force signature and acceptability).

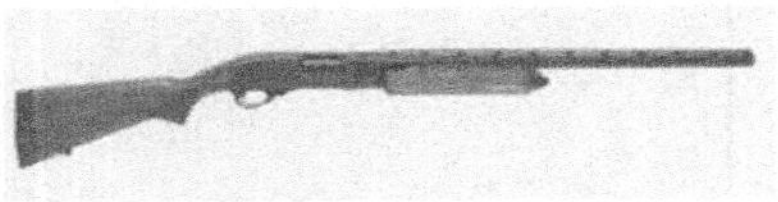

Figure 5-9. 12-Gauge Shotgun

M7 66-MILLIMETER LAUNCHER WITH M315 INSTALLATION KIT

5-12. The M7 (see *Figure 5-10*) is a 66-millimeter vehicle-mounted, NL, grenade-launching device that is mounted on a HMMWV. It is an indirect fire support system that can deliver the M98 distraction grenade that creates a flash-bang effect, L96A1 antiriot grenade, or M99 blunt trauma grenade that creates a sting-ball effect. The M315 installation kit is used to install an M7 discharger on the turret ring of appropriate HMMWV variants. An adjustable bracket allows the launch angle to be depressed for engaging targets at ranges of 50, 75, and 100 meters. The system enforces standoff distances and deters potential threats.

Figure 5-10. M7 Launcher

MODIFIED CROWD CONTROL MUNITION-GROUND EMPLACEMENT

5-13. Similar in nature to the claymore mine, the modified crowd control munition-ground emplacement (MCCM-GE) (see *Figure 5-11*) is an NL munition that can deliver six hundred .32-caliber rubber pellets and a flash-bang effect. The MCCM-GE will stop, confuse, disorient, and/or temporarily incapacitate area targets and/or personnel at ranges out to 30 meters.

WARNING
Avoid employing the MCCM-GE within 10 meters of personnel as it may cause serious bodily harm or death.

Figure 5-11. MCCM-GE

12-GAUGE, FIN-STABILIZED POINT, NONLETHAL CARTRIDGE (M1012)

5-14. The M1012 (see *Figure 5-12*) is a single projectile round made of hard rubber that is shaped like a bomblet and designed to be fired a at a single target. With a muzzle velocity of 500 feet per second, the M1012 has the effective range of no closer than 5 meters and no further than 30 meters. Engagement inside 5 meters could cause serious injury or death. Beyond 30 meters the kinetic energy dissipates to the point where the round becomes ineffective.

Figure 5-12. M1012

12-GAUGE, CROWD DISPERSAL, NONLETHAL CARTRIDGE (M1013)

5-15. The M1013 (see *Figure 5-13*) is a multiple projectile round with .23 caliber hard rubber pellets that is designed to be fired at and employed with the purpose of affecting multiple targets. With a muzzle velocity of 900 feet per second, the M1013 has an effective range of no closer than 5 meters and no further than 30 meters. Engagement inside 5 meters could cause serious injury or death. Beyond 30 meters, the kinetic energy dissipates to the point where the rubber pellets become ineffective.

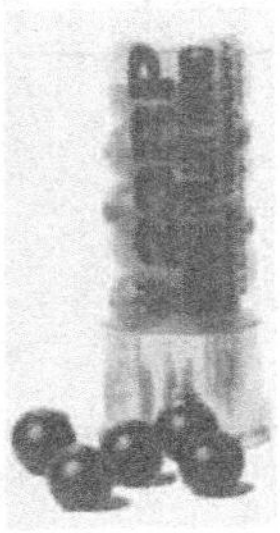

Figure 5-13. M1013

40-MILLIMETER SPONGE GRENADE (M1006)

5-16. The M1006 (see *Figure 5-14*) is a single projectile round made of pliable foam rubber with a hard plastic back. The M1006 is designed to be aimed and fired at a single target. With a muzzle velocity of 265 feet per second, the M1006 has an effective range of no closer than 10 meters and no further than 50 meters. Engagements inside 10 meters could cause serious injury or death. Beyond 50 meters, the kinetic energy dissipates to the point where the round becomes ineffective.

Figure 5-14. M1006

40-MILLIMETER CROWD DISPERSAL CARTRIDGE

5-17. The crowd dispersal cartridge (CDC) (see *Figure 5-15*) is a multiple projectile round with .48 caliber hard rubber balls that is designed to be fired and employed with the purpose of affecting multiple targets. With a muzzle velocity of 450 feet per second, the CDC has an effective range of no closer than 10 meters and no further than 30 meters. Engagement inside 10 meters could cause serious injury or death and beyond 30 meters the kinetic energy dissipates until the rubber balls become ineffective.

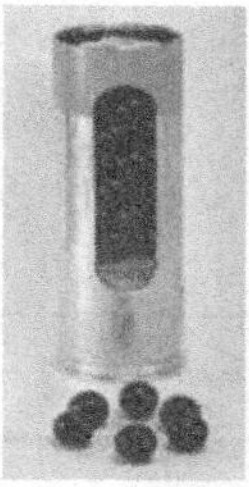

Figure 5-15. CDC

RUBBER BALL GRENADE (GG04)

5-18. The GG04 (see *Figure 5-16,* page 5-10) is a multiple projectile, flash-bang grenade with 100 .25-caliber hard rubber pellets. Each grenade has a fuse delay of 2.8 to 3 seconds with a flash measuring approximately 1 million CP and 180 decibels at 3.5 feet. At detonation, rubber pellets are dispatched at 360° with an effective range of 2 to 3 meters and a maximum engagement range of 15 to 20 meters. The rubber ball grenade is designed to be hand-thrown or muzzle-launched from a 12-gauge shotgun.

RUBBER BALL GRENADE LAUNCHING CUP AND AA30 12-GAUGE LAUNCH ROUND

5-19. The launch cup (see *Figure 5-17,* page 5-10) firmly attaches to the 12-gauge shotgun without any additional tools. It works in concert with the AA30 12-gauge launch round to propel the rubber ball grenade approximately 100 meters. The shooter simply inserts the full body and the safety lever of the grenade into the launching cup, pulls the pin, loads a 12-gauge round, and fires (approximately) at a 30° angle. The desired effect is to have the grenade explode approximately 7 to 10 feet above the target.

RELOADABLE FLASH-BANG GRENADE (M84)

5-20. The M84 (see *Figure 5-18,* page 5-10) is a reloadable flash-bang thrown grenade. Each grenade has a 1- to 2-second fuse delay, with a flash capability of 1.5 to 2.5 million CP and a bang of 168 to 175 decibels. Although it is part of the NLCS, use of the M84 is not recommended for use in crowd control situations. Specialty teams, such as security response teams, and search teams may use this device effectively.

Figure 5-16. GG04

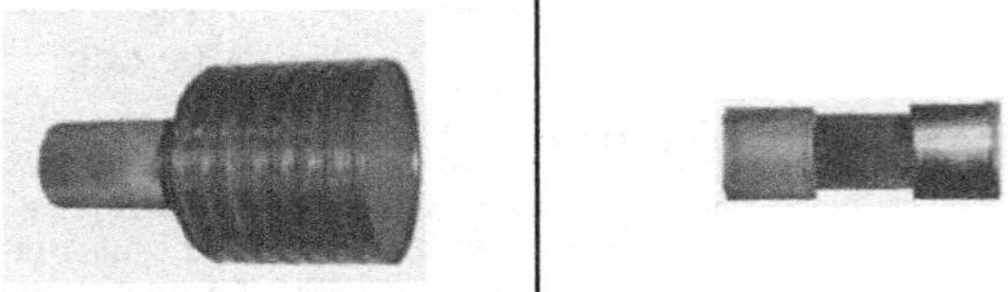

Figure 5-17. Launching Cup and AA30

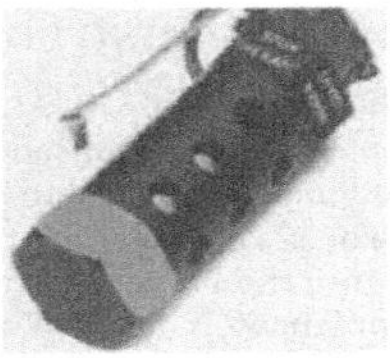

Figure 5-18. M-84

FLEX-CUFS

5-21. Flex-Cufs consist of a tough pliable plastic band with a self-locking mechanism in the center for each end. When the Flex-Cuf (see *Figure 5-19*) is threaded the restraint bands circle around the wrists or ankles, impeding movement and securing the individual. They are lightweight and disposable and require a cutting tool or other instrument to remove.

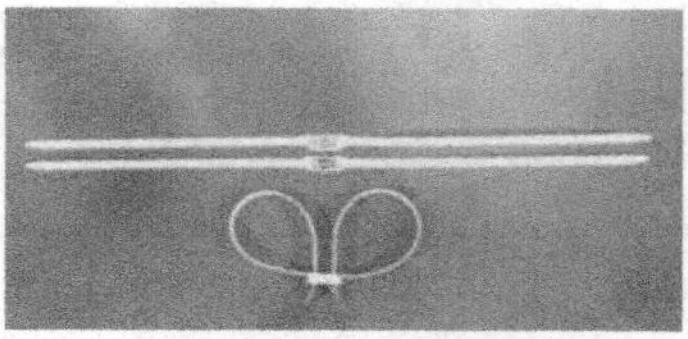

Figure 5-19. Flex-Cufs

INDIVIDUAL RIOT CONTROL AGENT DISPERSER (M36)

5-22. The M36 (see *Figure 5-20*) contains a Dibenz (F, -1 4-oxazepine [CR] solution. It can deliver 25 one-second bursts out to 12 feet. Individual disposable RCA dispersers are intended primarily for self-defense or to keep rioters out of arm's reach of soldiers conducting crowd control formations or soldiers engaged in missions where a noncombatant exists.

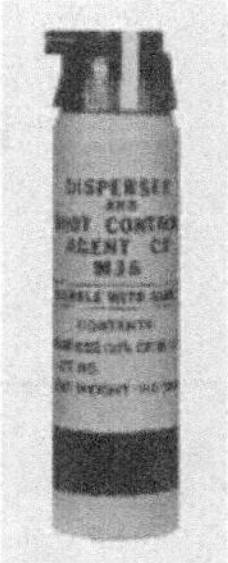

Figure 5-20. M36

MIDSIZE RIOT CONTROL AGENT DISPERSER (M37)

5-23. The M37 (see *Figure 5-21,* page 5-12) is the size of a standard fire extinguisher that uses compressed air to force the RCA out to a range of 30 feet. It has the capacity to employ 18 bursts for 3 seconds per charge. It is excellent for providing a wide coverage of RCAs onto a hostile crowd while maintaining excellent standoff capability. The M37 can be refilled and is rechargeable. It can be filled with CR solution (liquid agent) or CS1 (dry agent). For training purposes, CR can be substituted with water and CS1 can be substituted with talcum powder.

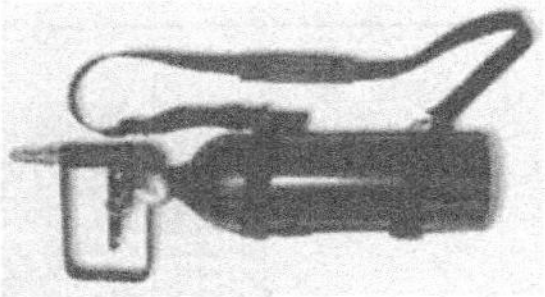

Figure 5-21. M37

SQUAD RIOT CONTROL AGENT DISPERSER (M33A1)

5-24. The M33A1 (see *Figure 5-22*) is designed to provide crowd control and protection at the squad level. It is capable of projecting a ballistic stream of RCAs beyond 25 feet in up to 25 half-second bursts. It consists of a frame and harness assembly, compressed-gas cylinder (agent container assembly), air pressure assembly, gun and hose assembly, multijet spray unit, and check valve assembly. The M33A1 can be refilled and is rechargeable. For training purposes, CR can be substituted with water and CS1 can be substituted with talcum powder.

Figure 5-22. M33A1

MISSION ENHANCERS

5-25. Mission enhancers aid the commander in various tactical situations. Mission enhancers include various types of devices used for illumination, vehicle denial, traffic control missions, and force protection enhancement.

PORTABLE BULLHORN

5-26. The portable bullhorn (see *Figure 5-23*) provides squad leaders with the capability to project their voices above the noise and commotion created by crowds and mobs. The megaphone has a built-in siren for alarm purposes that allows the operator to use the device as an ordinary megaphone and siren alarm during emergencies. The bullhorn is portable and easy to carry. The acoustic range of the megaphone for a clear voice output is ¾ to 1 mile. The power output of the megaphone is rated at 15 watts with a maximum rating of 20 watts. The dimensions of the megaphone are 14.5 inches with a 9-inch diameter bell. It weighs 3.5 pounds without batteries. Battery requirements for the megaphone are eight dry-cell size C batteries.

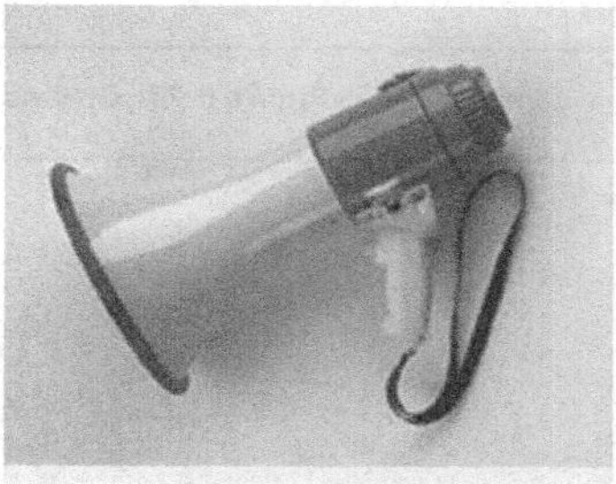

Figure 5-23. Portable Bullhorn

GROUND-MOUNTED BULLHORNS

5-27. Ground-mounted bullhorns (see *Figure 5-24,* page 5-14) are a critical communication enhancement device for conducting crowd control tactics. The bullhorn can facilitate communication with the crowd in conjunction with PSYOP support and assistance with the communication of commands to troops engaged in the crowd control process. The power output of the ground-mounted bullhorn is rated at 22 watts. The dimensions of the bullhorn are 13.75 by 9 by 14.75 inches, and it weighs 15 pounds.

INDIVIDUAL VOICE AMPLIFICATION SYSTEM (M7)

5-28. The M7 (see *Figure 5-25,* page 5-14) is a critical communication enhancement device for conducting crowd control tactics using RCAs while wearing the M40 protective mask. The M7 is fitted to the M40 protective mask to facilitate oral communication and increase the user's ability to communicate using radios and other devices.

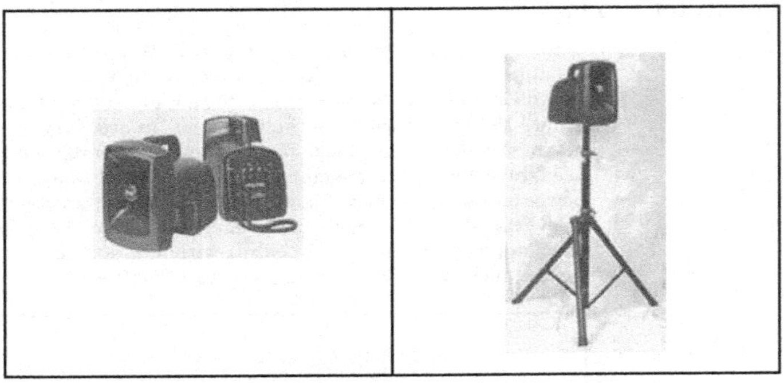

Figure 5-24. Ground-Mounted Bullhorns

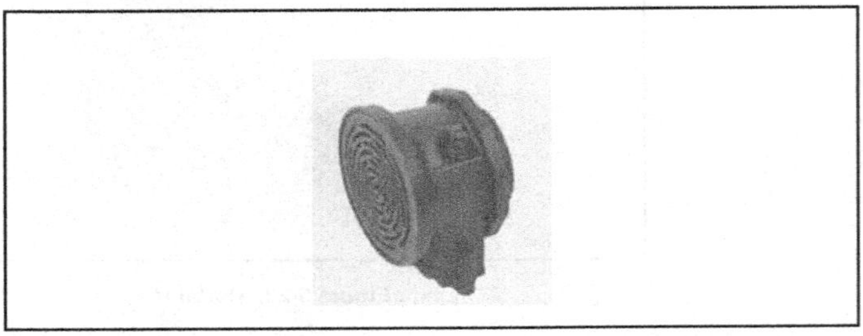

Figure 5-25. Individual Voice Amplification System

INDIVIDUAL HIGH-INTENSITY SEARCHLIGHT WITH HOLSTER

5-29. The individual high-intensity searchlight (see *Figure 5-26*) is used primarily for illumination in crowd control operations at night. An individual can use it for general illumination of the operational area to pinpoint agitators and threats, reduce the ability of the rioters to see troop formations and actions, enhance tactical deception and actions, and enhance tactical deception techniques for units conducting crowd control operations.

HIGH-INTENSITY LIGHT

5-30. A high-intensity light (see *Figure 5-27*) is intended for use in low light or night conditions. A high-intensity light can project a beam that will enable the user to identify an individual person up to 1,900 yards away. It has an adjustable beam spread of 1° to 15° and will run continuously at maximum power for up to 45 minutes. The internal power supply requires 4 hours for a full charge.

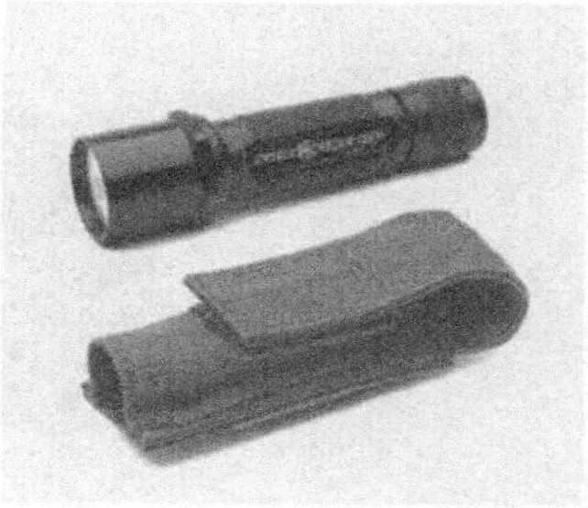

Figure 5-26. Individual High-Intensity Searchlight with Holster

Figure 5-27. High-Intensity Light

HIGH-INTENSITY LIGHT SYSTEM WITH 12-GAUGE SHOTGUN ATTACHMENT KIT

5-31. This 12-guage high-intensity light system (see *Figure 5-28,* page 5-16) is used to illuminate possible targets in crowd control situations where there is little or no light. It serves to blur or blind the aggressor's ability to see. It mounts easily to the end of the 12-gauge shotgun barrel.

40-MILLIMETER CARRYING POUCH

5-32. The 40-millimeter carrying pouch is made from durable canvas material, (see *Figure 5-29,* page 5-16) and can be easily affixed to load-bearing equipment (LBE) and/or a load-bearing vest (LBV) using Velcro straps. It is capable of carrying up to six 40-millimeter rounds.

Figure 5-28. 12-Gauge High-Intensity Light System With 12-Gauge Shotgun Attachment Kit

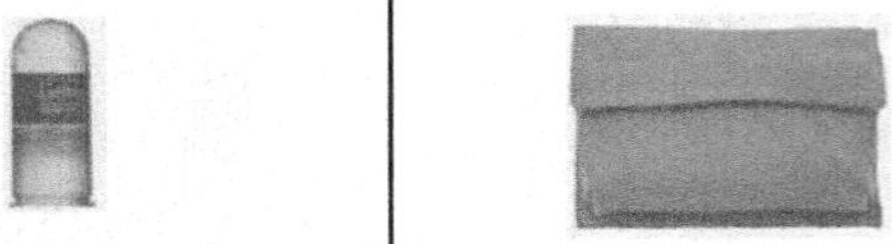

Figure 5-29. 40-Millimeter Carrying Pouch

12-GAUGE UTILITY POUCH

5-33. The 12-gauge utility pouch is made from durable canvas material (see *Figure 5-30*). It can be easily affixed to an LBE and/or LBV using metal clasps. The 12-gauge utility pouch is capable of carrying up to twenty-five 12-gauge shotgun rounds.

Figure 5-30. 12-Gauge Utility Pouch

12-GAUGE BUTTSTOCK CUFF

5-34. The 12-gauge buttstock cuff (see *Figure 5-31)* firmly attaches to the shotgun by sliding over the stock. Elastic straps aid in preventing the buttstock cuff from sliding or rolling. It holds up to seven 12-gauge cartridges, which give the soldier quick access to additional rounds.

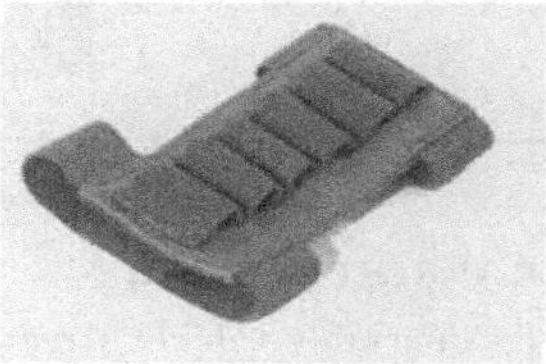

Figure 5-31. 12-Gauge Buttstock Cuff

DIVERSIONARY AND/OR RUBBER BALL GRENADE POUCHES

5-35. The diversionary and/or rubber ball grenade pouch (see *Figure 5-32*) is an ammunition-carrying pouch made of durable canvas material. It is partitioned in a manner to provide the soldier with a means of holding up to six rubber ball grenades or six flash-bang grenades in separate compartments.

Figure 5-32. Diversionary and/or Rubber Ball Grenade Pouch

CALTROPS

5-36. The caltrop (see *Figure 5-33,* page 5-18) is a field-expedient area denial system. Securing approximately 50 caltrops to 550-type military cord allows a soldier to quickly toss and recover the system. Caltrops are used to deny vehicle and pedestrian movement in designated areas. The four prongs are approximately 2 inches tall and will puncture tires or boots. A soldier can easily employ the caltrop by scattering it in a designated area hidden by camouflage.

Figure 5-33. Caltrop

PORTABLE VEHICLE ARREST BARRIER

5-37. A portable vehicle arrest barrier (PVAB) (see *Figure 5-34*) is a lightweight, portable, manually-emplaced and recoverable-barrier entrapment device that is used (on command) to create a barrier that will stop vehicles being driven by known or suspected hostile forces with minimal damage to the vehicle and its occupants. Two individuals can easily emplace it in approximately 4 to 8 hours. It has the capability of stopping a light truck (up to 7,500 pounds) at speeds of up to 45 miles per hour with a stopping distance of less than 200 feet. It is not to be used as a stand-alone system. The PVAB may be used at dismount points, checkpoints, and roadblocks to prevent unauthorized access by wheeled vehicles into or out of areas.

Figure 5-34. PVAB

TRAINING DEVICES

5-38. Training devices that commanders can use to effectively prepare their soldiers are included in the NLCS. Training devices allow soldiers to learn and build confidence in the development of personal defensive skills. These are critical pieces of equipment because they allow soldiers to train at almost full speed, which provides reinforcement of realistic motor skills. Training in this manner has proven to be more effective than just striking out at the air.

IMPACT TRAINING SUIT

5-39. Impact training suits (see *Figure 5-35*) are made of closed-cell shock-absorbent foam that allows soldiers to hone their riot baton skills through realistic hand-to-hand training. Impact training suits absorb strikes inflicted by the current expandable and wooden riot batons. They provide protection for the head, face, hands, and legs without significant degradation of the wearer's mobility.

Figure 5-35. Impact Training Suits

TRAINING STRIKE BAGS

5-40. Training strike bags (see *Figure 5-36*) give the soldier a durable, yet absorbing surface to strike or kick for training. It also helps develops proficiency in open-hand control and riot baton techniques.

Figure 5-36. Training Strike Bag

PRACTICE GRENADE BODY (GG05)

5-41. The GG05 is blue in color and is easily identified as a practice grenade (see *Figure 5-37*). It is designed to replicate the live GG04 rubber ball grenade. It is made of durable rubber and can be hand-thrown or muzzle-launched from the 12-gauge shotgun. It can be reloaded.

Figure 5-37. GG05

PRACTICE GRENADE FUSE

5-42. The practice grenade fuse (see *Figure 5-38*) is identified by its blue spoon and brown stripe. It easily screws into the body of the GG05. Once used, unscrew the practice grenade fuse from the body and reload it with a new fuse.

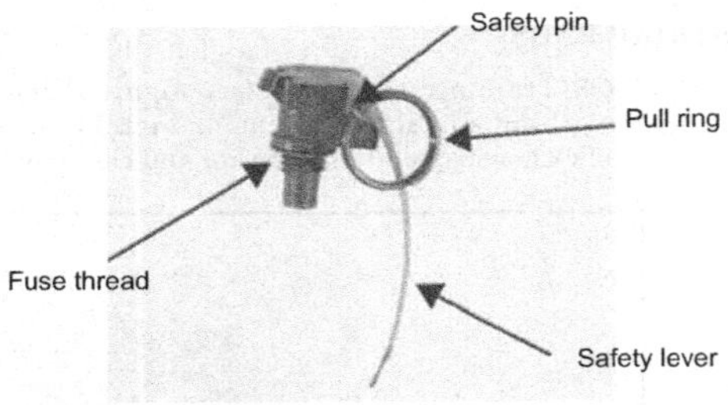

Figure 5-38. Practice Grenade Fuse

CORE CAPABILITIES

5-43. Core capabilities are those fundamental competencies that enable the commander to achieve the desired operational outcome. In the case of NLW, this includes providing a flexible means of response in order to protect friendly forces and/or influence the actions of potential adversaries and noncombatants. With the availability of the NLCS (see *Figure 5-39*) these goals can be achieved without resorting to lethal force and in a manner that will minimize collateral damage. The core capabilities associated with NL effects fall into the following two major categories:

NOTE: There are nine platoon equivalents within a light infantry battalion.

PLT Organization	Personnel and Weapons	Nonlethal Weapons & Equipment
PLT	7 personnel (5 M16/M4 and 2 M60/ M240)	3 riot face shields 3 sets of shin guards 3 sets of full-body restraints 3 individual voice amplification systems 3 individual RCA dispensers with carriers 1 portable bullhorn 1 ground-mounted bullhorn 1 high-intensity searchlight 1 set of caltrops 3 ballistic riot face shields 3 sets of ballistic shin guards 3 ballistic riot body shields 1 riot baton training suit 1 riot baton strike pad
SQD SQD SQD	3 personnel (3 M16/M4)	3 riot face shields 3 sets of shin guards 3 riot body shields 3 expandable riot batons with carriers 9 sets of cinch-strap restraints 3 individual voice amplification systems 3 individual RCA dispensers with carriers 3 individual high-intensity search lights with carriers (see *Figure 5-26*)
TM TM TM TM TM TM	24 personnel (12 M16/M4 and 6 M16/ M4 with M203 6 SAW)	24 riot face shields 24 sets of shin guards 100 sets of disposable flex-cuffs 24 individual voice amplification systems 24 individual RCA dispensers with carriers 6 expandable riot batons with carriers 11 grenade pouches 12 forty-millimeter grenade pouches 6 twelve-gauge ammunition pouches 6 twelve-gauge bitstock ammunition carriers

Figure 5-39. NLCS Distribution

- **Counterpersonnel.** NL counterpersonnel capabilities enable the application of military force with a reduced risk of fatalities or serious casualties among noncombatants or, in some instances, among enemy forces. There are several specific NL counterpersonnel capabilities to be explored. These include the means to influence the behavior and activities of a potentially hostile crowd, and the capability to bring a mob engaged in a riot under control. While there are many similarities in these two scenarios, each involves unique challenges, which may require radically different solutions. For more information, see *FM 3-22.40*.
- **Countermaterial.** NL countermateriel capabilities enhance operations by rendering equipment and facilities unusable without complete destruction. An NL countermateriel capabilities enable the

application of military force to defuse potentially volatile situations under circumstances in which more destructive conventional military means might prove counterproductive. For example, pre-emptive strikes against troublesome and aggressive nations may be politically unacceptable when only conventional weapons with the high risk of casualties are involved. With NL countermateriel capabilities, the enemy's ability to threaten its neighbors could be curtailed with less political risk by attacking only their weapons of war and their supporting infrastructure. For more information, see *FM 3-22.40.*

ADDITIONAL CAPABILITIES

5-44. NLW are only NL when fired within their parameters.

DANGER

If NLW are fired outside their parameters, they may cause serious injury or death.

5-45. If a situation turns from NL to lethal, an NL gunner only needs to change his point of aim and distance to inflict serious bodily harm and/or death to the threat. The closer the soldier is to the target, the greater the kinetic energy and the more severe the effects are to the threat.

EMPLOYMENT CONSIDERATIONS

5-46. NL munitions and equipment are not completely NL; they are NL by intent. The term NL does not guarantee zero mortality or nonpermanent damage. NLW can add flexibility to combat operations and enhance force protection by providing an environment in which friendly troops can engage threatening targets with a reduced risk of noncombatant casualties and collateral damage.

5-47. If the tactical situation dictates a NL response, a soldier who levels a weapon loaded with NL munitions must be trained in distance to the target and appropriate points of aim. Based on the soldier's understanding and training on the characteristics and capabilities of the round, they can easily discern and engage targets to the desired effect. If the tactical situation dictates a lethal response, soldiers can easily adjust the point of aim to a more vulnerable target location on the body and move in closer to the target. NL munitions can be used as a lethal response if the situation dictates and the intent of the soldier is to inflict death.

NONLETHAL WEAPONS CAPABILITIES IN FORMATIONS

5-48. The mix of NLW within the formation is based on METT-TC. For example, the first rank of the formation will have riot shields, riot batons, and their personal weapon (slung across their back [left to right] with the butt up and muzzle down or holstered). The second row will have a mix of shotguns and M203s. Shotguns and M203s may be used as area denial or point target weapons depending on the specific munitions used. The shotgun provides cover for the slower loading M203. The commander may move these weapons within the formation to meet the needs of the mission and to create distance between the formation and the crowd (see *Appendix D*).

Chapter 6

Civil Disturbance Formations

Control force formations, when properly employed and executed against a crowd, are some of the most practical methods of crowd control. The riot baton has been the weapon of choice for the control force. With the development of NLW, the commander now has more options at his disposal. Before NLW, the only other choices were chemical irritants, MWDs, and lethal force. Now commanders can create a standoff distance with NLW that reduces the risk of close fighting with the mob. During any crowd control operation, the commander must protect the control force with lethal overwatch. This overwatch may be in the form of marksmen and observer teams (sniper with observer) and a response force with lethal capabilities.

OVERVIEW

6-1. With advances in NLW, it is possible to create an effective standoff distance of 15 to 100 meters between the crowd and the control force formation. It is now possible to advance and disperse the crowd without coming into direct contact with it. Some NLW may also be used to assist in the identification and apprehension of instigators.

6-2. Control force formations are used to disperse, contain, and/or block a crowd. These formations are more effective in urban areas than in open areas, such as parks or fields. When control force operations are employed in urban areas, it is easier to split a crowd into smaller segments, isolate instigators, or funnel the crowd into the desired location by using buildings and other man-made structures. Commanders must be mindful of the fact that well-organized crowds may attempt to move to the flanks of a formation and/or get behind the formation to gain a tactical advantage.

6-3. Commanders must be aware of the limitations of control force formations. They are not the answer to all civil disturbance situations. Do not expose the formation to sniper fire and/or unnecessary violence simply for a show of force. Control force formations should be used when it is decided that a crowd poses a threat and must be dispersed or moved to a specified area. Before dividing a large crowd, the commander must consider that this might not solve the problem. It may worsen the problem by creating smaller elements that may engage the control force in small-mob tactics, such as sniping, looting, burning, and attempting to envelop the control force. Commanders must anticipate that breakaway crowds may engage in small-mob tactics. These tactics can be defeated by area control measures, such as building searches, saturation patrolling, and other tactics. After dispersing the crowd, the control force must not assume that there will be an immediate return to peaceful activities. The use of formations is only part of the total dispersal effort.

6-4. Commanders must assess and secure the areas of operations (AOs). This may be accomplished with the use of helicopter overflights and by visually securing rooftops, high elevations, and advantage points. Designated marksmen teams may be used to secure the rooftops and provide lethal overwatch for the control formation. The control force commander must be aware that designated marksmen teams are positioned on rooftops.

6-5. Crowd control formations and their support teams have more capabilities than just crowd dispersal. They also have the capability to apprehend and detain certain members of the crowd that the commander feels may instigate the crowd to further violence. Commanders must analyze the threat, determine the mission of the control force, and decide which control force formations will accomplish the mission according to METT-TC.

6-6. During CONUS operations, there are usually no enemy forces. However, commanders must analyze and prepare for any potential opposition. The formations discussed in this chapter are guidelines and may be adapted to fit any mission or situation. Whatever the modification, soldiers must stay in position and on line. Only through training and rehearsals will soldiers become proficient in basic formation movements and be able to adjust to changing situations. Soldiers must also be prepared to employ NLW. *Figure 6-1* is used as a key to identify soldiers within crowd control formations.

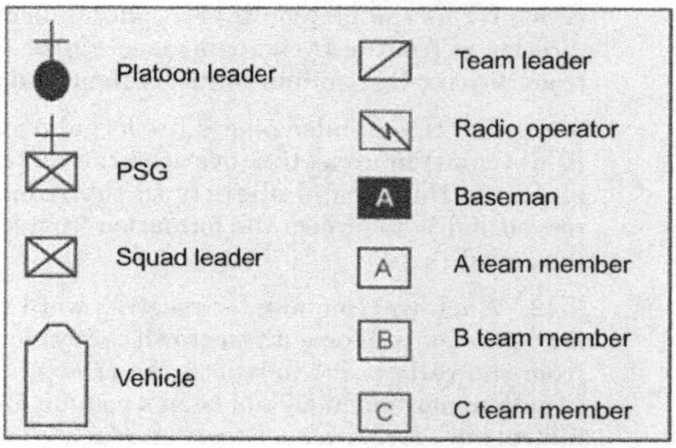

Figure 6-1. Symbols for a Formation Element

ACTIONS BEFORE MOVEMENT

6-7. Before moving the control force to the civil disturbance area and into a potentially hostile urban environment, the commander and his staff must review the preincident plans and expand them (as needed) based on current information. Because of the employment of forces against ambiguous threats,

IPB becomes critical. Information is the key to developing an appropriate response to civil disturbances.

6-8. Civil disturbance training must be part of the annual unit training plan. If a soldier is deployed to an area where civil disturbances are a recurring event, sustainment training must occur more frequently. Before moving the control force to the rally point, an intense rehearsal must be conducted by the leadership and the control force. A top-down review of the rules of engagement (ROE/rules for the use of force (RUF) must be accomplished. The ROE/RUF must be part of the unit training plan and each soldier must be trained to standard (see *Chapter 8*). The use of NLW (type specific) must be granted to the on-site commander from the approval authority.

6-9. Information is the key to developing plans for appropriate responses to a civil disturbance, and IPB is crucial to the success of the mission. During the planning phase (see *Chapter 2*), the information provided by the IPB will aid the staff in selecting various routes and rally points. The selection of ingress and egress routes must include a variety of ways in and out of the area. Direct and indirect routes and those with cover and concealment are necessary. The rally point must be cleared and secured by an advanced party capable of controlling the area.

RALLY POINT ACTIONS

6-10. Once at the rally point, the commander makes contact with the local police, civil authorities, or military officer. He then assesses the situation, decides the next COA, and issues orders, as appropriate. The platoon sergeant (PSG) forms the platoon into a column formation and places team leaders directly in front of their teams (see *Figure 6-2,* page 6-4). Team and squad leaders make last minute checks of their soldiers and await orders.

6-11. The commander places his lethal overwatch (designated marksmen [DM] teams) in areas that overlook the control force and the mob. They may be to the flanks and slightly to the front of the formation. It is their responsibility to protect the formation from lethal fire by constantly scanning the crowd.

6-12. When the commander receives word that DM teams are in place, he forms the control force into a crowd control formation and moves them quickly from the rally point to where the crowd is assembled. Depending on the situation, movement should be in a column formation or one of the three basic control force formations. Reserve forces are left at the rally point until ordered forward as additional lethal overwatch or to reinforce the formation.

CONTROL FORCE FORMATIONS

6-13. Control force formations, when properly employed and executed against a crowd of limited size, are one of the most practical methods of crowd control. Experience has indicated that the most frequently used control force

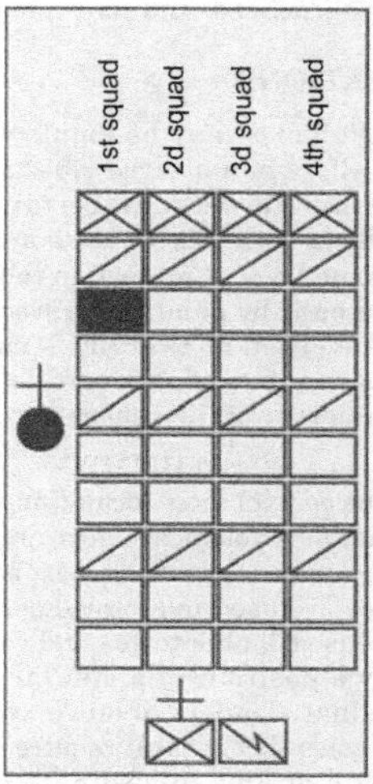

Figure 6-2. Column Formation

formations are the line, echelon, and wedge. The two lesser-used formations are the diamond and circle. Descriptions of all the formations are as follows:

- **Line formation.** The line is the basic formation and it is used more often because of its offensive and defensive applications. As an offensive formation, the line is used to push or drive crowds straight back, across an open area, or up a city street. As a defensive formation, the line is used to hold the crowd or to deny access to areas.
- **Echelon formation.** The echelon is an offensive formation used to turn or divert groups in open or built-up areas and to move crowds away from buildings, fences, and walls.
- **Wedge formation.** The wedge is an offensive formation that is used to penetrate and split crowds into smaller groups.
- **Diamond formation.** The diamond, when used as an offensive formation, is used to enter a crowd and is the formation of choice for extraction teams. As a defensive formation, the diamond is used when all-around security is required in open areas. The decision to use this formation is based on the conformation of the crowd.
- **Circular formation.** The circular formation is used for the same purpose as the diamond formation. The decision to use this formation is based on the conformation of the crowd.

6-14. There are many suitable variations of control force formations that may be employed, but appropriate commands and signals must be created to execute the formations. Because of their somewhat complicated nature and the coordination required for these formations, new variations must be trained, practiced, and rehearsed before they are used in a civil disturbance.

VEHICLES AND FORMATIONS

6-15. Vehicles may be employed with troops in control force formations, especially when a large riotous crowd is on hand. Although vehicles add strength to formations, certain precautions must be taken. Covering the windshield with sturdy, close mesh fencing and the standard safety glass will offer some level of protection to the occupants. Shields and/or mobile barriers may be built by mounting a wooden or metal frame strung with barbed wire across the front of a vehicle. Members of the formation should walk as near to the front corners of each vehicle as possible to keep rioters from attacking the sides and rear of the vehicles.

6-16. Up-armored HMMWVs, ASVs, or armored vehicles may be employed with the control force formation when determined by METT-TC. They present a strong psychological effect and offer protection for the occupants. Wheeled armored vehicles are best because they do not deface the pavement. Whenever vehicles are used in a crowd control formation, the commander must ensure that he is still able to see and control the formation. This may require him to occupy a position in a similar vehicle behind the formation or command subordinate leaders to move away from the front line and into line-of-sight (LOS) control. This will require additional hand-and-arm signals to allow the commander and subordinate leaders to communicate. Commanders can effectively use vehicles as primary communication tools by using horns and lights as signals to organize movement or actions. However vehicles are used, it takes a good measure of training and practice to be effective.

ELEMENTS OF A CONTROL FORCE FORMATION

6-17. Four elements make up the basic crowd control formation. They are as follows:

- **Base element.** The base element is made up of two ranks. The first rank is shield holders, while the second rank contains the NLW. This is the front line of the formation.
- **Support element.** The support element forms up in a column formation behind the base element and has a variety of uses. It may be used to replace base element members as needed, provide lateral or direct support, or perform extraction team operations. The support element helps the base element by performing the following three essential tasks:
  - **General support.** The general-support element is formed from an uncommitted squad in the platoon (generally the 2d squad). When a company is tasked as the control force element, one of the platoons becomes the general-support element. The element is in a column formation centered on and behind the main formation. From this formation, the general-support element can move to lateral or direct support, as needed.
  - **Lateral support.** The lateral-support element is used to protect the flanks of the formation. This is done by moving a set number of teams forward from the general-support element or by using the

end teams in the formation; however, this will make your formation smaller. Once in position, these teams become part of the formation, with their riot shields facing the flanks of the formation.

  - **Direct support.** When direct support is ordered, the general-support element moves forward as they form their own formation. While moving forward, the squad and team leaders from the control force formation step back and allow the riot shield holders from the support element to step behind riot shield holders of the original formation. This allows the support riot shield holders to form up between and behind the riot shield holders of the original formation. Now there are two lines of riot shield holders between the crowd and NLW firers. This formation is the strongest of the formations and requires more planning and practice to master.

- **Command element.** The command element contains several different members. A general configuration for the command element is the platoon leader, the PSG, a radio operator, a video recorder operator (if required), and an interpreter (if required). This element does not have a fixed location within the formation and moves about as needed.
- **Lethal overwatch element.** The lethal overwatch element is a team formed from reserve security forces.

6-18. The reserve support element is not part of the control force formation until it is brought forward from the rally point to join the formation. It remains until released by the formation commander. Lethal overwatch teams are formed from the reserve force. Once lethal overwatch teams are deployed and in position, they are under the control of the formation commander. Communication between the lethal overwatch team and the formation commander is a priority because he approves target selection and engagement.

WEAPON POSITIONS

6-19. At the commander's discretion, members of the control force formation may carry their weapons in any of the positions prescribed in *FM 3-21.5*. However, the primary method of carrying weapons in the control force formation is at the safe-port position (see *Figure 6-3*). It allows the soldier to control both ends of the weapon while moving in and out of the formation and advancing on the crowd.

6-20. High-port position (see *Figure 6-3*) is a position of complete readiness. It is used whenever troops are in contact with a crowd that is showing resistance or not withdrawing. This is a tiring position and is hard to maintain for extended periods. Commanders must rest their troops at every opportunity by using less tiring positions.

Safe-port arms | High-port arms

Figure 6-3. Weapons Positions

COMMANDS

6-21. Commands are given to the control force formation orally or with hand-and-arm signals. Oral commands for the on-guard position are given in one count. All other commands are given in two counts, a preparatory command followed by a command of execution. However, oral commands cannot be relied on completely, so commanders must plan to use hand-and-arm signals too. When vehicles are part of the formation, radios become an alternate method for relaying commands. To improve communication, commanders may use hand-and-arm signals (see *Figure 6-4,* page 6-8) with oral commands. The basic commands used in control force formations are outlined in *Table 6-1,* page 6-9. These commands can be added to or combined to form more complex commands, therefore forming more complex formations.

MOVEMENTS

6-22. Control force formations move in the same manner as regular formations. This allows the commander to more precisely maneuver the formation by commanding the number of steps it should move and in what direction, such as "Five Steps, Forward March." The normal rate of march for entering and leaving a control force formation is double-timing. The half-step march is used when the formation is in direct contact with the crowd. This slows the formation down and allows for better command and control. The squad and team leaders echo preparatory commands and provide cadence while in control force formations.

INTERVAL AND DISTANCE

6-23. Interval is the lateral space between elements. Distance is the space between elements in a column. The usual interval and distance between soldiers in control force formations is 30 inches. The interval and distance can be adjusted based on METT-TC. If the crowd has to be physically pushed back, a close interval is preferred. This allows the riot shields to be overlapped, creating a stronger wall. During peaceful demonstrations, the interval could increase to double arm width, allowing the formation to cover more area and still control the crowd. The interval and distance could change several times during the course of operations before the crowd is dispersed.

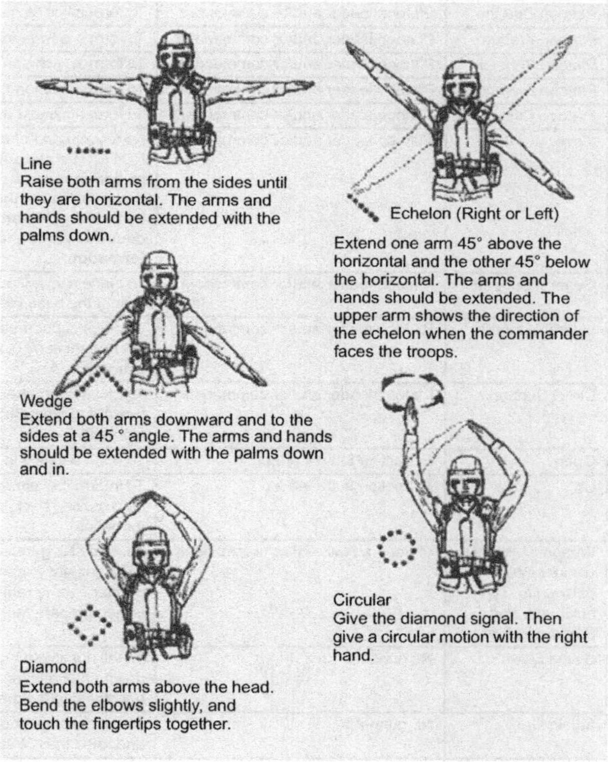

Figure 6-4. Hand-and-Arm Signals

Table 6-1. Basic Commands

Commands for	Command	Given by	Purpose
Formations	Platoon On Line	Platoon leader and/or commander	To form a line formation.
	Platoon Wedge	Platoon leader and/or commander	To form a wedge formation.
	Platoon Echelon	Platoon leader and/or commander	To form an echelon formation.
	Platoon Diamond	Platoon leader and/or commander	To form a diamond formation.
	Platoon Circle	Platoon leader and/or commander	To form a circular formation.
	Move	Platoon leader and/or commander	As a command of execution. **NOTE: The platoon leader and/ or commander also identifies the location for the formation by pointing his arm to the desired location for the formation.**
Support	General Support	Platoon leader and/or commander	To place a specified unit in the rear of the base element.
	Lateral Support	Platoon leader and/or commander	To place support elements on the left or right flank of the base element.
	Direct Support	Platoon leader and/or commander	To move support elements forward to strengthen the base element.
	Open	Extraction team leader	To open a space in the formation.
	Up	Extraction team leader	To inform the extraction team to start moving back to the main formation.
Fire	Weapon System Gunners, Number of Rounds, Type of Round, and Prepare to Fire	Platoon leader and/or commander	To ready NL gunners to fire weapons; for example, M203 gunners, three rounds, area target, prepare to fire.
	Shield Down	NL gunners	To tell riot shield holders to get down on one knee and lower the riot shield to the ground.
	Shield Up	NL gunners	To tell riot shield holders to stand and raise the riot shield.
Weapons	Port Arms	Platoon leader and/or commander	To use when not in contact with the crowd.
	High Port	Platoon leader and/or commander	To tell NL gunners what position to hold their weapons in while the formation is in physical contact with the crowd. This allows for rapid targeting and firing of NL munitions.

EXTRACTION TEAM OPERATIONS

6-24. The extraction team is a squad (see *Figure 6-5*) that is generally formed from the support element, but could also be formed from reserve security forces. Once the team is formed, it is OPCON to the formation commander until the team's mission is complete and it returns to the rear of the formation. This team provides the riot control formation with the means to employ NL and lethal cover forward of the formation. It also provides the formation with the ability to remove and detain rioters downed by munitions.

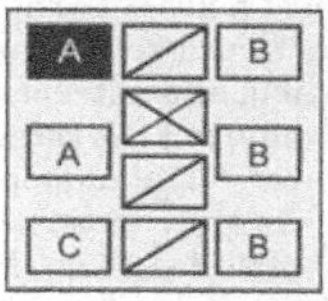

Figure 6-5. Extraction Team Formation

The extraction team—

- Extracts vehicles or personnel that are in immediate danger from the crowd.
- Detains and escorts downed rioters to the rear of the formation.
- Restrains and searches subjects.
- Covers confined or congested areas where a full riot control formation cannot be inserted.

6-25. The extraction team is usually a squad. While the recommended distance for deploying the extraction team from the control force formation is no farther than 10 meters, this may change based on METT-TC. The control formation protects the extraction team and provides immediate support, if necessary.

6-26. The extraction team may deploy from anywhere within the formation. When given orders to deploy, the squad leader of the extraction team gives the formation commands, identifies at least two personnel in the squad to conduct the search and apprehension of the subject, and leads the squad to the base line of the formation. He then extends his arm between the two riot shield holders and commands, "Open," while tapping the riot shield holders on the side of the shoulder. The two riot shield holders take one step backward and one step to the right or left. This clears a path for the extraction team. The extraction team leader then states the number of personnel leaving the formation, such as "Ten security personnel leaving." As the last extraction team member exits the formation he states, "Last Man." The two riot shield holders then return to their original position in the formation. The adjacent base element squad and/or team leaders pass the number of personnel in the team through the formation.

6-27. As the extraction team approaches an instigator or incapacitated demonstrator, the riot shield holders envelop the target and face in the direction of the nearest threat. NL gunners on the left and right cover their respective areas. The search and apprehension team immediately controls and restrains the target. To avoid confusion, one team member concentrates on controlling the subject while another team member applies a restraining system. When the search and apprehension team has the subject under control, the extraction squad leader sounds off with, "Up." Upon hearing the "Up" command, it is repeated by all extraction team members. Team members then grab a shoulder of the team member in front of them to backtrack into the formation. As the extraction team approaches the formation, the nearest base element leader will extend his arm and command, "Open," while tapping the riot shield holders on the side of the shoulder. The riot shield holders move one step backward and to the right or left to open the formation. As he enters the formation, the extraction team leader sounds off with the number of personnel entering the formation, such as "Ten security personnel and one demonstrator." The adjacent base element leader counts the number of personnel as they enter the formation. When the last man enters the formation, he states, "Last Man." The riot shield holders then return to their original places in the formation. The adjacent squad and team leaders in the base element pass information through formation that the extraction team has returned.

LETHAL OVERWATCH TEAMS

6-28. During a NL engagement, the use of a DM team provides confidence and safety to those facing the mob. If a lethal threat is presented, the DM team in an overwatch position (armed with a standard infantry rifle that is mounted with a high-powered scope) can scan the crowd, identify agitators and riot leaders for apprehension, or fire lethal fire if so ordered or warranted. It is also ideally suited for flank and countersniper operations.

6-29. The DM team sets the security overwatch and provides real-time intelligence and reporting that is vital to mission success. It provides coverage during the entire approach to the crowd, and its ability to select positions and provide cover fire (if and when warranted) is critical to the safety of the control force. Successful execution requires training, practice, and rehearsal with the control force formation.

6-30. The DM team is organized from a standard three- or four-man team. Each individual on the team is equipped with a standard-issue weapon. One individual has an infantry rifle with a scope, another has a set of binoculars, and two other individuals provide security for the team. Each team is equipped with a radio for communication with the HQ element of the control force formation.

6-31. The teams must have a visual advantage over the crowd to provide lethal protection to the formation. This is best accomplished by placing them on nearby rooftops, in the upper floors of buildings, and on hilltops. They must have the optical equipment to identify mob leaders, instigators, and/or individuals with weapons. Team integrity must be maintained. In organizations with three-man teams, one individual becomes the marksman, another the observer, and another the security. When teams have a 4th member, that individual becomes the video camera operator. This will help identify individuals for later apprehension and provides a record of events.

6-32. Two teams should be deployed to cover the flanks of the formation. Each team should be deployed to a position that is in front of the formation and in an overwatch position. As the formation moves forward, the teams leapfrog forward from the formation.

6-33. The reserve forces, located at the rally point, can provide another lethal force of some size that can be brought forward, as needed. Each soldier in the control force formation should have a sidearm (9-millimeter pistol) or an M16 or M4 rifle. The rifle is slung diagonally across his back with the rifle butt over his left shoulder and the muzzle below his right hip. Lethal munitions are in a separate ammunition pouch that is isolated from NL munitions. This is to prevent the mixing of NL and lethal ammunition.

6-34. Once the lethal overwatch teams are deployed, they are under the OPCON of the formation commander. The lethal overwatch element must have direct and secure communications with the formation commander. The number of teams deployed is based on METT-TC. Teams will leapfrog to keep ahead of the advancing control force formation with one team in the overwatch position at all times.

RESERVE FORCES

6-35. A large reserve of soldiers should be maintained during civil disturbance operations. Knowing that a large reserve force is available provides confidence and safety within the control force and helps prevent them from overreacting to provocative acts with disorderly and criminal elements in the crowd. When determining the number of reserve forces required, it should be resolved in favor of a large number. Lethal overwatch teams, forces to augment the control force, and apprehension teams come from the reserve forces.

6-36. Apprehension teams are a must when it is likely that a large number of people will be apprehended. These teams provide an organized response to what could be a chaotic situation. The teams can be organized at the team or squad level, depending on the number of apprehensions expected. These teams usually operate behind the base element. During peaceful demonstrations, these teams can be sent forward to remove demonstrators who refuse to disperse.

6-37. Each team consists of an apprehension, movement, and security element (added if the control force cannot provide security); a recorder; and a video or still photographer. The apprehension element is an officer or NCO and a civilian police officer (CONUS or OCONUS) or US marshal. It decides who will be apprehended, informs the subject of the reasons, supervises the handling of the subject, and ensures that the apprehension is properly conducted and documented. The apprehension team is responsible for the overall apprehension process, limiting the number of individuals appearing in court to document the apprehension. The movement element helps the apprehension element handle the subject. They move, restrain, and search the individuals being apprehended under the supervision of the apprehension element. The security element prevents the crowd from interfering with the mission. The recorder documents the apprehension on *DD Form 2708*. When possible, the video and/or still photographer recorder photographs each subject with the apprehension element. This provides the apprehension element with the necessary documentation for use in court. One individual may need to conduct many of the duties of an apprehension team based on METT-TC.

6-38. The video and/or still photographer record the situation before, during, and after the apprehensions. These images may also be used as pictorial documentation for use in court. It is possible that the presence of video and still cameras actively recording the situation can cause crowd members to be less prone to unruly or criminal acts.

6-39. The behavior of the selected individual to be apprehended is observed to decide if the subject is violent, cooperative, or passive. Unless there is good reason to believe otherwise, the subject must be considered dangerous. The mood of the crowd is observed to estimate what it might do. Is it only curious onlookers or does it support the individual targeted for apprehension? Will it interfere?

6-40. When apprehending an individual, the actions must be kept as simple as the situation allows. The element of surprise can give the apprehension team an advantage over the apprehended individual. It is best not to use surprise if the individual might panic and injure someone. Select an area for the apprehension that is to the advantage of the team and not the individual. Avoid apprehensions in crowded areas as these areas provide many avenues of escape and contain supporters of the apprehended individual.

6-41. All verbal commands given to the apprehended individual must convincing authority with short, clear, and distinct commands. Voice, attitude, and deportment must convey an expectation that the apprehended individual will comply.

6-42. The individual should be searched immediately; however, circumstances (violence or sympathizers for the individual being apprehended) may dictate temporarily postponing the search until the individual can be moved to a safer area. Flex-Cufs or hand irons may be applied.

SQUAD FORMATIONS

6-43. Civil disturbance training is a collective task. Using team and squad concepts, team leaders instruct their teams and squad leaders instruct their squads. Safety is key when using force on force training for the control force formation. Soldiers must understand that serious injuries may occur if they do not follow guidelines set by the commander.

6-44. The smallest formation is a squad formation, which is used in back of the main formation to cover side streets. Squad members must know their positions regardless of which control force formation they are in, squad to company.

6-45. Training must enforce soldiers' understanding of the need to stay informed for individual protection. Individual soldiers are much more vulnerable to attack when they break ranks and chase after crowd members. When an individual soldier breaks ranks he not only puts himself in danger, but the entire formation.

6-46. When moving a squad into a crowd control formation from a column formation, the squad leader takes one or more steps to the right of the squad and faces it. He gives the command and hand-and-arm signal for the formation he wants. Then he indicates where he wants the formation located by pointing in that general direction. The A team leader commands, "Follow Me." He places the baseman at the position indicated by the squad leader. The squad forms on the baseman position.

LINE FORMATION

6-47. The command for forming a squad in a line formation is "Squad Line Formation, Move." On the command of execution, the A team leader leads the baseman to the place indicated by the squad leader. The squad members align themselves in sequence with the baseman at normal intervals (see *Figure 6-6*). Based on the situation, the squad leader may designate a specific interval. He does this when he gives the preparatory command. If no interval is specified, the squad automatically forms using the normal interval.

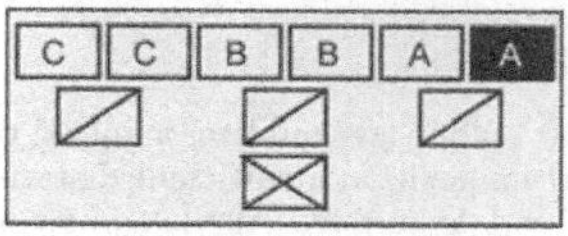

Figure 6-6. Squad Line Formation

ECHELON FORMATION

6-48. The command for having a squad form in an echelon formation is "Squad Echelon Right (or Left), Move" (see *Figure 6-7*). On the command of execution, the A team leader places the baseman at the location indicated by the squad leader. The squad members align themselves in sequence with the baseman, one pace to the side and one pace to the rear.

WEDGE FORMATION

6-49. The command for moving a squad formation into a wedge formation is "Squad Wedge, Move." On the command of execution, the A team leader moves the baseman to the place indicated by the squad leader. The A team lines up to the left of the baseman, one pace to the left and one pace to the rear of each preceding man. B team members align themselves with the baseman, one pace to the right and one pace to the rear of each preceding man. C team members align themselves in the same way to the left of the baseman (see *Figure 6-8*).

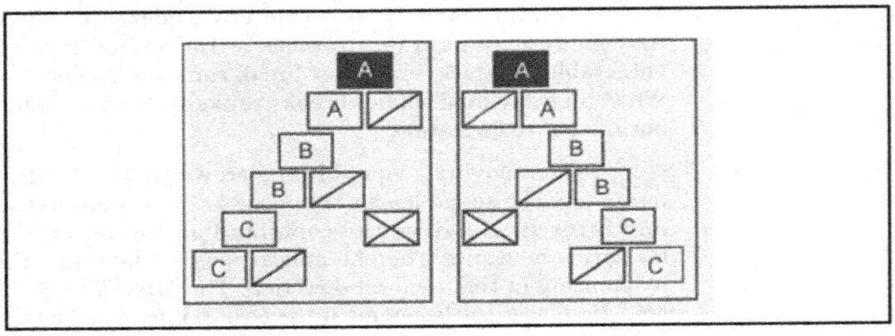

Figure 6-7. Squad Echelons Left and Right Formations

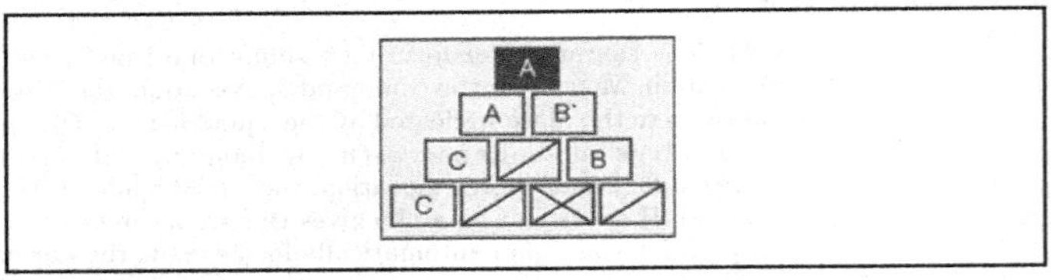

Figure 6-8. Squad Wedge Formation

ASSEMBLY

6-50. When assembling a squad from the line formation, the squad leader takes a position a sufficient distance to the rear of the squad and commands, "Squad Assemble." At the same time, he raises his right hand in the air and makes a circular motion. Squad members, who hear the command, automatically come to the port arms position. The A team leader and the baseman do an about-face movement.

6-51. The other squad members face toward the baseman. On the command of execution "Move," the squad leader points to the place where he wants the squad to assemble. The A team leader double-times to the designated spot, and the other members of the squad follow him. Team leaders fall into the file in their designated positions as they move to the AA. To assemble from a squad echelon formation, the steps are the same as a line formation.

6-52. To assemble from a squad wedge formation, the squad leader takes the same steps as for the line and echelon formations. B team members do a half left face, and the A team and C team members do a half right face. The C team pauses at the baseman position and allows the B team to clear the formation before moving to the AA.

DIAMOND AND CIRCLE FORMATIONS

6-53. The diamond and circle formations are used during extraction team operations. They are small formations used to penetrate the crowd or cover small areas. The decision on which formation to use is based on METT-TC and the ability of the squad to perform.

6-54. Commands for forming a squad in a diamond or circle formation is "Squad Diamond (or Circle), Move." On the command of execution, the A team leader moves the baseman to the place indicated by the squad leader. The A team lines up to the left of the baseman, covering the 8 to 12 o'clock positions. B team members align themselves to the right of the baseman, covering the 12 to 4 o'clock positions. C team members align themselves covering the 4 to 8 o'clock positions (see *Figure 6-9*).

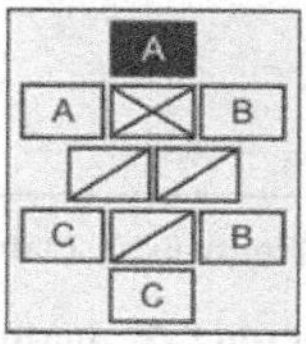

Figure 6-9. Squad Diamond or Circle Formation

6-55. To assemble from a squad diamond or circle formation, the squad leader takes the same steps as he would for a line formation. The A team leads the squad to the AA followed by the B team and then the C team. Teams fall into a column formation as they file back to the AA.

THREE-SQUAD PLATOON FORMATIONS

6-56. The platoon HQ consists of the following individuals: the platoon leader, the PSG, and a messenger or radio operator. When directed by the platoon leader, other individuals such as the selected marksman, interpreter, RCA disperser operators, and firefighters may augment the platoon HQ. In forming all control force formations from the column, the platoon leader moves to the right of the platoon and faces them when giving his commands. He gives the hand-and-arm signal and the preparatory command. On the command of execution "Move," he points to the approximate location where he wants the platoon to form. The platoon leader pauses between the preparatory command and the command of execution to allow each squad leader to issue a preparatory command to his squad.

Platoon Line Formation

6-57. The command for forming a platoon line is "Platoon on Line, Move." Immediately following the platoon leader's preparatory command, the squad leaders of the 1st and 3d squads give the command, "Follow Me." The leader of the 2d squad gives the command, "Stand Fast." On the platoon leader's command of execution, the 1st and 3d squads move forward to the point designated by the platoon leader. The 1st team member of the 1st squad is the baseman for the platoon formation. The 1st squad forms a squad line to the left of the baseman while the 3d squad forms a squad line to the right of the baseman and dresses with the 1st squad. After the 1st and 3d squads have cleared the platoon, the 2d squad leader gives the command, "Follow Me." The 2d squad splits and the A and B teams form a line to the right of the 3d squad.

The C team (with the squad leader) forms a line to the left of the 1st squad (see *Figure 6-10*).

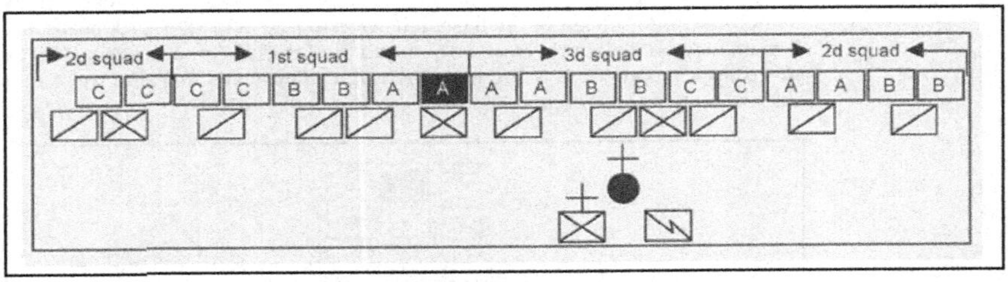

Figure 6-10. Platoon Line Formation

Assembling From a Platoon Line Formation

6-58. When assembling the platoon from the platoon line formation, the platoon leader and his HQ personnel take a position to the rear of the platoon. He then gives the command, "Platoon Assemble," while raising his right arm and making a circular motion above his head. Platoon members automatically assume the port arms position. The 1st squad leader and the baseman do an about face. The 3d squad leader moves to the head of his squad. All other members do an about-face movement toward the baseman. On the platoon leader's command of execution, the platoon leader points to where he wants the platoon to assemble. The 1st squad leader gives the command, "Follow Me," and moves his squad toward the AA. The squad moves to make a single file line to the baseman position followed by the C team of the 2d squad. As the 1st squad clears the baseman position, the 2d squad leader comes to a halt at the baseman position. The 3d squad leader gives the command, "Follow Me," and leads his squad to the AA followed by the A and B teams of the 2d squad. The A team leader of the 2d squad comes to a halt at the baseman position. Once the 3d squad has cleared the baseman position, the 2d squad leader gives the command, "Follow Me," and leads his squad to the AA. The 2d squad reassembles into their original column formation as they move to the AA. As the 1st squad comes to a halt (facing the crowd) in the AA, the 3d squad dresses with the 1st squad and leaves space for the 2d squad to return to its place in the formation. Once the platoon is formed into a column formation, HQ personnel take their position in the formation. The PSG then gives the command, "Countercolumn March," and moves the platoon to the rally point.

Platoon Line Formation With General Support

6-59. Whenever the command for establishing a crowd control formation contains the phrase "In Support" (without modification), it means that the support element is to remain in general support.

6-60. The command for forming a platoon line with the 2d squad in general support is "Platoon Line, 2d Squads in Support, Move." The 1st and 3d squads of the platoon line with general support execute the line, while the 2d squad remains in the column (see *Figure 6-11*). To assemble a platoon with a squad in general support, the procedure is the same as the platoon line, except that the 1st and 3d squads dress with the 2d squad.

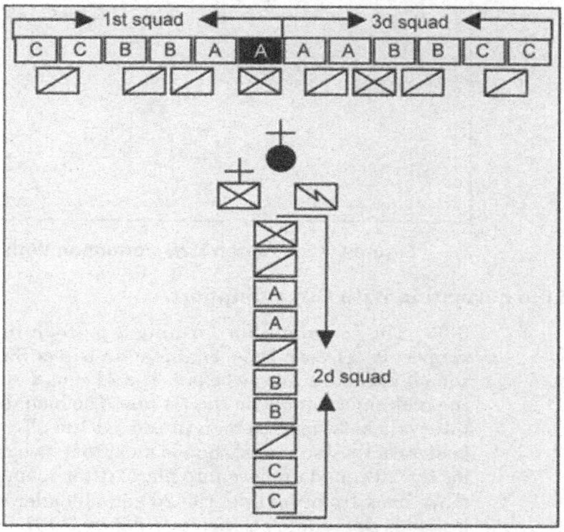

Figure 6-11. Platoon Line Formation With General Support

Platoon Line Formation With Lateral Support

6-61. The command for forming a platoon line with one squad in lateral support is "Platoon Line, 2d Squad in Lateral Support, Move." The 1st and 3d squads execute the line as before, and the 2d squad stands fast. After the line has been formed, the 2d squad leader gives the command, "Follow Me." The A and B teams move to the right flank while the squad leader and the C team move to the left flank of the formation. The 2d squad forms columns behind the last men on the line formation. If no direction is given, the 2d squad splits to cover both flanks. A squad may be used to cover one flank by giving the command, "2d Squad Right (or Left) Lateral Support" (see *Figure 6-12*).

6-62. The 2d squad may be moved from general to lateral support at any time by the platoon leader. The platoon leader commands, "2d Squad, Lateral Support, Move." To have the 2d squad join the line from either support position, the platoon leader commands, "2d Squad, Extend the Line, Move." The 2d squad leader commands, "Follow Me." On the command of execution, the squads set up individual lines, closing and dressing on the existing line. To assemble the platoon from a platoon line with lateral support, follow the same steps as for the platoon line.

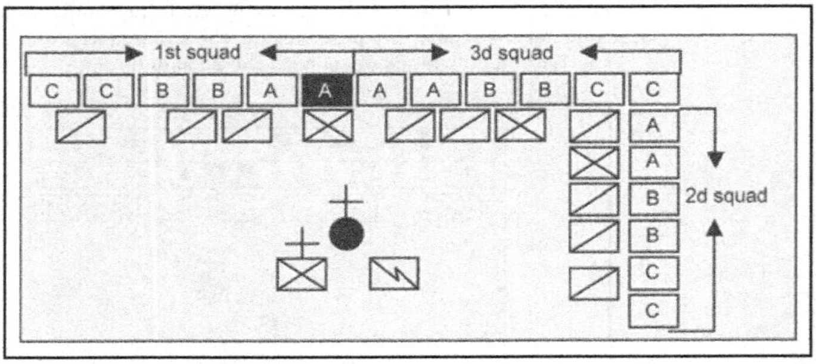

Figure 6-12. Platoon Line Formation With Lateral Support (Right)

Platoon Line Formation With Direct Support

6-63. The command for forming a platoon line with one squad in direct support is "Platoon Line, 2d Squad in Direct Support, Move." The 1st and 3d squads execute a line as before. The 2d squad executes a squad line directly to the rear and centered on the 1st line. The men in the supporting line cover the intervals between the men in the 1st line. To do this, the squad and team leaders of the 1st and 3d squads must take two steps to the rear to allow space for the 2d squad to move into place. After seeing the squad and team leaders move back from the line, the 2d squad leader commands, "Follow Me," and leads his squad into position (see *Figure 6-13*).

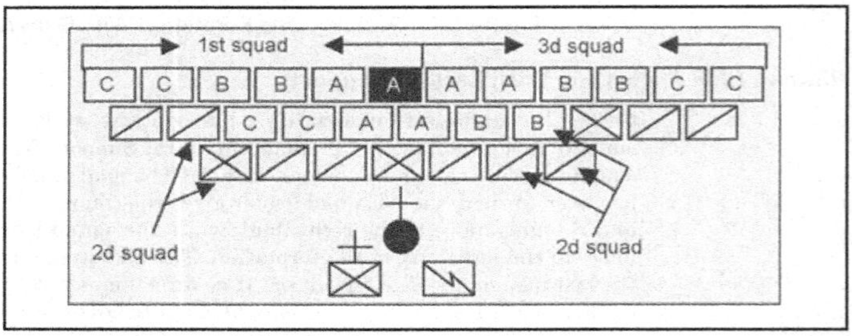

Figure 6-13. Platoon Line Formation With Direct Support

6-64. To assemble a platoon with direct support, the procedure is the same as the platoon line, except that the 2d squad moves to the AA first. To assemble the support squads from any position to general support, the platoon leader commands, "2d Squad in Support, Move." The 2d squad then returns to a column behind the line formed by the other two squads.

Platoon Echelon Right and Left Formations

6-65. The command for forming a platoon echelon right is "Platoon Echelon Right, Move." After the platoon leader's preparatory command, the squad leader of the 1st squad commands, "Follow Me." The squad leaders of the 2d and 3d squads command, "Stand Fast." On the command of execution, the 1st squad moves out and executes an echelon right at the location designated by the platoon leader. As the 1st squad clears the column, the 3d and 2d squads extend the echelon. The 2d squad performs support to this formation in the same ways it supports a line formation.

6-66. The command for forming a platoon echelon left is "Platoon Echelon Left, Move." The platoon echelon right is formed in the same manner as the echelon left (see *Figure 6-14*). *Figures 6-15; 6-16,* page 6-22; and *6-17,* page 6-23 show support formations in a left echelon.

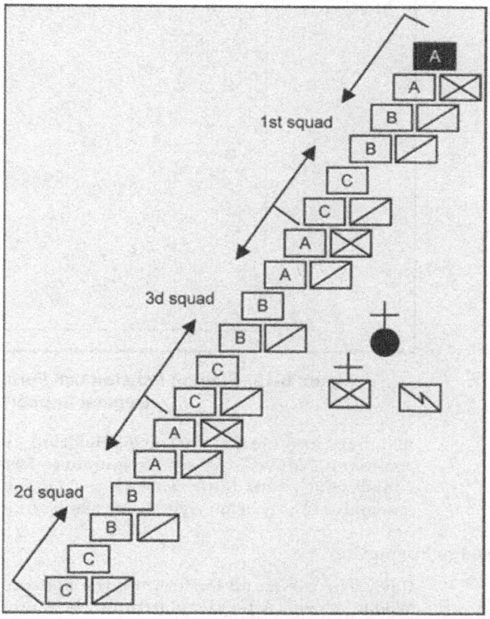

Figure 6-14. Platoon Echelon Left Formation

6-67. To assemble the platoon, the commander takes his position to the rear of the formation with the platoon HQ. On the preparatory command, "Platoon Assemble," the 1st squad leader does an about-face movement while all other

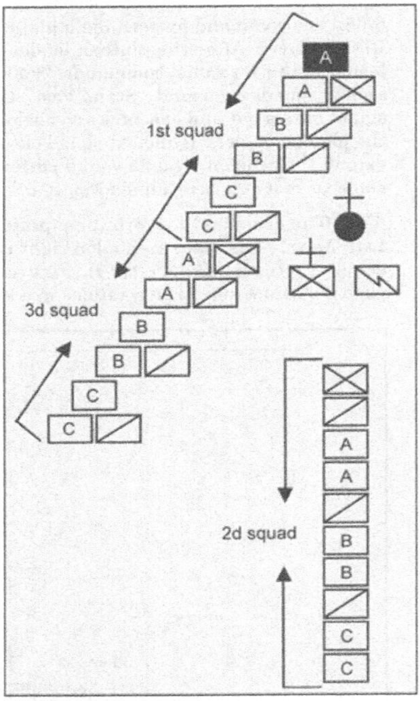

Figure 6-15. Platoon Echelon Left Formation With 2d Squad in General Support

members assume the port arms position and face the baseman. On the command, "Move," the 1st squad moves first, followed by the 3d and 2d squads. Each squad dresses with the 1st squad as it reaches the AA. The assembly of the echelon right formation is the same as the echelon left.

Platoon Wedge Formation

6-68. The command for forming the platoon wedge formation is "Platoon Wedge, Move." After the platoon leader's preparatory command, the squad leaders of the 1st and 3d squads command, "Follow Me." At the same time, the 2d squad leader commands, "Stand Fast." On the platoon leader's command of execution, the 1st and 4th squads move directly to the front. The 1st squad executes an echelon left on the baseman while the 3d squad executes an echelon right. When the last man of the 1st and 4th squads clear the 2d squad, the 2d squad leader commands, "Follow Me," and moves out to the left and right. The 2d squad splits and the A and B teams go to the right and the C

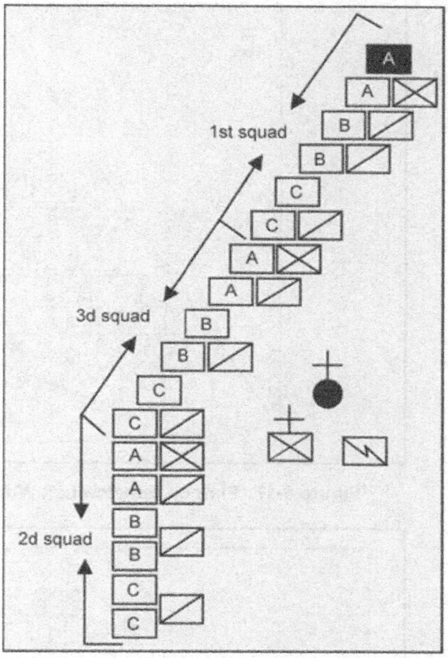

Figure 6-16. Platoon Echelon Left Formation With 2d Squad in Lateral Support

team goes to the left to extend the formation (see *Figure 6-18*). Assembly of the wedge formation is the same as the platoon line.

Platoon Wedge Formation With General Support

6-69. The command for forming a platoon wedge formation with support is "Platoon Wedge, 2d Squad in Support, Move." The 1st and 3d squads execute the wedge while the 2d squad remains in a column formation (see *Figure 6-19,* page 6-24).

Platoon Wedge Formation With Lateral Support

6-70. The command for forming a platoon wedge formation with lateral support is "Platoon Wedge, 2d Squad in Lateral Support, Move." The 1st and 3d squads execute the wedge while the 2d squad stands fast. After the 1st and 3d squads have cleared the column, the 2d squad leader commands, "Follow Me." If no direction is given, the 2d squad splits to cover both flanks. A squad

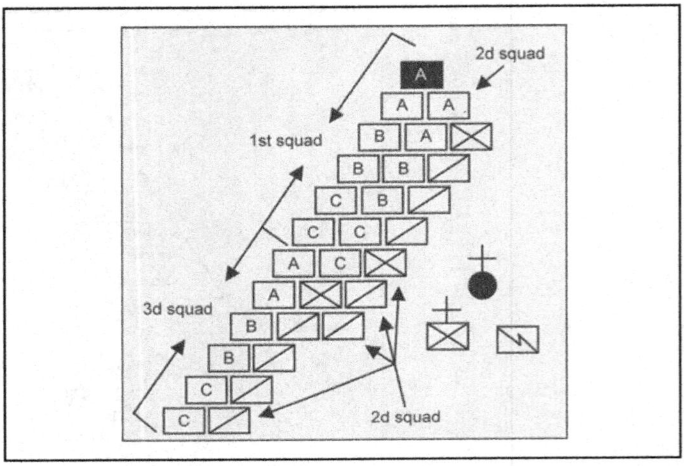

Figure 6-17. Platoon Echelon Left With 2d Squad in Direct Support

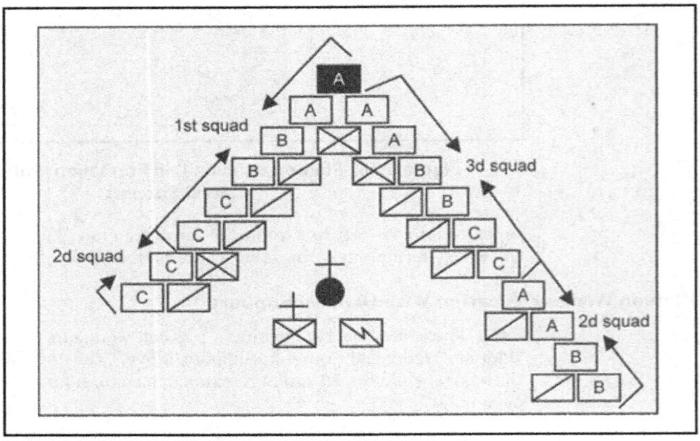

Figure 6-18. Platoon Wedge Formation

may be used to cover one flank by commanding, "2d Squad Right (or Left) Lateral Support" (see *Figures 6-20* and *6-21*, page 6-25).

6-71. To move the 2d squad from general support to lateral support, the platoon leader commands, "2d Squad, Lateral Support, Move." To have the 2d squad join the wedge from general or lateral support, the platoon leader

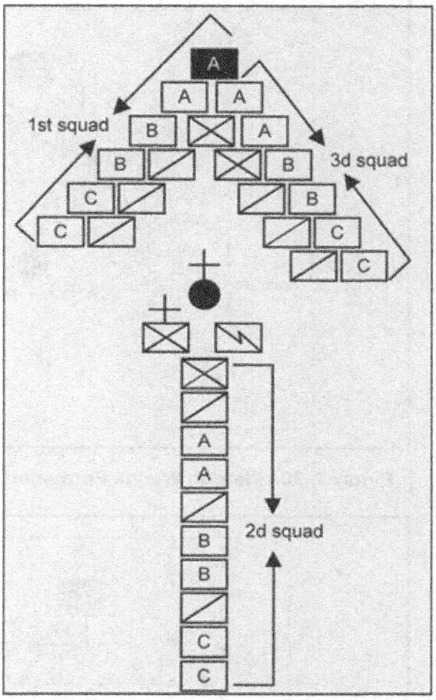

Figure 6-19. Platoon Wedge Formation With 2d Squad in General Support

commands, “2d Squad, Extend the Wedge, Move.” To extend the wedge, the 2d squad splits and the teams move to the left and right, respectively.

Platoon Wedge Formation With Direct Support

6-72. The command for forming a platoon wedge formation with the 2d squad in direct support is “Platoon Wedge, 2d Squad in Direct Support, Move.” The 1st and 3d squads execute a wedge. The 2d squad waits for the squad and/or team leaders to move to the rear, as in the line formation, before executing a wedge directly behind and centered on the main formation. The individuals in the supporting wedge cover the intervals between individuals in the leading wedge (see *Figure 6-22,* page 6-26).

Platoon Diamond and Circle Formations

6-73. These formations are used in open areas or where the crowd may be able to envelop the formation. The diamond and circle formations are primarily defensive in nature and difficult to maneuver. The decision about which formation to use is based on the ability of the platoon to perform them.

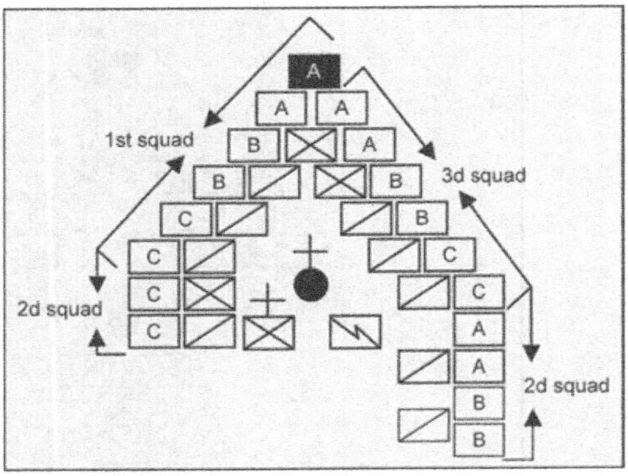

Figure 6-20. Platoon Wedge Formation With 2d Squad in Lateral Support

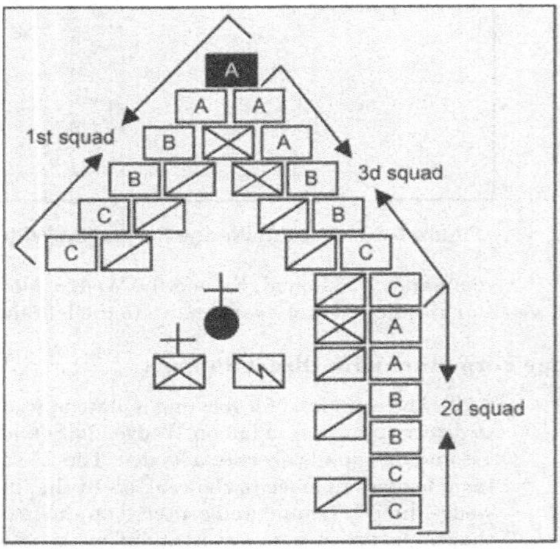

Figure 6-21. Platoon Wedge Formation With 2d Squad in Right Lateral Support

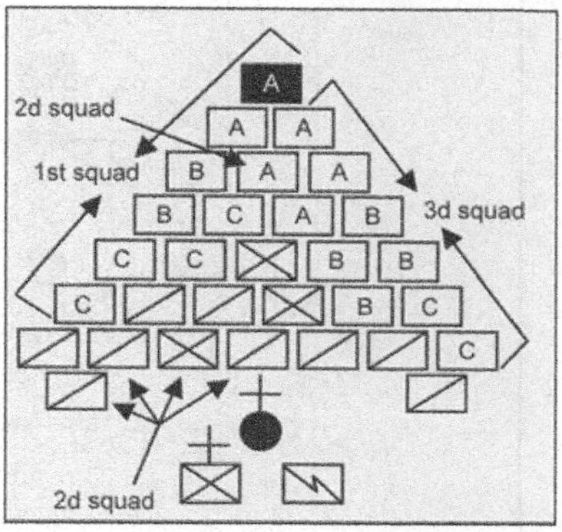

Figure 6-22. Platoon Wedge Formation With 2d Squad in Direct Support

The command for having a platoon form in a diamond or circle formation is "Platoon Diamond (or Circle), Move." On the command of execution, the 1st squad leader moves the baseman to the place indicated by the squad leader. The baseman becomes the 12 o'clock position while the 1st squad lines up to the left of the baseman covering the 8 to 12 o'clock positions. The 3d squad aligns itself to the right of the baseman, covering the 12 to 4 o'clock positions and the 2d squad aligns itself, covering the 4 to 8 o'clock positions (see *Figures 6-23* and *6-24*).

6-74. To assemble from a platoon diamond or circle formation, the HQ element takes the same steps as that of a line formation. The 1st squad leads the platoon back to the AA followed by the 2d and 3d squads. Squads fall into a column formation as they file back to the AA.

RELIEF ELEMENTS IN PLACE

6-75. During crowd control operations, the platoon leader may rotate his squads to give them a rest. He is not limited to using only those squads specified in the preceding examples as the lead elements of the formations. He can replace the base elements by commanding direct support to the 1st squad and then directing the 1st squad into general support. The support element (the 2d squad) then passes through the 1st squad. Once the 2d squad is formed directly behind the 1st squad, the 1st squad leader commands, "Right Face." The members of the 1st squad do a right-face movement and the 2d squad steps quickly though the 1st squad and forms the base formation. The 1st squad leader then commands, "Follow Me," and leads his squad to the general support position. This procedure is often necessary when protective masks are used and chemical irritants employed. The 1st squad then replaces the 3d squad in the same manner. Platoons participating in company operations are relieved in the same manner.

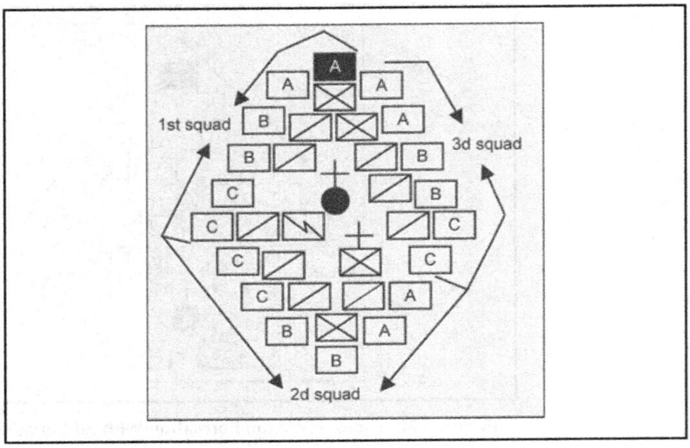

Figure 6-23. Platoon Diamond Formation

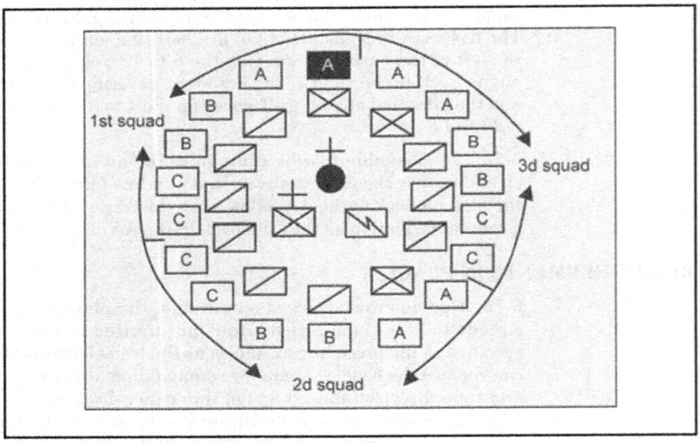

Figure 6-24. Platoon Circle Formation

FOUR-SQUAD PLATOON FORMATIONS

6-76. When forming control force formations from the column, the platoon leader moves to the right of the platoon and faces it when giving his commands. He gives the hand-and-arm signal and the preparatory command. On the command of execution, "Move," he points to the approximate location where he wants the platoon to form. The platoon leader pauses between the preparatory command and the command of execution to allow each squad leader to issue a preparatory command to his squad.

Platoon Line Formation

6-77. The command for forming a platoon line formation is "Platoon Line, Move." Immediately following the platoon leader's preparatory command, the squad leaders of the 1st and 4th squads command, "Follow Me." The squad leaders of the 2d and 3d squads command, "Stand Fast." On the platoon leader's command of execution, the 1st and 4th squads move forward to the point designated by the platoon leader. The 1st A team member of the 1st squad is the baseman for the platoon formation. The 1st squad forms a squad line to the left of the baseman and the 4th squad forms a squad line to the right of the baseman with all lines dressed to the right. After the 1st and 4th squads have cleared the platoon, the 2d and 3d squad leaders command, "Follow Me." The 2d squad forms a squad line to the left of the 1st squad. The 3d squad forms a squad line to the right of the 4th squad. All squads dress with the 4th squad (see *Figure 6-25*).

Platoon Line Formation Assembly

6-78. When assembling from the platoon line formation, the platoon leader and his HQ personnel take a position to the rear of the platoon. He then gives the command, "Platoon Assemble," while raising his right arm and making a circular motion above his head. Platoon members automatically assume the port arms position. The 1st squad leader and the baseman do an about-face movement. All other platoon members do an about-face movement toward the baseman. On the platoon leader's command of execution, he points to where he wants the platoon to assemble. He commands, "Follow Me," and moves his squad toward the AA. The 1st squad files to the baseman position, followed by the 2d squad. As the 2d squad clears the baseman position, the 4th squad leader then commands, "Follow Me," and leads his squad to the AA, followed by the 3d squad. As the 1st squad comes to a halt facing the crowd in the AA, the 2d squad assumes its position to the right of the 1st squad. As the 4th squad assembles, it dresses on the 1st squad, leaving space for the 3d squad to return to its place in the formation. Once the platoon is formed into a column formation, HQ personnel take their position in the formation. The PSG then commands, "Countercolumn, March," and moves the platoon to the rally point.

Platoon Line Formation With General Support

6-79. Whenever the command for establishing a crowd control formation contains the phrase "In Support Without Modification," it means that the support element is to remain in general support. The command for forming a platoon line with two squads in general support is "Platoon Line, 2d and 3d Squads in Support, Move." The 1st and 4th squads execute the line, while the 2d and 3d squads remain in the column (see *Figure 6-26,* page 6-30). To assemble a platoon with two squads in general support, the procedure is the same as that of the platoon line, except that the 1st and 4th squads dress on with the 2d and 3d squads.

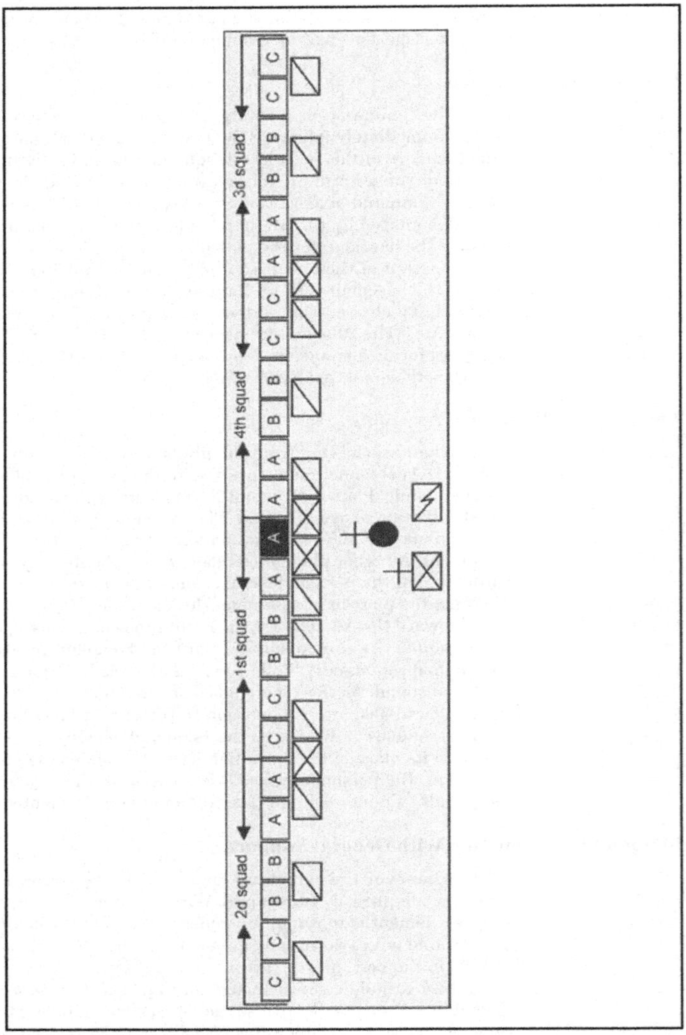

Figure 6-25. Platoon Line Formation With Four Squads

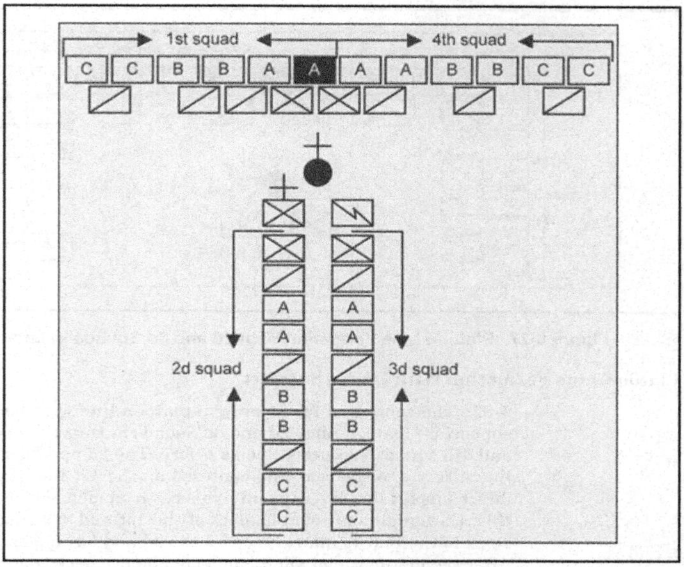

Figure 6-26. Platoon Line Formation With d and 3d Squads in General Support

Platoon Line Formation With Lateral Support

6-80. The command for forming a platoon line formation with two squads in lateral support is "Platoon Line, 2d and 3d Squads in Lateral Support, Move." The 4th and 1st squads execute the line as before, and the 2d and 3d squads stand fast. After the line has been formed, the squad leaders of the 2d and 3d squads command, "Follow Me." The 2d squad forms a column behind the last man of the 1st squad. The 3d squad forms a column behind the last man of the 4th squad (see *Figure 6-27*).

6-81. Squads may be moved from general support to lateral support at any time by the platoon leader. He commands, "2d and 3d Squads, Lateral Support, Move." To have the 2d and 3d squads join the line from either support position, the platoon leader commands, "2d and 3d Squads, Extend the Line, Move." The 2d and 3d squad leaders command, "Follow Me." On the command of execution, the squads set up individual lines, closing and dressing with the existing line. To assemble the platoon from a platoon line with lateral support, the procedure is the same as that of the platoon line, except lateral support squads face forward until they move to the AA.

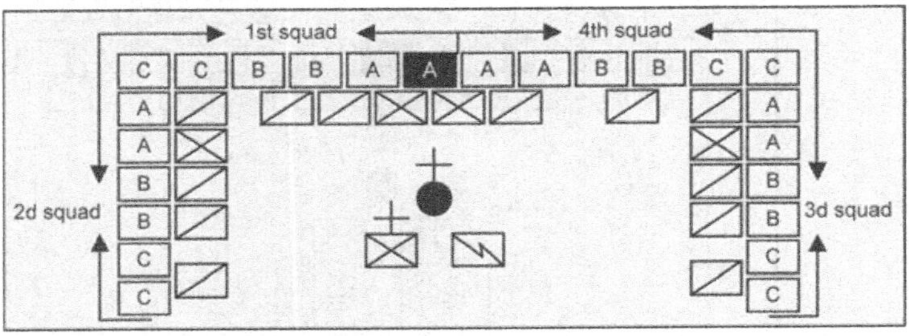

Figure 6-27. Platoon Line Formation With 2d and 3d Squads in Lateral Support

Platoon Line Formation With Direct Support

6-82. The command for forming a platoon line with two squads in direct support is "Platoon Line, 2d and 3d Squads in Direct Support, Move." The 1st and 4th squads execute a line as before. The 2d and 3d squads execute a 2d line directly to the rear and centered on the 1st line. The members in the direct support line cover the intervals between members in the baseline. To do this, the squad and team leaders of the 1st and 4th squads must take two steps to the rear to allow space for the 2d and 3d squads to move into place. After seeing the squad and team leaders move back from the line, 2d and 3d squad leaders command, "Follow Me," and lead their squads into position (see *Figure 6-28*).

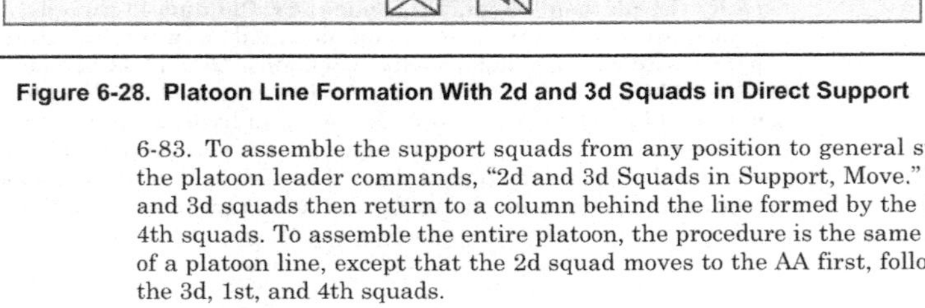

Figure 6-28. Platoon Line Formation With 2d and 3d Squads in Direct Support

6-83. To assemble the support squads from any position to general support, the platoon leader commands, "2d and 3d Squads in Support, Move." The 2d and 3d squads then return to a column behind the line formed by the 1st and 4th squads. To assemble the entire platoon, the procedure is the same as that of a platoon line, except that the 2d squad moves to the AA first, followed by the 3d, 1st, and 4th squads.

Platoon Line Formation With One Support Squad

6-84. The command for forming a platoon line formation with one squad in support is "Platoon Line, 3d Squad in Support, Move." The 1st squad moves out and executes a squad line. The 2d and 4th squads form a line on the side of the left and rights sides of the 1st squad. The 3d squad remains in column formation. The support squad may be used in lateral support on one or both sides of the formation or in direct support of any segment of the formation. To assemble the platoon, the procedure is the same as that of a platoon line, except that all squads dress with the 3d squad.

Platoon Echelon Right and Left Formations

6-85. The command for forming a platoon echelon right formation is "Platoon Echelon Right, Move." After the platoon leader's preparatory command, the squad leader of the 1st squad commands, "Follow Me." The squad leaders of the 2d, 3d, and 4th squads command, "Stand Fast." On the command of execution, the 1st squad moves out and executes an echelon right at the location designated by the platoon leader. As each squad clears the column, the next successive squad moves out and extends the echelon that has been formed. To assemble the platoon, the procedure is the same as that of a platoon line, except that all squads file back in order.

6-86. The command for forming a platoon echelon left formation is "Platoon Echelon Left, Move." The platoon echelon left is formed in the same manner as the echelon right, but in reverse order (see *Figure 6-29*). To assemble the platoon, the procedure is the same as that of an echelon right. See *Figures 6-30,* page 6-34*;* *6-31,* page 6-35*;* and *6-32,* page 6-36, for examples of echelon formations with support.

Platoon Wedge Formation

6-87. The command for forming the platoon wedge formation is "Platoon Wedge, Move." After the platoon leader's preparatory command, the 1st and 4th squad leaders command, "Follow Me." At the same time, the 2d and 3d squad leaders command, "Stand Fast." On the platoon leader's command of execution, the 1st and 4th squads move directly to the front. When the last members of the 1st and 4th squads clear the 2d and 3d squads, the 2d and 3d squad leaders command, "Follow Me," and move out to the left and right. The 1st squad executes an echelon left on the baseman and the 4th squad executes an echelon right. The 2d squad forms an echelon left on the last man of the 1st squad. The 3d squad forms an echelon right on the last man of the 4th squad. This completes the wedge formation (see *Figure 6-33,* page 6-36). To assemble the platoon, the procedure is the same as that of a platoon line.

Platoon Wedge Formation With General Support

6-88. The command for forming a platoon wedge formation with two squads in support is "Platoon Wedge, 2d and 3d Squads in Support, Move." The 1st and 4th squads execute the wedge while the 2d and 3d squads remain in column formation (see *Figure 6-34,* page 6-37).

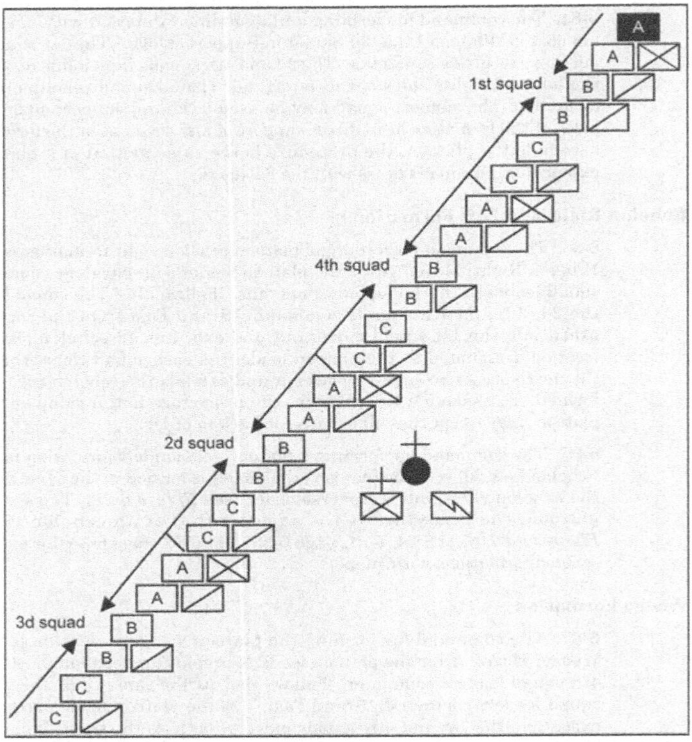

Figure 6-29. Platoon Echelon Left Formation With Four Squads

Platoon Wedge Formation With Lateral Support

6-89. The command for forming a platoon wedge formation with lateral support is "Platoon Wedge, 2d and 3d Squads in Lateral Support, Move." The 1st and 4th squads execute the wedge while the 2d and 3d squads stand fast. After the 1st and 4th squads have cleared the column, the 2d and 3d squad leaders command, "Follow Me," and move out to the left and right. The 2d squad forms a column formation behind the last man of the 1st squad. The 3d squad forms a column formation behind the last man of the 4th squad (see *Figure 6-35,* page 6-38).

6-90. To move the 2d and 3d squads from general to lateral support, the platoon leader commands, "2d and 3d Squads, Lateral Support, Move." To have the 2d and 3d squads join the wedge from general or lateral support, the

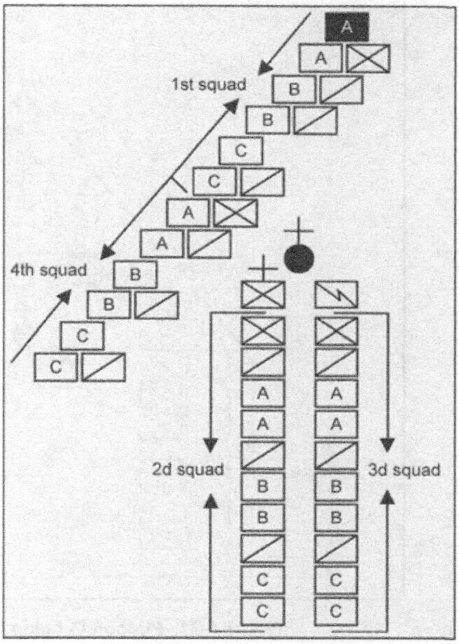

Figure 6-30. Platoon Echelon Left Formation With 2d and 3d Squads in General Support

platoon leader commands, "2d and 3d Squads, Extend the Wedge, Move." The 2d and 3d squad leaders command, "Follow Me," and move out to the left and right to extend the wedge on the 1st and 4th squads.

Platoon Wedge Formation With Direct Support

6-91. The command for forming a platoon wedge with two squads in direct support is "Platoon Wedge, 2d and 3d Squads in Direct Support, Move." The 1st and 4th squads execute a wedge. The 2d and 3d squads execute a wedge and close in on the leading wedge. The members in the supporting wedge cover the intervals between members in the leading wedge. To do this, the squad and team leaders of the 1st and 4th squads must take two steps to the rear to allow space for the 2d and 3d squads to move into place. After seeing the squad and team leaders move back from the line, the 2d and 3d squad leaders command, "Follow Me," and lead their squads into position (See *Figure 6-36,* page 6-38). To move the 2d and 3d squads back to general support, the platoon leader commands, "2d and 3d Squads Support, Move." Then the 2d and 3d squads move to their general support positions. To

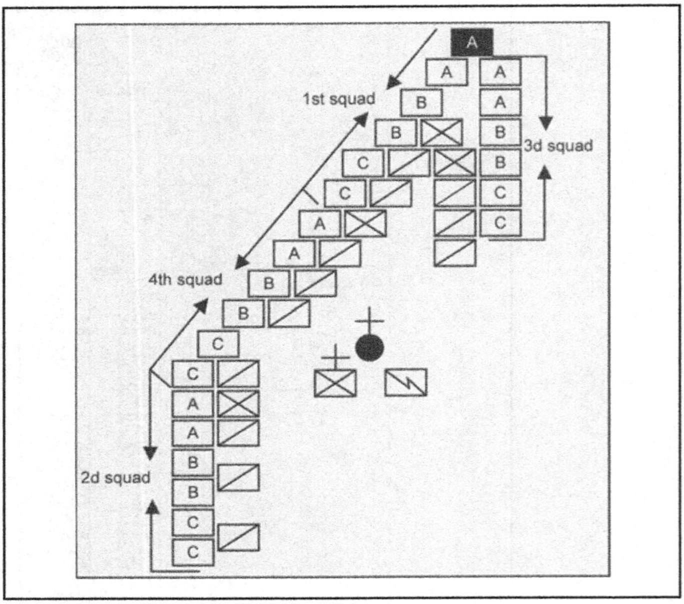

Figure 6-31. Platoon Echelon Left Formation With 2d Squad in Lateral Support

assemble a platoon from a platoon wedge with direct support, the procedure is the same as that of a platoon wedge, except that the 2d squad assembles first, followed by the 3d, 1st, and 4th squads.

Platoon Wedge Formation With One Support Squad

6-92. The command for forming a platoon wedge formation with one squad in support is "Platoon Wedge, 3d Squad in Support, Move." The 1st squad moves out and executes a squad wedge. The 2d and 4th squads form echelons left and right on the 1st squad. The 3d squad remains in the column formation.

Platoon Diamond and Circle Formations

6-93. The platoon diamond and circle formations are used in open areas or where the crowd may be able to envelop the formation. They are primarily defensive in nature and hard to maneuver. The decision concerning which formation to use is based on the ability of the platoon to perform them. The command for having a platoon form in a diamond or circle formation is "Platoon Diamond (or Circle), Move." On the command of execution, the 1st squad leader moves the baseman to the location indicated by the squad leader. The baseman becomes the 12 o'clock position. The 1st squad lines up to the

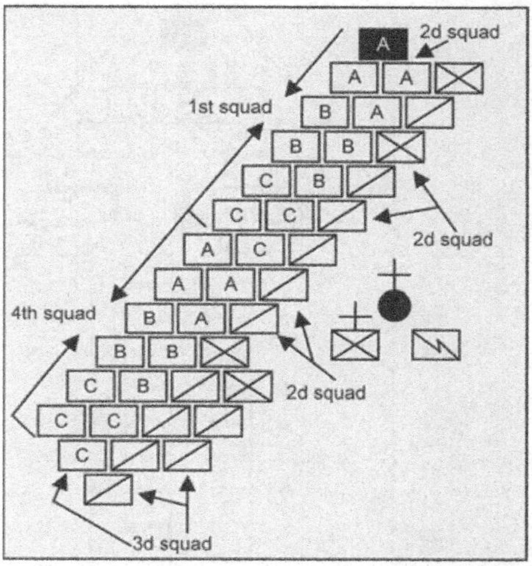

Figure 6-32. Platoon Echelon Left Formation With 2d and 3d Squads in Direct Support

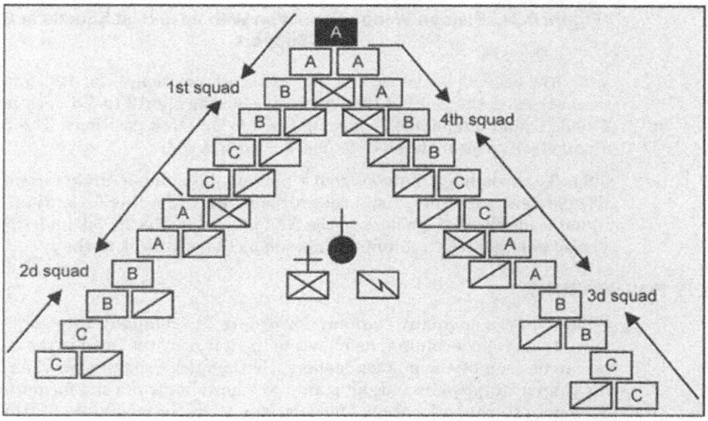

Figure 6-33. Platoon Wedge Formation With Four Squads

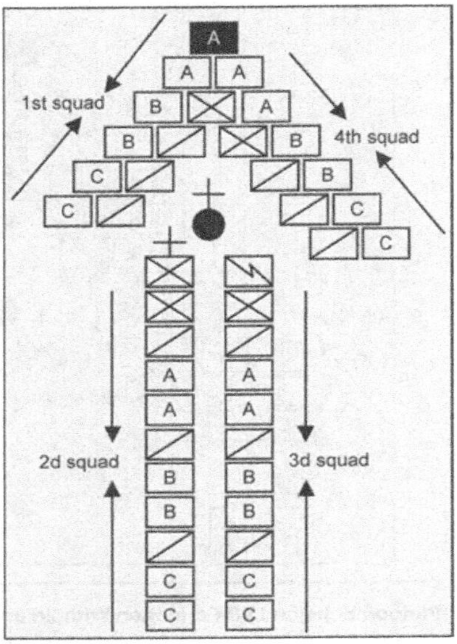

Figure 6-34. Platoon Wedge Formation With 2d and 3d Squads in General Support

left of the baseman covering the 9 to 12 o'clock positions. The 4th squad aligns themselves to the right of the baseman covering the 12 to 3 o'clock positions. The 3d squad aligns itself covering the 3 to 6 o'clock positions. The 2d squad aligns itself covering the 6 to 9 o'clock positions.

6-94. To assemble a platoon from a platoon diamond or circle formation, the HQ element uses the same procedures as that of a line formation. The 1st squad leads the platoon back to the AA followed by the 2d, 3d, and 4th squads. The squads fall into a column formation as they file back to the AA.

COMPANY FORMATIONS

6-95. When a company commander orders his company into control force formations from a column, he moves to the left near the head of the column so he can be seen by the platoon leaders. He then faces the company. As he gives his preparatory command, he points to where he wants the formation. If he does not indicate a location, the company forms immediately in front of the lead platoon.

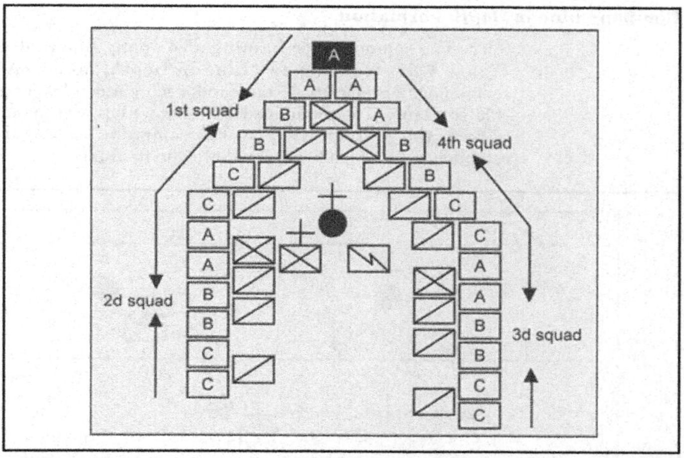

Figure 6-35. Platoon Wedge Formation With 2d and 3d Squads in Lateral Support

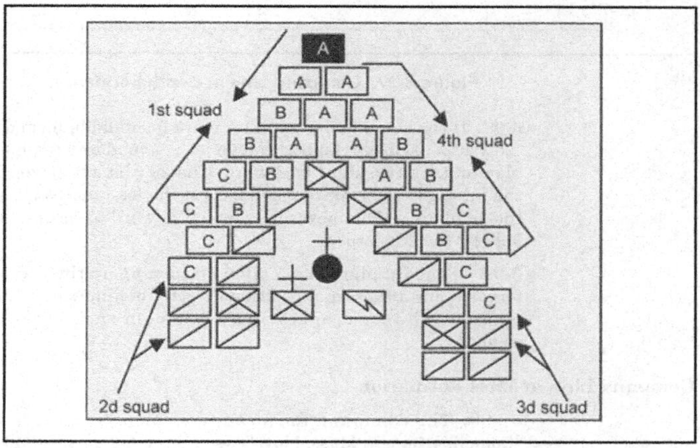

Figure 6-36. Platoon Wedge Formation With 2d and 3d Squads in Direct Support

Relieving Elements in Place

6-96. During crowd control operations, the platoon leader may rotate his squads to give them a rest. See *paragraph 6-76,* page 6-27.

Company Line in Depth Formation

6-97. The command for forming a company line in depth formation (see *Figure 6-37*) is "Company Line in Depth, Move." After the company commander's preparatory command, each platoon leader gives the command, "Platoon Line." The squad leaders follow with their commands to their squads to form the platoon line. On the command of execution, each platoon establishes a platoon line immediately to its front.

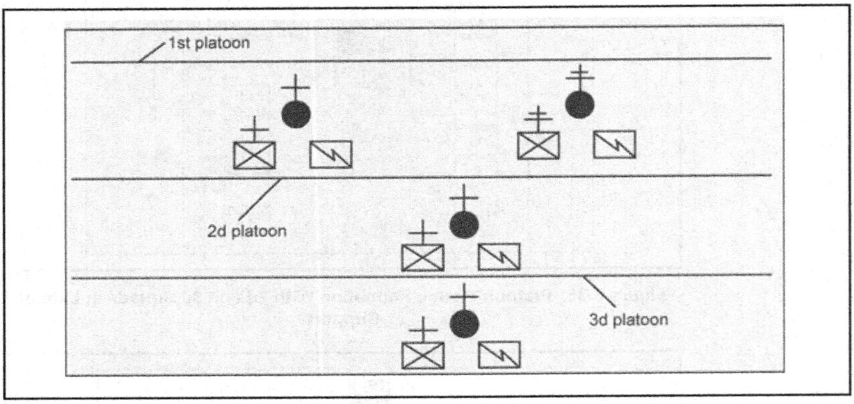

Figure 6-37. Company Line in Depth Formation

6-98. If the commander wants a more formidable formation, he commands, "2d Platoon, Direct Support, Move." The squad and team leaders from the 1st platoon take two steps backward. The 2d platoon moves forward and covers the intervals between the members of the lead platoon. The platoon leader of the 3d platoon then moves his platoon forward to occupy the position formerly held by the 2d platoon.

6-99. If the 3d platoon is called on to support the company line, it moves forward and takes up a position directly behind the 2d platoon in the same manner. Platoon leaders and PSGs help each other in controlling the company.

Company Line in Mass Formation

6-100. The command for forming a company line in mass formation is "Company Line in Mass, Move" (see *Figure 6-38,* page 6-40). The company line in mass is the same as a company line with both platoons in direct support. Instead of forming it in stages, the company commander indicates his desire for this formation in his initial command. The platoons each form a line, and the 2d and 3d platoons close on the 1st platoon without further command.

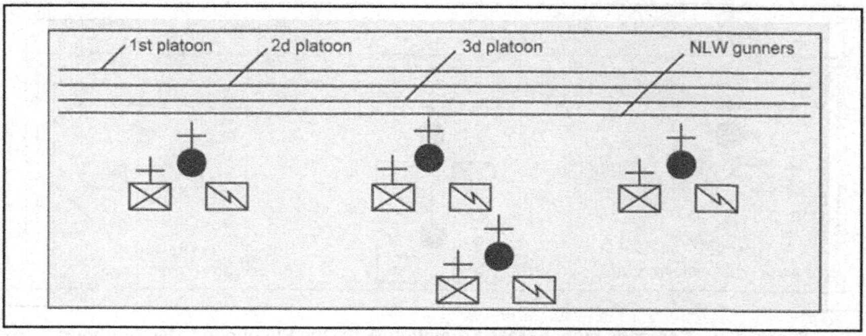

Figure 6-38. Company Line in Mass Formation

Company Line Formation With Support

6-101. In company formations, the 1st platoon usually forms the lead element and the 2d and 3d platoons provide support (see *Figures 6-39; 6-40;* and *6-41*). The support platoons can be employed in the same manner as the support squads in platoon formations. One platoon can be relieved from the lead element by another platoon while in formation by having a support platoon pass through the lead platoon.

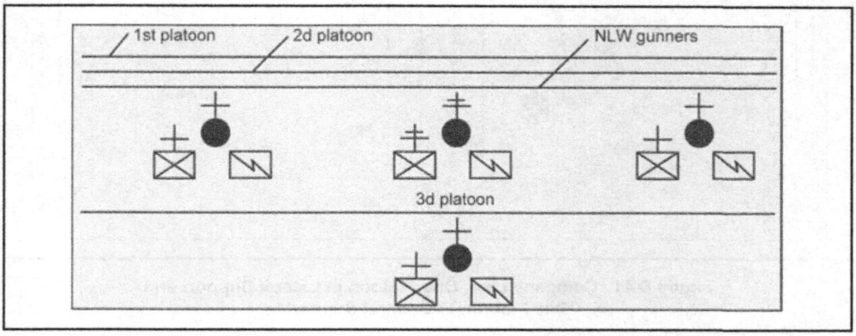

Figure 6-39. Company Line in Mass Formation With One Platoon in Depth

Company Echelons and Wedge Formations

6-102. Company echelons (see *Figures 6-42,* page 6-42*; 6-43,* page 6-43*; 6-44,* page 6-43*;* and *6-45,* page 6-44) and company wedge formations (see *Figures 6-46,* page 6-45; *6-47,* page 6-46; *6-48,* page 6-46; and *6-49,* page 6-47) are formed in the same manner and with the same variations as the company line formation. A company assembles from a crowd control formation in the same manner as a platoon or a squad does. However, due to the size of company

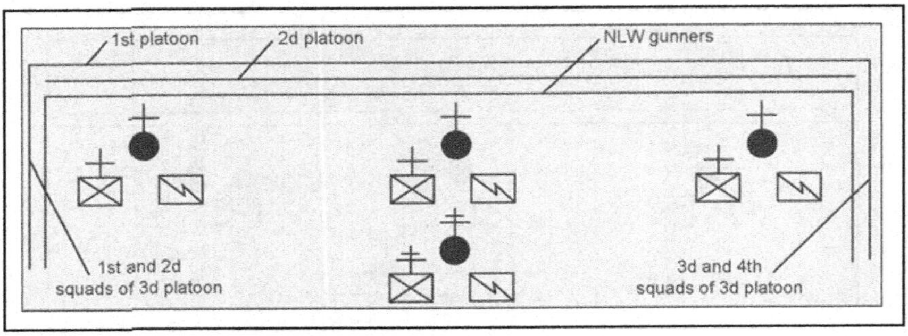

Figure 6-40. Company Line in Mass Formation With One Platoon in Lateral Support

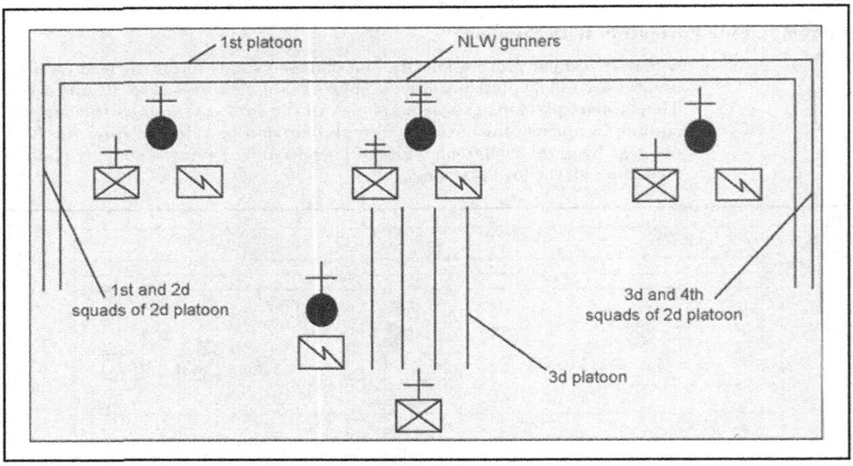

Figure 6-41. Company Line, One Platoon in Lateral Support and One Platoon in General Support

formations, the commander must consider the area that the company is operating in when he assembles the company. If he follows the guidelines of platoon assembly, he will have no difficulty. When assembling the company, the commander takes a position to the rear of the formation and commands, "Company Assemble." The platoon leaders take charge of their respective platoons. Usually, the 3d platoon assembles first, just behind the formation. After the 3d platoon is assembled, the 2d platoon assembles in front of the 3d platoon and the 1st platoon assembles in front of the 2d. All platoons assemble facing the commander. The 2d and 3d platoons then dress with the 1st platoon. As each platoon is assembled, they are ordered to port arms by the platoon leader to await further orders from the commander. If four platoons are used, the 4th platoon assembles first.

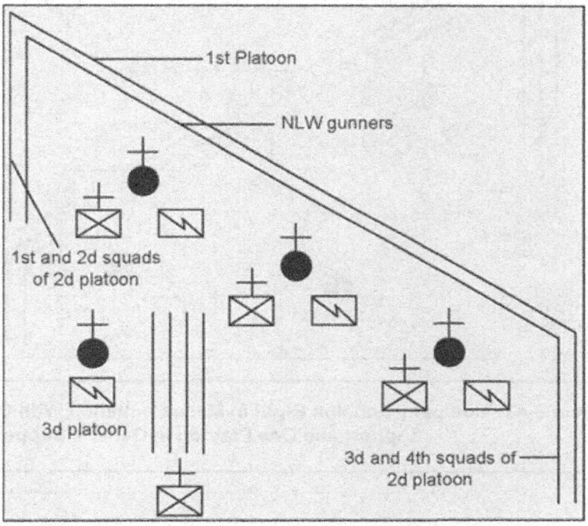

Figure 6-42. Company Echelon Right Formation With One Platoon in Lateral Support and One Platoon in General Support

Company and Platoon Formations With Vehicles

6-103. The commands for vehicles and foot troops are the same as those for foot troops alone. Because of the additional noise and distances involved, the company commander gives the hand-and-arm signal for the line. The command and the signal apply to the lead platoon only. The other platoons remain in general support until further directed (see *Figures 6-50,* page 6-48; *6-51,* page 6-49; *6-52,* page 50; *6-53,* page 6-50; *6-54,* page 6-51; *6-55,* page 6-52; and *6-56,* page 6-53).

6-104. The motor section moves out with vehicle number two moving to the right. Vehicle number two passes the lead vehicle and establishes the position for the center of the line. At the same time, the 3d and 4th vehicles swing out to the left and right. They form to the left and right and slightly to the rear of vehicle number two. The 1st and 4th squads of the lead platoon then move forward and form lines to the left and right, on the front end of vehicle number two. As soon as the troops are in position, the 3d and 4th vehicles close in on the squads and tighten the formation. The 2d and 3d squads of the lead platoon then form a line to the left and right of the 3d and 4th vehicles to complete the company line. The command vehicle (vehicle number one) takes

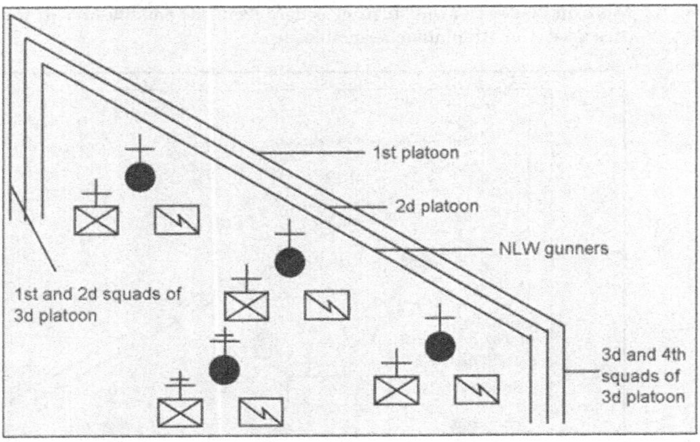

Figure 6-43. Company Echelon Right in Mass Formation With One Platoon in Lateral Support and One Platoon in General Support

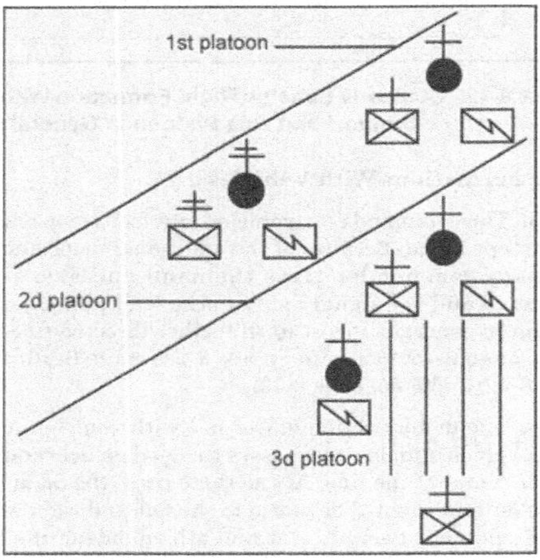

Figure 6-44. Company Echelon Right in Mass Formation With One Platoon in Lateral Support

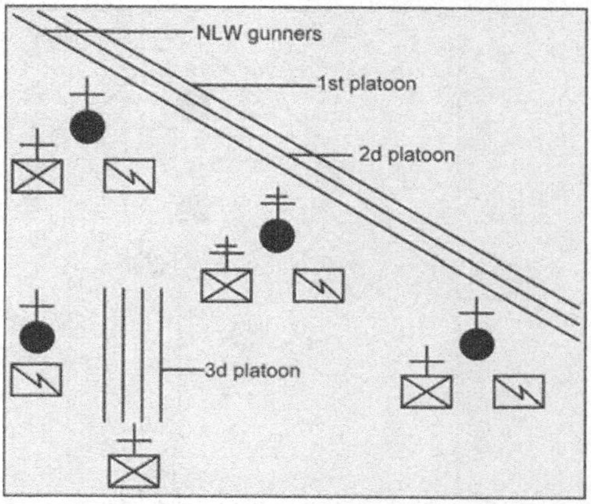

Figure 6-45. Company Echelon Right in Mass Formation With One Platoon in General Support

up a position behind the line where the commander can direct and control the unit. The 2d and 3d platoons move forward and are used in general, lateral, or direct support. In some instances, foot troops move into position first and vehicles join the formation, as needed.

6-105. To form a company echelon or a company wedge with vehicles, the company follows the same procedures used for company formations without vehicles. Vehicles are moved into the appropriate positions using procedures similar to those used for a company line.

6-106. To assemble troops and vehicles, the command vehicle moves to a position behind the formation. The commander faces the formation and gives the vehicle section the signal for assembly. Immediately, the vehicles return to their proper positions in column formation behind the command vehicle while the troops stand fast. The second in command then assembles the troops in the usual manner.

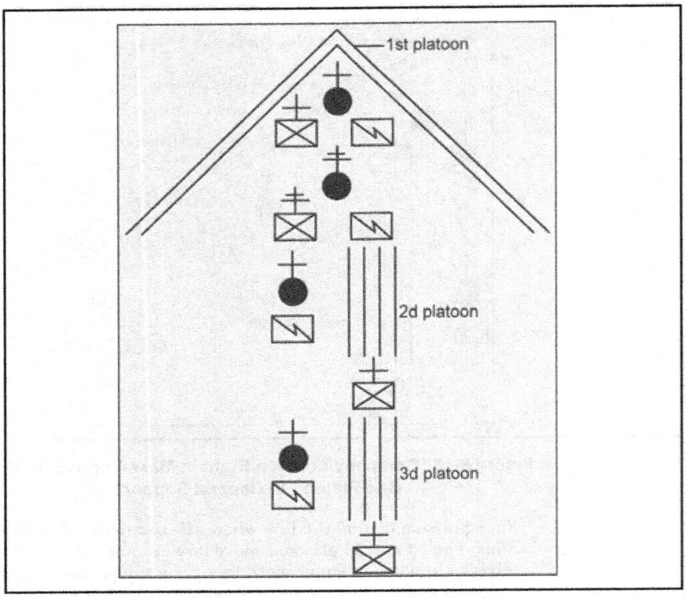

Figure 6-46. Company Wedge Formation With Two Platoons in General Support

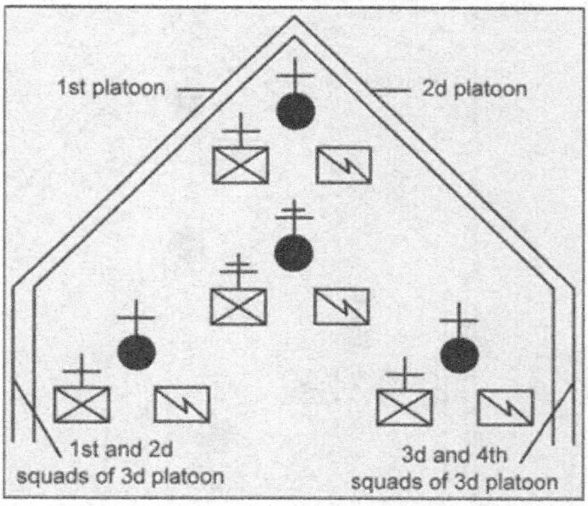

Figure 6-47. Company Wedge Formation With One Platoon in Lateral Support

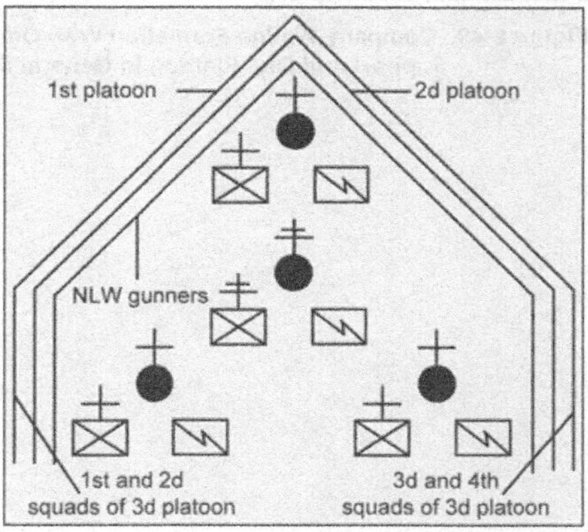

Figure 6-48. Company Wedge Formation With One Platoon in Lateral Support and One Platoon in General Support

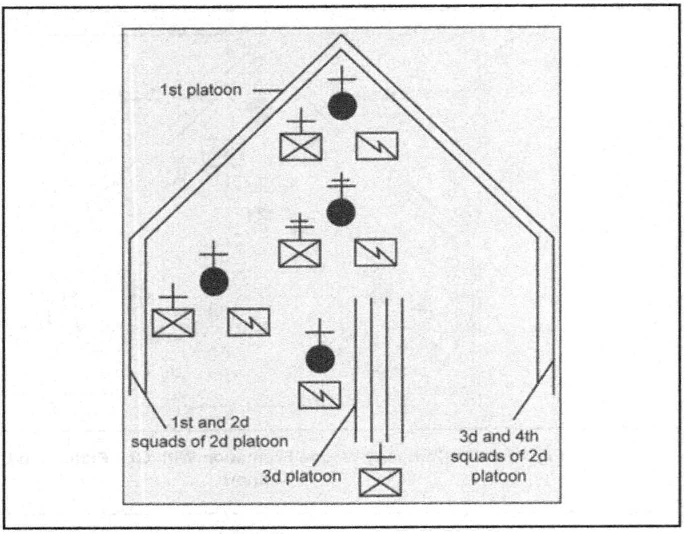

Figure 6-49. Company Wedge Formation With One Platoon in Lateral Support and One Platoon in General Support

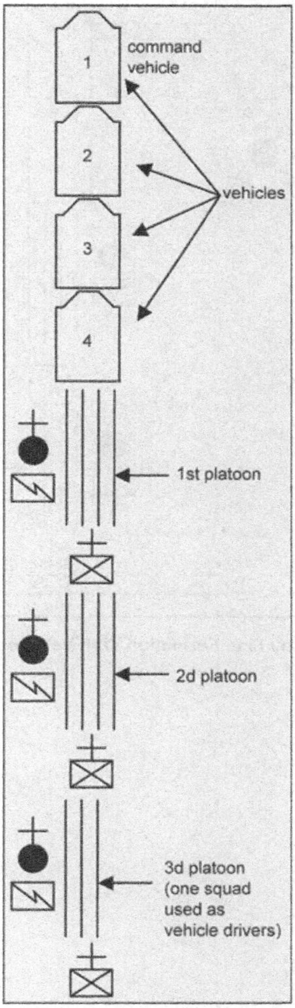

Figure 6-50. Company Column Formation With Vehicles

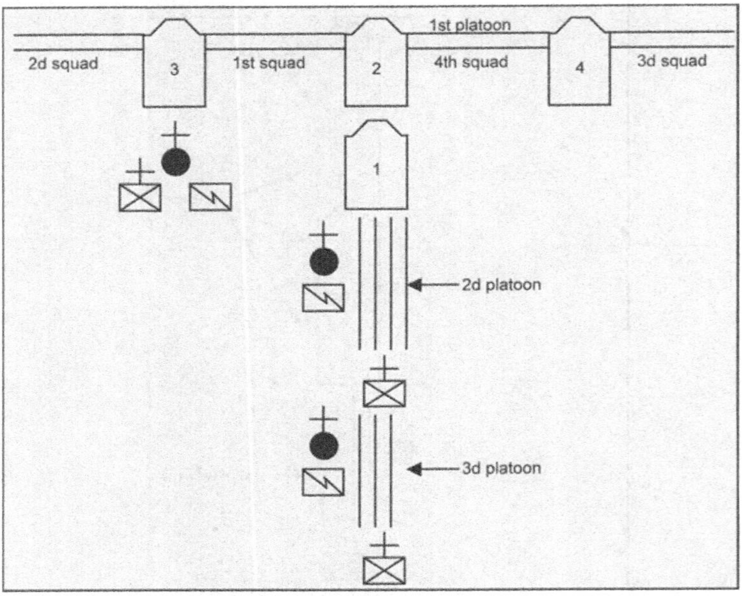

Figure 6-51. Company Line Formation With Vehicles and Two Platoons in General Support

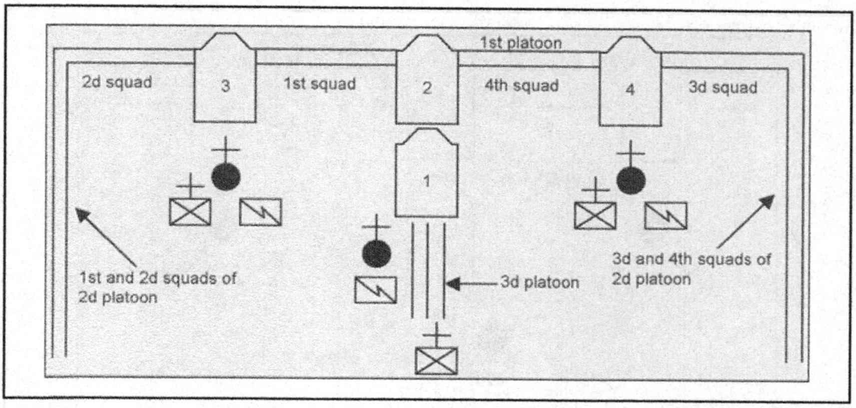

Figure 6-52. Company Line Formation With Vehicles, One Platoon in Lateral Support, and One Platoon in General Support

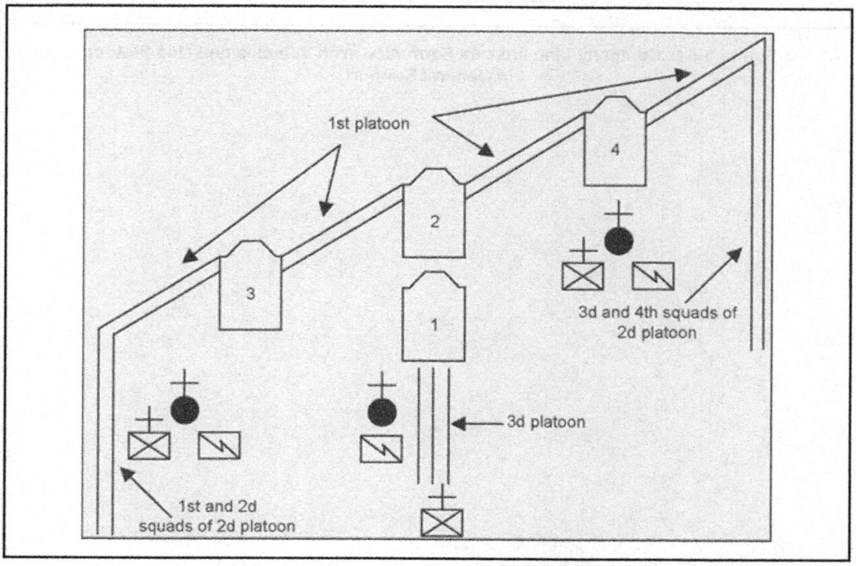

Figure 6-53. Company Echelon Right Formation With Vehicles, One Platoon in Lateral Support, and One Platoon in General Support

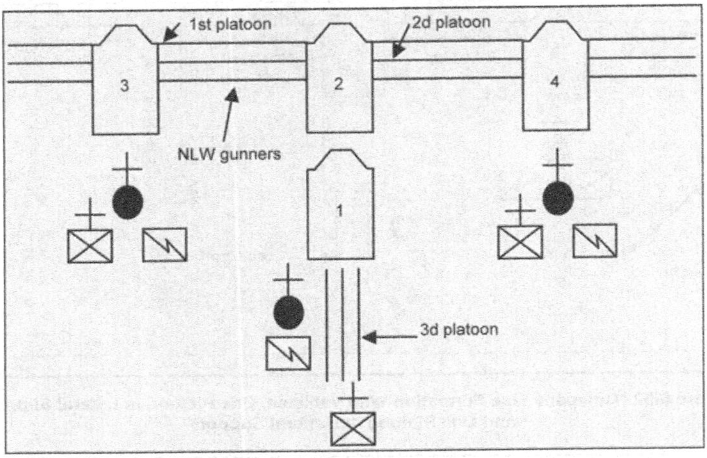

Figure 6-54. Company Line in Mass Formation With Vehicles and One Platoon in General Support

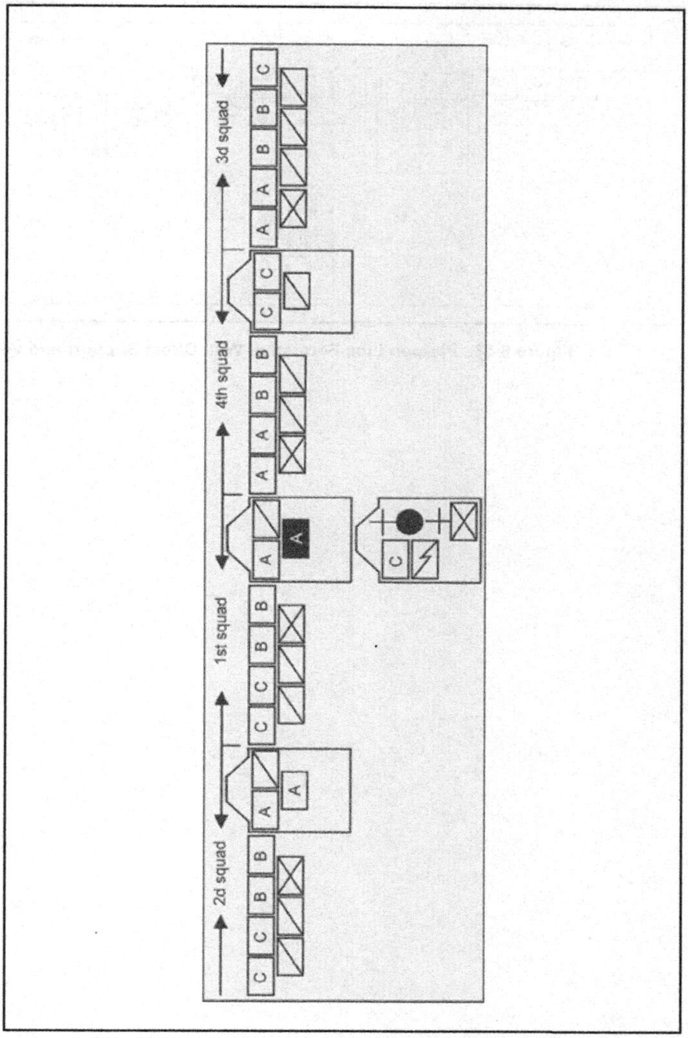

Figure 6-55. Platoon Line Formation With Vehicles

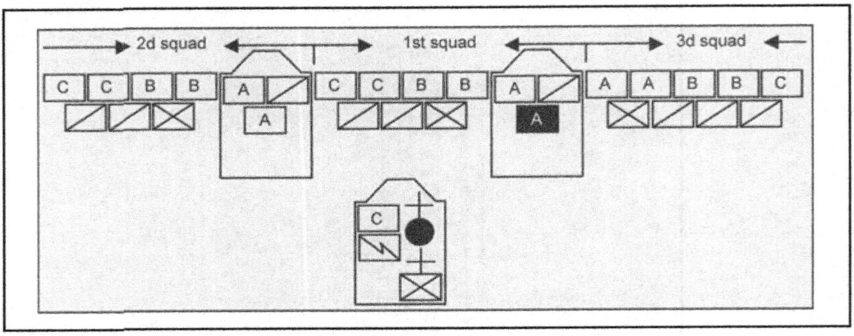

Figure 6-56. Platoon Line Formation With Direct Support and Vehicles

Chapter 7

Civil Disturbance Operations in Confinement Facilities

Civil disturbance doctrine is usually written for quelling disturbances that occur in large open areas where the size of the responding force or the type of formation employed is not hampered by a lack of space. However, in the military confinement environment, inmate disturbances often occur in small, closed areas, such as an inmate's housing unit, a gymnasium, a dining facility, or a chapel where maneuver space and equipment limit the size of the element that may be employed. The layout of each facility is different and presents its own unique problems when employing forces. This chapter covers the use of small military police teams, forced cell move teams (FCMTs), and the formations used to move an unruly inmate from one cell to another. Small generic riot control formations are discussed to provide the commander with a basic employment option that he can tailor to his facility to enable him to regain control of the affected area. The use of NL munitions and RCAs are incorporated into this discussion.

CROWD DYNAMICS WITHIN THE CONFINEMENT FACILITY

7-1. In confinement facility disturbances, inmates employ a number of tactics to resist control or to achieve their goals. Tactics may be unplanned or planned and nonviolent or violent. When a disturbance carries many purposes, it is likely that their tactics are well planned.

NONVIOLENT TACTICS

7-2. Nonviolent tactics range from name-calling to building barricades. Inmates may attempt to distract control force members by shouting at or ridiculing them or by using abusive language, obscene remarks, taunts, and jeers. The inmates' goals are to anger and demoralize the control force. They also want authorities to take actions that may later be exploited as acts of brutality.

VIOLENT TACTICS

7-3. Violent crowd tactics used by inmates are often extremely destructive and can include physical attacks on cadre, other inmates, and property. Use of violent tactics is limited only by the attitudes and ingenuity of the inmates, the training of their leaders, and the materials available to them. Inmates often commit violence with crude homemade weapons. If unplanned violence occurs, inmates will use mops, brooms, chairs, beds, or whatever else is on hand as weapons of violence. During planned violent disturbances, inmates can easily conceal makeshift weapons or tools to use against the control force. Inmates often erect barricades to impede troop movement and/or to prevent the control force from entering certain areas or buildings. Inmates can be expected to vent their emotions on individuals (cadre or inmates), the control force, their equipment, and/or the institution's property. Inmates direct dangerous objects like carts, barrels, liquids, and burning shoe polish at the control force.

GRIEVANCE PROTEST

7-4. A grievance protest may be organized as a riot. However, under normal circumstances, a riot of this type is not of an extremely violent nature. It may turn violent when leaders attempt to exploit any successes or weaknesses found in the control force.

UNORGANIZED RIOTS

7-5. Unorganized riots are characterized as being spontaneous in nature. However, they could be exploited and diverted by leaders at any subsequent stage of the riot, turning it into a different type. They may begin as an isolated assault against authority figures, or acts of civil disobedience. Under determined leadership, an unorganized riot could change to an organized riot.

MULTIPLE RIOTS

7-6. Multiple riots may happen when a disturbance occurs in one area of the confinement facility, causing others to riot in their areas. To control this situation with only one riot control force available, the force commander should first subdue the most violent riot. At the same time, he should employ some of his control forces to contain the other riots until the main control force is able to move to those locations to subdue them.

CADRE BEHAVIOR

7-7. The cadre is susceptible to crowd behavior. They become emotionally stimulated during tense confrontation with unruly and violent inmates. Commanders must be aware of the mood and attitude of the crowd and its effect on the control force cadre. To counteract the effects of crowd behavior on the cadre, the commanders must institute rigorous training and firm and effective leadership. This training must include a complete awareness and understanding of the use of force and the commander's intent. These are all necessary to offset the effect of crowd contagion upon the control force. Even with the best training and leadership, control force members must exercise individual and collective restraint.

PLANNING

7-8. Preplanning is the preparation and training conducted before a crisis occurs. The purpose of preplanning is to have plans and standing operating procedures (SOPs) in place so the cadre may react to an emergency and successfully contain and neutralize the situation. The preplanning process includes plan development, intelligence gathering, and training.

NOTE: Intelligence gathering is essential for the successful containment, neutralization, and/or prevention of a riot.

7-9. Once the control force team has been alerted to a disturbance within the confinement facility, leaders further develop preplans to fit the situation much the same way that DOD develops war plans.

7-10. During the planning process, leaders must be aware of the potential risks involved in quelling disturbances within a confinement facility. Careful planning minimizes collateral damage and risks to control force members and hostages.

7-11. The use of NLW and NL munitions must also be considered when developing plans. There must be strict accountability and control of RCAs, their employment, and other appropriate NL means (such as high-pressure water hoses). When using RCAs, plans must be flexible enough to accommodate changes in the situation and weather because it impacts the effectiveness of some agents. Planners must also consider the—

- Location of the disturbance.
- Estimated number of rioters.
- Accessibility of weapons, tools, or cleaning supplies.
- Inmate's military training.
- Hostage situation.

7-12. Based on an analysis of these factors, the commander makes an estimate of the situation. The estimate must be as thorough as time permits. Using the estimate, the commander considers COAs, selects RCAs, and determines the need for engineer support. In choosing a COA, consider the—

- Number of hostages.
- Number of inmates holding the hostages.
- Leader.
- Attitude and demeanor of the rioters.
- Location of the riot.
  - Inside areas of the confinement facility, such as the dining facility, work site, or housing wing.
  - Outside areas of the confinement facility, such as in exercise or work areas.
- Accessibility of weapons or material to make weapons.
- Accessibility of chemicals, such as cleaning supplies, cooking oils, and shoe polish.

USE OF CHEMICAL IRRITANTS

7-13. The use of chemical irritants can be a valuable NL tool for control force leadership to consider during the planning phase. These chemical irritants can drive a threat from an established, enclosed position or deny the rioters access to a certain area without long-lasting effects to those involved. The proper use of chemical irritants may prevent the control force from having to enter a dangerous area; however, improper use can cause injury, death, or property damage.

USE-OF-FORCE GUIDELINES

7-14. The installation commander authorizes the use of chemical irritants after the control force leader determines the type and dosage. The SOP outlines the procedures for securing authorization and provides guidance to help the control force leader make his decisions. The confinement facility commandant should inform the installation commander of the situation, since the installation commander is ultimately responsible. Team members must be aware of how chemical irritants affect personnel and plan accordingly. Inmates not involved must be removed from the area and the chemical cloud's path. The fire department and emergency medical services (EMS) should be on standby during operations involving chemicals.

PEPPER IRRITANT

7-15. The pepper irritant receives its name from varieties of the capsicum plant that is it made from. The active ingredient, Oleoresin Capsicum (OC), is extracted from the plant and micropulverized to make the irritant. It is neither a chemical nor a gas, but an all-natural organic substance. It is effective against individuals who are emotionally disturbed or under the influence of drugs or alcohol. It is effective against domestic and wild animals. OC is often used to divert the threat from making or continuing an assault. OC does not cause permanent damage. Its effects last 30 to 40 minutes after individuals are placed in fresh air.

7-16. OC can have various effects on people. Some effects are—

- Swelling of mucous membranes and the upper respiratory system; however, it does not shut down the system.
- Burning (intense) and discoloration (bright red) of exposed skin.
- Dilating the capillaries.
- Swelling of the eyelids.
- Burning and involuntary closing of the eyes.
- Coughing uncontrollably.
- Gagging.
- Gasping for air.
- Losing strength and coordination (temporarily).

OC has some disadvantages. They are that—

- The canister depressurizes over time if not checked regularly.
- Inmates may gain access to OC and use it against SRT members.
- The canister must be shaken on a regular basis.

DISSEMINATION OF CHEMICAL IRRITANTS

7-17. There are six methods for disseminating chemical irritants.

Pyrotechnic Method

7-18. In the pyrotechnic method of dissemination, the chemical irritant is placed in a canister with an inert material that is ignited when the device is activated. The agent is then carried into the air on the smoke particles of the inert material.

7-19. There are some disadvantages when using a pyrotechnic device indoors. These disadvantages are that the device—

- May cause a fire.
- Has a very slow saturation time (40 to 45 seconds to burn completely).
- Can be thrown back by the threat, even while burning.

Bursting Method

7-20. In the bursting method of dissemination, the irritant and an inert powder are placed in a saw-toothed canister. After a small detonation, the canister splits and expels the irritant in a cloud.

7-21. The bursting method has its advantages and disadvantages. Some advantages are that—

- There is no risk of fire.
- There is no risk of the canister being thrown back.
- The contents disseminate quickly.
- The canister is effective, easy to carry, and easy to control.

Some disadvantages are that—

- There is a slight possibility of fragmentation when detonated.
- The fuse head may separate from the canister.
- Only 50 to 90 percent of the irritant may be disseminated.
- The direction of the cloud formation depends on where the device is detonated.

Aerosol Method

7-22. In the aerosol method of dissemination, the irritant is suspended in an inert liquid that is located in the rear portion of the device. When the projectile penetrates the structure, the rear portion splits open and the irritant is disseminated in a mist. The aerosol method is most appropriate for tactical operations when chemical irritants are used before entry.

7-23. The aerosol method has its advantages and disadvantages. Some advantages are that it—

- Is easily carried.
- Is an excellent deployment system (uses 40-millimeter, 37-millimeter, or 12-gauge ammunition).
- Disperses the contents quickly.
- Is easily controlled and accurate (fin-stabilized).

NOTE: Practice firing is always necessary.

Some disadvantages are the possibility of—

- Injury to individuals if it is fired from close range.
- Damage to the structure if it strikes a weak portion.

Projectiles Method

7-24. Chemical-containing projectiles can be very effective when used correctly. A soldier must know which projectile to use in each unique situation.

7-25. **Thirty-Seven-Millimeter Projectile.** A 37-millimeter projectile can incapacitate anyone in an enclosed area of 4,500 cubic feet. It can penetrate the following:

- A sheet of ¾-inch-thick plywood at 25 yards.
- An automobile windshield at 33 yards.
- A hollow-core door at 58 yards.

7-26. **Forty-Millimeter Projectile.** A 40-millimeter projectile is more effective at greater ranges than a 37-millimeter projectile because of its rifled bore. It can incapacitate anyone in an enclosed area of 4,500 cubic feet. It can penetrate the following:

- A 1-inch-thick plywood at 10 yards.
- An automobile windshield at 55 yards.
- A hollow-core door at 109 yards.

7-27. **Twelve-Gauge Projectile.** A 12-gauge projectile can be loaded into a shotgun with no modifications to the weapon. It can be fired as a single round, or it can be magazine-fed. The 12-gauge projectile can incapacitate anyone in an enclosed area of 1,000 cubic feet. The angle at which the projectile strikes the barrier can affect its penetration capability. It can penetrate the following:

- A sheet of ¾-inch plywood at 33 yards.
- An automobile windshield at 33 yards.
- A hollow-core door at 100 yards.

Fogging Method

7-28. The irritant is suspended in a fogging solution and dispersed with a fogging machine. It is not recommended for tactical operations.

7-29. The fogging method has its advantages and disadvantages. Some advantages are that it—

- Produces enough chemical irritant to cover an enclosed area of 100,000 cubic feet in 26 seconds.
- Disseminates quickly.
- Provides high-volume capabilities.
- Is effective for riot control or other crowd dispersal situations.

Some disadvantages are that it is—

- Noisy.
- Cumbersome.

Multipurpose Grenade Method

7-30. A multipurpose grenade can be hand-thrown or launched from a shotgun with an adapter. It has an extended shelf life of 6 years and an adjustable fuse delay of 2 to 5 seconds. The irritant is located in the cylinder portion of the device and is forced out the bottom of the grenade upon detonation.

> **WARNING**
> **A multipurpose grenade deployed incorrectly could result in injury. Safety rules should be observed at all times.**

RECORD OF EVENTS

7-31. Reporting procedures for serious incidents should be included in the plans. A record of events must be initiated to provide a basis for the preparation and submission of a formal report to higher HQ. As a minimum, include the—

- Time the incident was reported and by whom.
- Time the incident was reported to the facility commander.
- Time the control force was assembled.
- Time the control force entered the facility.
- Weather conditions as they relate to the use of RCAs.
- Number of cadre and inmates injured or killed, including how they were injured or killed and the medical attention given to them.
- Time the operation was completed.
- Time the riot control force restored order.

TRAINING

7-32. Guard units and associated teams must train on a regular basis in the five basic riot control formations. There must be a continuous training program established, to include—

- Interpersonal communication.
- Self-defense.
- Use of force.
- Use of a riot baton.
- Fatal areas of the body to avoid when using the riot baton.
- First aid.
- Emergency plans.
- Intelligence and counterintelligence techniques.
- The use of RCAs and the various methods of dispersing them.
- The use of NLW.
- The use and effects of high-pressure water.
- Riot control formations.

EQUIPMENT

7-33. The recommended basic riot gear used by the confinement riot control force is as follows:

- Riot baton.
- Riot shield.
- Helmet with face shield.
- Groin protector.
- Flak vest.
- Leather gloves.
- Shin protection.

RIOT BATON

7-34. The riot baton is an invaluable weapon in crowd control situations. A riot baton in the hands of a well-trained control force member is the most appropriate weapon, except with extremely violent crowds. The riot baton is not meant to replace NL munitions, but is instead used in situations in which NL munitions are not needed. Losing a riot baton to the crowd does not present a serious threat.

7-35. Use of the riot baton is based on the commander's appraisal of the situation and his choice of force option. Control forces committed with the riot baton must have teams equipped with RCAs, NLW, and lethal coverage. If the confrontation increases in intensity, the commander has these force options available and may rapidly engage the control force.

7-36. A riot baton is a versatile offensive weapon that varies in length from 26 to 42 inches with the optimum length being 36 inches. It is made of hickory and does not shatter or break easily. Both ends of the riot baton are rounded to prevent unnecessary injury. It is approximately 1¼ inches in diameter. Each riot baton has a hole drilled at the grip end with a leather thong threaded through the hole. The thong helps secure the riot baton to the user's hand.

RIOT SHIELD

7-37. The riot shield (see *Figure 7-1*) is an excellent piece of defensive equipment. It is used by control force personnel in the front and on the flanks of the formation.

7-38. The riot shield has its advantages and disadvantages. Some advantages are that—

- OC pepper spray or high-pressure water can be used to back the control force.
- It provides protection from thrown objects.
- It absorbs some of the impact of rioters hurling themselves at the control force.

The disadvantages of the riot shield are that it—

- Severely limits the use of the riot baton.
- May crack if hit with a heavy object.

Figure 7-1. Riot Shields

RIOT CONTROL FORMATIONS

7-39. The five basic formations for civil disturbance operations are the line, wedge, echelon, diamond, and circle. In the correctional environment the basic formations for disturbances are the line, wedge, or echelon with apprehension teams in support. Generally, the diamond and circle formations are not used in the correctional environment. The minimum size force that the commander should consider is an augmented platoon, with a two-to-one advantage over the rioters.

7-40. Correctional facility cadre must be proficient in all riot control formations. These are usually trained at the squad level and above, but performed by a platoon or company-sized unit. Squad line, echelon left, echelon right, and wedge formations form the basis for platoon and company formations. Each squad must be adept in the basic formations before practicing in platoon-sized or larger formations. For more detailed information on formations, see *STPs 19-95B1-SM* and *19-95B24-SM-TG*. In the correctional environment, the support platoon plays a key role by supporting the control formation with apprehension and equipment teams. See *Chapter 6* for an in-depth discussion on civil disturbance formations.

APPREHENSION TEAMS

7-41. Apprehension teams give the control force the ability to extract inmates. Inmates identified as leaders or agitators and inmates that are injured or no longer want to resist can be extracted from the crowd. Apprehension teams can also be used to enter confined areas where the formation cannot operate.

7-42. The apprehension team will consist of eight members and one leader. Each team will have four shield holders, two nonlethal gunners and two apprehension/search members. The squad leader controls the movement of the team and assists the team in exiting the control formation. He maintains communications with the formation and coordinates the team's operations with the formation. The squad leader also ensures that the team uses the minimum amount of force necessary to extract inmates.

7-43. The apprehension team may deploy through the middle of the formation or from around either of the flanks.

7-44. The apprehension team should be deployed within a distance of approximately 10 meters so that the formation can still provide immediate support. If the apprehension team is deployed farther than 10 meters, the formation must advance to close the gap.

7-45. The team will exit the formation in two columns; shield holders in the first and second ranks, nonlethal gunners in the third rank, apprehension/search members in the fourth rank and the leader last. Team members grip the shoulder or equipment of the member in front of them. The shield holders will envelop the inmate that is to be extracted and form a protective barrier. Once the shield holders have enveloped the inmate, the nonlethal gunners will cover their sides with their weapons while the apprehension/search members control and restrain the inmate. When the team has the inmate under control, the squad leader will give the order for the team to withdraw. The team withdraws in reverse order (apprehension/search members with the inmate, followed by nonlethal gunners and shield holders). Team members then move backwards, face the crowd and grip the shoulder or equipment of the member in front of them.

TEAM EQUIPMENT

7-46. The apprehension team uses the following equipment:

- Flex-Cufs.
- Restraint cutters.
- Pepper spray.
- Helmets with face shields.
- Groin protectors.
- Flak vests.
- Elbow pads.
- Leather gloves.
- Shin protectors.
- Knee pads.
- A video camera.

TEAM COMPOSITION

7-47. The apprehension team is composed of five team members. This team consists of four members and a squad leader. Team member one is the team leader.

Squad Leader

7-48. The squad leader assists the team in exiting the control formation and ensures that the team uses the minimum force necessary to subdue inmates. He is also responsible for signaling the PSG if more help is needed.

Team Member One (Team Leader)

7-49. Team member one's responsibility is to make initial contact with the inmate and use pepper spray and apply Flex-Cufs, as needed. He also carries a set of restraint cutters. Once the apprehension team has returned to the relative safety of the control formation, team member one searches the inmate.

7-50. Team member one receives the command to move forward and secure the inmate from the squad leader. This command is usually given by a signal and verbal command. Team member one then notifies the main force commander where his team exits the control force. Once the team is in position to exit the main formation, team member one signals the member of the formation in front of him. This is usually a prearranged signal, such as a tap on the right shoulder. The control force opens and the team exits the formation and makes contact with the assigned inmate. Team members then subdue and control the inmate. If pepper spray is needed, team member five shouts, "Pepper." This gives the other team members time to react to the command and spray the inmate. Team member one then applies Flex-Cufs to the inmate's wrists and ankles. Depending on the mission, the inmate may be left for follow-on security forces or carried behind the control force to be secured by security forces.

7-51. Should the mission require that the inmate be moved, team member one shouts, "Prepare to Move," this allows each member to grab an arm and position the inmate toward the control force formation. When the team is ready to move, team member two shouts, "Ready," alerting team member one that the team is ready for the last command. Team member one then commands, "Move." The team enters the control formation with the inmate. The inmate is searched by team member one and turned over to the security team.

Team Member Two

7-52. Team member two stands directly behind team member five with his hands on team member five's waist. He protects the left side of team member five by moving to team member five's left and helps to secure the inmate after team member one makes initial contact. He repeats team member five's commands and controls the inmate's head after the inmate is on the ground.

Team Member Three

7-53. Team member three stands behind team member two with his hands on team member two's waist. He protects the right side of team member five by moving to team member five's right and helps to secure the inmate after initial contact.

Team Member Four

7-54. Team member four stands behind team member three with his hands on team member three's waist. His task is to help subdue the inmate in any way possible. He usually concentrates on the inmate's body below the knees. He acts as the eyes and ears for the team during control of the inmate.

Team Member Five

7-55. Team member five gives commands to control the inmate.

FORCED CELL MOVE TEAMS

7-56. The intent of a forced cell move is to move an unruly and/or uncooperative inmate from one cell to another. This is a difficult task and must not be taken lightly. A forced cell move must be thoroughly planned, rehearsed, and resourced to ensure the safety of the team members and the inmate.

7-57. A number of reasons may compel the commander to consider a forced cell move. The inmate may refuse to eat, take medication, accept medical attention, get a haircut, keep himself clean, work, or move or rotate cells when ordered to. These are some of the reasons that the commander may order a forced cell move.

TEAM COMPOSITION AND EQUIPMENT

7-58. The FCMT is composed of six corrections cadre and two support cadre from military police and medical sections specially trained in forced cell movement. Primarily, the team consists of the following personnel:

- One officer in charge (OIC) (E7 or above).
- Five corrections cadre.
- One military police investigator.
- One medic.

7-59. The FCMT uses the following standard equipment:

- Helmets with face shields.
- Groin protectors.
- Flak vests.
- Elbow pads.
- Leather gloves.
- Shin protectors.
- Knee pads.
- A pinning shield.
- Restraints (hand or leg irons or Flex-Cufs).
- A video camera.

ACTIONS PRIOR TO TEAM EMPLOYMENT

7-60. The OIC gives an operations order (OPORD) to team member five (the team leader) of the FCMT. As a minimum, the OPORD should answer the following questions about the incident leading up to the move:

- Did the inmate fail to comply with orders?
- Did the inmate assault cadre or other inmates?
- Is the objective of the move to force a shower or haircut?
- Is OC pepper spray authorized for use?
- Has the inmate been moved before?

- What is the condition of the area where the inmate is currently located?
- Has the inmate tested positive for human immunodeficiency virus (HIV)?
- Does the inmate have a weapon or the materials needed to make one?
- Is the area barricaded?
- Has he smeared himself with body fluids or other slippery substances?
- Can you approach the inmate without risk to him and/or others?

7-61. The team is assembled for inspection by the OIC, and the team members are required to brief the OIC on their individual duties and responsibilities. A line formation is the customary formation for this inspection. It should take place out of site of the inmates.

7-62. Each soldier takes one step forward and briefs the OIC as shown in the following example: *"I am (rank and name), team member one. My mission is to pin the inmate using the minimum amount of force necessary. I carry the pinning shield."* He then faces so the camera can record the number printed on the back of his flak jacket and holds up the pinning shield [see *Figure 7-2*]).

Figure 7-2. Recording Team Members' Duty Position

INDIVIDUAL TEAM MEMBER RESPONSIBILITIES

7-63. Each individual team member has his own responsibilities during a forced cell move. Each team member must do his designated job while working as a team.

Team Member One

7-64. Team member one is the eyes and ears of the team. His responsibility is to carry the pinning shield up to the cell door and describe the scene with as much detail as possible (especially including anything that may cause harm to the team). The pinning shield is used to protect the team from body fluids thrown at the them by the inmate.

7-65. Team member one describes the scene in the cell to his teammates by turning his head to the right and speaking in a loud voice so that all team members can hear him. A slap on the right shoulder by team member two alerts him that all team members have heard him. If he receives a slap on the left shoulder from team member two, he repeats the situation report (SITREP) in a loud, clear voice. The decision to enter the inmate's area with or without the pinning shield is team member one's decision, unless the inmate is known to have a weapon. Should he decide to enter without the pinning shield, team member one then enters with both hands held approximately chest high. When team member one is unable to determine if the inmate has a weapon, the pinning shield is used. Charging through the open cell door, team member one drives the inmate to the bunk or ground with the shield. Once pinned, team member one secures the inmate's head and ensures that it is not twisted or turned in relation to its natural form. He may have to use the various pressure points on the inmate's head to help subdue him.

Team Member Two

7-66. Team member two steps forward and briefs the OIC as follows: *"I am (rank and name), team member two. My mission is to secure the inmate's right arm using the minimum amount of force necessary. I carry one set of hand irons."* He then holds the hand irons up for video camera taping (see *Figure 7-3*) and says, *"I also carry one canister of OC pepper spray to be used at the direction of the team OIC."* He then holds up the canister for video camera taping and faces so the camera can record the number on the back of his flak jacket (see *Figure 7-2,* page 7-13.)

Figure 7-3. Recording Team Members' Equipment

7-67. Team member two lines up directly behind team member one. He keeps contact with team member one until he can secure the inmate's strong arm. Team member two taps team member one on his right or left shoulder after being tapped by team member three. Once inside the cell, team member two secures the prisoners strong hand with wrist restraints and announces, "*Hands secure,*" to team member five.

Team Member Three

7-68. Team member three steps forward and briefs the OIC as follows: *"I am (rank and name), team member three. My mission is to secure the inmate's left arm using the minimum amount of force necessary."* He then faces so the camera can record the number on the back of his flak jacket (see *Figure 7-4*).

Team member three lines up directly behind team member two. Team member three keeps contact with team member two until he can grasp a portion of the inmate's body. He then concentrates on securing the inmate's weak hand and assisting team member two in applying hand restraints. Team member three taps team member two on his right or left shoulder after being tapped by team member four.

Figure 7-4. Recording Number on Flak Jacket

Team Member Four

7-69. Team member four steps forward and briefs the OIC as follows: "*I am (rank and name), team member four. My mission is to secure the inmate's right leg using the minimum amount of force necessary. I carry one set of leg irons.*" He holds the leg irons up for the camera to record (see *Figure 7-2,* page 7-13). He then faces so the camera can record the number on the back of his flak jacket (see *Figure 7-2*). Team member four lines up directly behind team member three. He taps team member three on his right or left shoulder after being tapped by team member five. Team member four keeps contact with team member three until he can secure a portion of the inmate's body. Team member four concentrates on securing the right leg of the inmate. It is his responsibility to apply the leg irons. Once the legs are secure, he announces, "*Legs secure,*" to team member five.

Team Member Five (Team Leader)

7-70. Team member five steps forward and briefs the OIC as follows: "*I am (rank and name), team member five. My mission is to secure the inmate's left leg using the minimum amount of force necessary. I carry a restraint key (hold up for the camera to view). I am also the team leader for this mission and it is my responsibility to ensure that the team uses the minimum amount of force necessary to subdue the inmate.*" He then faces so the camera can record the number on the back of his flak jacket (see *Figure 7-4*).

7-71. Team member five taps team member four on the right shoulder when he has heard the SITREP from team member one. If he does not hear the report or needs to hear it again, he taps team member four on the left shoulder. Team member five keeps in contact with team member four until he can secure a portion of the inmate's body. He secures the left leg and assists team member four in applying the leg irons. Once he has heard that the hands and legs are secure, team member five signals for the team to conduct an equipment check. The team members check the equipment in place.

NOTE: At no time is the inmate to be released by anyone.

Medic

7-72. The medic briefs the OIC in the following manner: *"I am (rank and name), the medic. My mission is to treat all injuries that may occur during the move."* He follows the instructions given to him by the OIC/NCOIC. His mission is to observe the inmate for signs of injury before the forced cell move and assess the inmate's condition during and after the move.

Military Police Investigator

7-73. The military police investigator (MPI) is responsible for videotaping the entire forced cell move. This includes the OIC/NCOIC briefing the OPORD, each team member briefing his responsibilities to the OIC, and any medical treatment rendered to the inmate.

ACTIONS DURING THE FORCED CELL MOVE

7-74. There are many actions performed during a forced cell move. Each member of the FCMT must complete his task successfully to accomplish the forced cell move.

Officer in Charge or Noncommissioned Officer in Charge

7-75. The OIC, an E7 or above, is responsible for ensuring that the team uses the minimum amount of force necessary to subdue the inmate. He is also responsible for ensuring that all necessary precautions have been taken to ensure the safety of the team and the inmate. The OIC gives the inmate the initial orders to comply and informs him of the consequences of noncompliance. He does this by placing himself at the front of the inmate's cell (see *Figure 7-5*). The orders are not given until it can be enforced by the FCMT. The following prepared statement is read to the inmate by the OIC: *"I am (rank and name), OIC of this FCMT. At this time I am ordering you to lie face down on your bunk (or other designated location) with your head to the rear of the cell. Cross your hands behind your back and cross your feet as the team enters the cell and uses the minimum amount of force necessary to place you in restraints and move you to (give location). Failure on your part to comply with these instructions or to resist the FCMT will result in the use of force (OC pepper spray if applicable). Do you understand these instructions? If so, lie down on the bunk."*

7-76. After reading the instructions to the inmate, the OIC faces the camera and states whether the inmate complied. He then waits 60 seconds and motions for the team to position itself in front of the inmate's cell (see *Figure 7-6*). The OIC directs the opening of the cell when the team is positioned at the entrance to the cell and prepared to enter. He positions himself so he can observe team entry to and departure from the cell. If OC spray is authorized, the OIC tells the team when to spray the inmate with OC spray. This task is done in silence. Communication between team members should be done through hand-and-arm signals, unless noted otherwise.

Figure 7-5. OIC Reading Compliance Statement to Inmate

Figure 7-6. OIC and FCMT

Team Member One

7-77. Once the team is positioned in front of the inmate's cell, team member one starts his report. At a minimum, he must answer the following questions and report them to the team:

- Does the inmate have a weapon, if so, what kind?
- Does the inmate appear dry?
- Is the inmate clothed or wearing extra clothing?
- Has the inmate smeared anything on his body?
- Can you see the inmate's hands?
- Is the cell barricaded or visibly booby-trapped?

- Is the cell dry?
- Are cell furnishings intact?

7-78. Based on the answers to the above questions, team member one must decide whether to enter the cell with or without the shield (see *Figure 7-7*). Team member one's ultimate goal is to pin the inmate on the bunk or floor.

Figure 7-7. FCMT Preparing to Enter a Cell

Team Member Two

7-79. Team member two keeps contact with team member one until he can secure the inmate's strong arm. If he cannot immediately secure the inmate's arm, he secures a portion of the inmate's body and progresses to the inmate's strong arm. Once the inmate's hands are secure, he announces, "*Hands secure,*" to team member five (see *Figure 7-8).*

Team Member Three

7-80. Team member three keeps contact with team member two until he can grasp a portion of the inmate's body. Team member three concentrates on securing the inmate's weak hand and assists team member two in applying Flex-Cufs. Team member three maintains control of the inmate's hands and looks at team member five for further instructions (see *Figure 7-8*).

Team Member Four

7-81. Team member four keeps contact with team member three until he can secure a portion of the inmate's body. Team member four concentrates on securing the inmate's right leg. He applies leg irons and when the inmate's legs are secure, he announces, "*Legs secure,*" (see *Figure 7-8*).

Team Member Five (Team Leader)

7-82. Team member five keeps contact with team member four until he secures a portion of the inmate's body. Team member five concentrates on securing the inmate's left leg and assists with the application of leg irons. When he is informed that the hands and legs are secure, the team searches the inmate for weapons and contraband. Once the search is complete, team member five commands, "Prepare to Lift."

Figure 7-8. FCMT Securing an Inmate

Movement of the Inmate

7-83. Team member one positions his hands to control the inmate's head during the move, turning the inmate's head toward the wall. Team members two and three position themselves on the inmate's right and left sides. They grasp the wrist or forearm of the team member opposite them to form a platform for the inmate to lie on. Team member four positions himself on the inmate's right side even with the inmate's knees. This places him in position to control the inmate's legs. Team member five's position is on the inmate's left side and enables him to control the inmate's lower legs and feet. When in position, team member five ensures that all team members are in place before commanding, "Ready, Lift."

7-84. On the command "Lift," all team members simultaneously stand with the inmate. They should sandwich the inmate between them to facilitate control. Team member five then commands, "Prepare to Turn, Ready—Turn." The inmate is always turned toward the bunk. If a bunk is not available, team member five should specify which direction to turn the inmate. The inmate is then removed from the cell (see *Figure 7-9*, page 7-20).

7-85. Once the inmate is on the tier, team member five commands, "Prepare to Lower, Team—Lower." The team simultaneously lowers the inmate to the floor with the team keeping positive control of the inmate. The medic asks the inmate, "*Do you have any injuries at this time?*" (see *Figure 7-10,* page 7-20). If the inmate has no injuries, the medic looks at the camera and states, "*Inmate (inmate name) appears to have sustained no injuries.*" If the inmate has sustained injuries, the medic treats those injuries and completes a sworn statement. If OC was employed, the medic decontaminates the inmate using water and paper towels. The medic instructs the inmate to tilt his head to the side and flush his eyes with water for approximately 5 minutes. The medic then dries the inmate's face.

Figure 7-9. Moving the Inmate

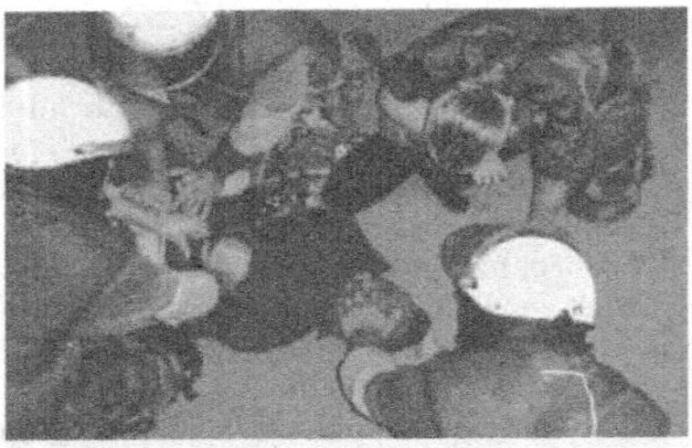

Figure 7-10. Treating Injuries

Placement of the Inmate in a Cell

7-86. Team member five orders an equipment check before returning the inmate to a cell. He then commands, "Prepare to Lift, Team—Lift." Team members position themselves exactly where they were when they removed the inmate from the cell. Next, team members simultaneously lift the inmate and return him to the predetermined cell. They move the inmate into the cell head first with his face toward the bunk. Once inside the cell, team member five commands the team to lower the inmate onto the bunk (see *Figure 7-11,* page 7-20). Team members then remove the inmate's restraints, keeping positive control of the inmate at all times. Team members use a nonverbal signal to notify team member five when the restraints are removed. He signals another equipment check. Team members then account for all equipment in place and prepare to exit the cell. The team exits the cell in the same manner as they entered, but in reverse order.

Figure 7-11. Placing Inmate in Cell

7-87. Team member four moves into position where he can control both of the inmate's legs while team member five moves toward the cell door. When team member four is ready to exit the cell, team member three moves to control the inmate's legs, team member two moves to where he can control both of the inmates arms, and team member one maintains control of the inmate's head. Team member four maintains contact with team member three at all times. When team member three is ready to exit the cell, team member two moves into a position where he can control the inmate's lower body and team member one moves into position to control the inmate's upper body. Using a prearranged, nonverbal signal, the team pulls team members one and two from the cell. The FCMT is prepared to reenter the cell if the inmate attempts to attack team members one and two as they exit the cell. The FCMT remains ready to reenter until the cell door is secured.

Chapter 8

Civil Disturbance Training

Training requires planning and preparation. Unit commanders structure their training program according to the mission essential task list (METL) (see *FM 7-1)* and expected contingencies. These training plans include prior training and the experience and proficiency of their unit. Leaders agree that the operational tempo and expected contingencies for the next 10 to 15 years will likely involve low-intensity military operations other than war (MOOTW). Therefore, civil disturbance training must be included in the commander's annual training program and must cover all aspects of that type of operation. To effectively function as a member of a control force, individual soldiers must be trained in civil disturbance operations. Training must be intensive, realistic, and sustained. Any lack of training (from basic soldier skills to leadership skills) can have an adverse effect to the point of endangering lives, especially the lives of soldiers. Leader involvement in the training process is essential for mission success in crowd control operations.

UNIT TRAINING STRATEGY DEVELOPMENT

8-1. Developing a unit training strategy begins with the mission analysis portion of the MDMP and identifying the GRM. Missions that require soldiers and units to enforce treaties or accords, protect the lives of civilians in uncertain or hostile environments (such as NEOs), or provide large-scale humanitarian assistance demand some sort of graduated-response criteria to maintain order and prevent uncertain environments from becoming hostile.

8-2. When developing training strategies, planners should focus training and evaluation on the echelon level above the target element (for example TF) or the unit identified as needing training.

8-3. Planners must agree on the intent of the GRM. The GRM is for most a training and rehearsal tool. It provides leaders with likely vignettes that can be incorporated into COA analysis, predeployment training, and rehearsals. The GRM becomes a handy reference during situations that require graduated responses.

TRAINING PLAN DEVELOPMENT

8-4. In developing a training plan for civil disturbances, many considerations should be addressed in terms of available resources. Considerations include the number of personnel to train, the equipment needed and its availability, the time available to train to standard, the location of where training will be executed, and the incorporation of SOPs (based on the theater of operations

[CONUS or OCONUS]). A personnel assessment is necessary to determine the training and experience level of soldiers. It is critical to identify and use soldiers with previous real-world experience and leaders that have graduated from the Interservice Nonlethal Individual Weapons Instructor Course (INIWIC) as primary instructors.

8-5. Commanders should structure the unit training program based on the METL, expected contingencies, prior training, and the proficiency of the unit. In developing a training program, use the seven-step approach. For in-depth instructions on training development, see *FMs 7-0* and *7-1*.

8-6. The objective is to train and evaluate units in the performance of civil disturbance procedures. Proficiency in performing civil disturbance procedures is reflected by a unit's ability to quickly and deliberately execute the disturbance control procedures required to prevent loss of life and property or to regain control of a civil disturbance in its assigned area of responsibility.

8-7. Training should focus on, but is not limited to—

- Instructors.
- Individual and collective drills.
- Leaders.
- Equipment.
- Weapons.
- Munitions.
- Live-fire range exercises.
- Qualification and familiarization.
- Soldier training exercises (STXs).
- Culminating exercises.

8-8. There are several successful methods to evaluate trained leader skills and trained soldier skills using scenarios, vignettes, lanes, or STXs. Flexibility is an important key in training plan development due to time restraints.

RANGE SETUP

8-9. The key to setting up a successful range for NL munitions is to ensure that they are in compliance with installation safety regulations. Kinetic energy is the force that pushes the projectiles down range. In most cases, kinetic energy will only carry these NL munitions projectiles 75 meters or less. This gives trainers flexibility as to the locations where a live-fire exercise can be conducted, especially in a field-expedient situation. *Figure 8-1* is an example for conducting a NL live fire range setup.

8-10. As illustrated in *Figure 8-1*, mark off 5-, 10-, 20-, 30-, 40-, and 50-meter distances from the targets with markers. Each of the NL munitions has minimum and maximum effective ranges within the distances of 5 to 50 meters. Impact power and accuracy are greater at close distances and depleted at farther distances. Targets can be anything from the standard E-type to more realistic three-dimensional manikins. Evenly divide your shooters into three groups as shown in *Figure 8-1*. Group one will initially be shield holders, group two will be shooters, and group three will be grenadiers.

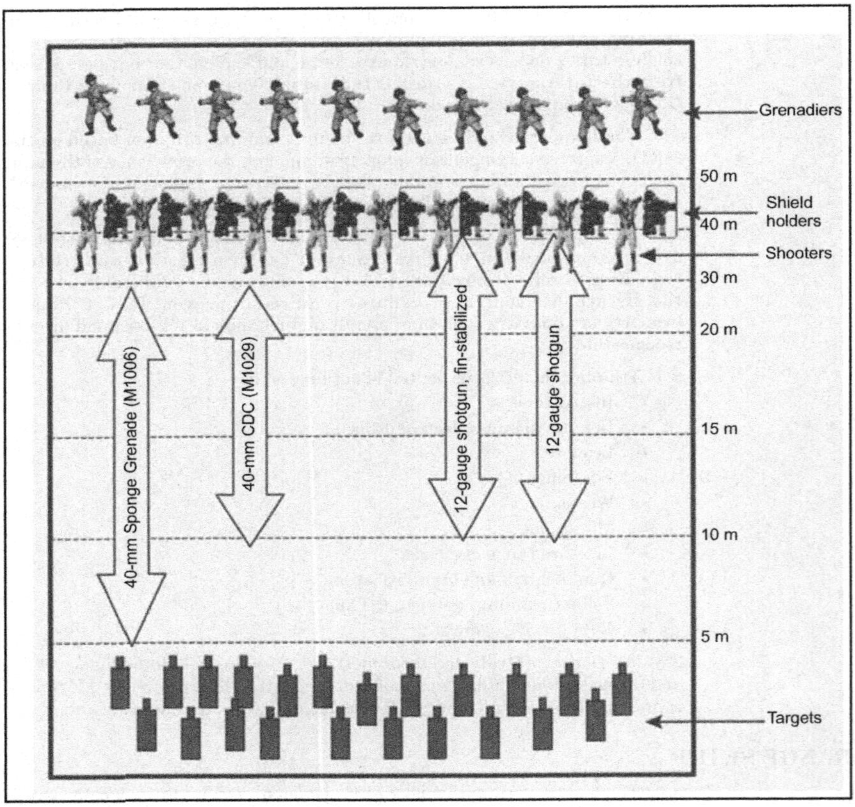

Figure 8-1. NL Munitions Range Setup

8-11. The formation should begin at the farthest effective range firing line (50 meters). After each volley of fire, the formation moves closer to the targets, but stays outside the closest effective range firing line (5 meters). For safety reasons, ensure that the formation stays online throughout each iteration.

8-12. When moving forward to each firing point, it is recommended that the formation always come to a halt before the fire command is given. It is also important to use the tap down drill method (see *Figure 8-1*) rather than having the shooters move forward and fire between the shield holders.

8-13. Range instructors can use verbal commands, such as, "Ready," or "Fire," for each volley. Shooters can be briefed to move forward, tap down the shield holder, and engage targets at the first explosion of distraction grenades thrown by grenadiers. In that instance, the command could simply be "Grenadiers Prepare to Throw, Throw."

8-14. The success of any civil disturbance operation is through effective communication, especially when incorporating the use of NL munitions. For effective training, use verbal commands in conjunction with effective, predetermined hand-and-arm signals during live-fire exercises.

8-15. Each iteration requires two phases of fire. During the first phase, half of the shooters will be firing NL, 40-millimeter rounds using an M203. The others will be firing NL, 12-gauge rounds using a 12-gauge shotgun. In the second phase, shooters switch weapons (safely). It is critical that line safeties are alert during this transition between phases.

8-16. Upon completion of each iteration, clear the weapons and the range. Establish a safe method and return all groups back to the farthest firing line. Once that is accomplished, group one rotates to become group three, group three rotates to become group two, and group two rotates to become group one for the next iteration. Using this method affords a safe and efficient way to get all shooters through the proficiency course. *Figure 8-2* is an example of how to fire each volley based on its characteristics, effectiveness, and capabilities.

Distance	Command	Subject	Actions
10 Meters	12-Gauge Shotgun, Two Point	Shotgun shooters	Load two M1013 rounds.
	M203, Two Point	M203 shooters	Load one M1006 round and have one readily available.
	Ready, Fire	Shotgun shooters	Tap down shields and engage designated targets.
		M203 shooters	Tap down shields, engage designated targets, reload, and reengage.
	Clear Range	Range safeties	Ensure that weapons are clear and safe.
	Safeties, Clear All Weapons	All shooters	Face to right, keep weapons pointed up and downrange.
	Shooters, Right Face, Move	All personnel	Step off shield holder, then shooters, and grenadiers on the command, "Move." Grenadiers reform at 50 meters.

Figure 8-2. Sample Live-Fire Exercise

Distance	Command	Subject	Actions
15 Meters	Shotgun, Two Point	Shotgun shooters	Load two M1013 rounds.
	M203, Two Point	M203 shooters	Load one M1006 round and have another available.
	Grenadiers, Prepare to Throw	Grenadiers	Make distraction grenade ready.
	Throw	Grenadiers	Throw grenade.
		Grenadiers	Throw distraction grenade.
	Ready, Fire	Shot gun shooters	Tap down shields and engage designated targets.
		M203 shooters	Tap down shields, fire at designated targets, reload with M1006 rounds, and engage designated target.
	Clear Range	Range safeties	Ensure that weapons are clear and safe.
	Three Steps Forward	All shooters and grenadiers	Echo preparatory command.
	Move	All shooters and grenadiers	Stay on line and step off three steps, counting each step aloud.
	Clear Range	Range safeties	Ensure that weapons are clear and safe.
	Seven Steps Forward	All shooters and grenadiers	Echo preparatory command.
	Move	All shooters and grenadiers	Stay on line and step off seven steps, counting each step aloud.

Figure 8-2. Sample Live-Fire Exercise (Continued)

Distance	Command	Subject	Actions
20 Meters	Shotgun, Three Point	Shotgun shooters	Load three M1013 rounds.
	M203, One Area, Two Point	M203 shooters	Load one M1026 round and have two M1006 rounds available.
	Grenadiers, Prepare to Throw	Grenadiers	Make distraction grenade ready.
	Throw	Grenadiers	Throw grenade.
		Grenadiers	Throw distraction grenade.
	Ready, Fire	Shotgun shooters	Tap down shields and engage designated targets.
		M203 shooters	Tap down shields, fire area at targets, reload with M1006 rounds, and engage designated target.
	Clear Range	Range safeties	Ensure that weapons are clear and safe.
	Shotgun, Two Point	Shotgun shooters	Load two M1013 rounds.
	M203, Two Point	M203 shooters	Load one M1006 round and have another available.
	Grenadiers, Prepare to Throw	Grenadiers	Make distraction grenade ready.
	Clear Range	Range safeties	Ensure that weapons are clear and safe.
	Three Steps Forward	All shooters and grenadiers	Echo preparatory command.
	Move	All shooters and grenadiers	Stay on line and step off three steps, counting each step aloud.

Figure 8-2. Sample Live-Fire Exercise (Continued)

Distance	Command	Subject	Actions
30 Meters	Shotgun, Two Area, Two Point	Shotgun shooters	Load two M1012 rounds and two M1013 rounds.
	M203, Two Area, Two Point	M203 shooters	Load one round M1026 and have one M1026 and two rounds M1006 available.
	Grenadiers, Prepare to Throw	Grenadiers	Make distraction grenade ready.
	Throw	Grenadiers	Throw grenade.
		Grenadiers	Throw distraction grenade.
	Ready, Fire	Shotgun shooters	Tap down shields, engage targets, and engage designated target with M1013 rounds.
		M203 shooters	Tap down shields, fire area at target, reload with M1006 rounds, and engage designated target.
	Clear Range	Range safeties	Ensure that weapons are clear and safe.
	Seven Steps Forward	All shooters and grenadiers	Echo preparatory command.
	Move	All shooters and grenadiers	Stay on line and step off seven steps, counting each step aloud.

Figure 8-2. Sample Live-Fire Exercise (Continued)

Distance	Command	Subject	Actions
40 Meters	Shotgun, Two Area, One Point	Shotgun shooters	Load two M1012 rounds and one round M1013.
	M203, One Area, One Point	M203 shooters	Load one M1026 round and have one M1006 round available.
	Grenadiers, Prepare to Throw	Grenadiers	Make distraction grenade ready.
	Throw	Grenadiers	Throw grenade.
		Grenadiers	Throw distraction grenade.
	Ready, Fire	Shotgun shooters	Tap down shields and engage targets with M1013 rounds.
		M203 shooters	Tap down shields, fire area at target, reload with M1006 rounds, and engage designated target.
	Clear Range	Range safeties	Ensure that weapons are clear and safe.
	Seven Steps Forward	All shooters and grenadiers	Echo preparatory command.
	Move	All shooters and grenadiers	Stay on line and step off seven steps, counting each step aloud.
50 Meters	Shotgun, One Area	Shotgun shooters	Load one M1012 round.
	M203, One Area, One Point	M203 shooters	Load one M1026 round and have one M1006 round available.
	Grenadiers, Prepare to Throw	Grenadiers	Make distraction grenade ready.
	Throw	Grenadiers	Throw grenade
		Grenadiers	Throw distraction grenade.
	Ready, Fire	Shotgun shooters	Tap down shields, engage targets, and engage designated target with M1013 rounds.
		M203 shooters	Tap down shields, fire area at the target, reload with M1006 rounds, and engage the designated target.
	Clear Range	Range safeties	Ensure that weapons are clear and safe.
	Seven Steps Forward	All shooters and grenadiers	Echo preparatory command.
	Move	All shooters and grenadiers	Stay on line and step off seven steps, counting each step aloud.

Figure 8-2. Sample Live-Fire Exercise (Continued)

NOTE: Shield holders are referred to as group one, shooters are referred to as group two, grenadiers are referred to as group three.

8-17. Currently, there are no qualification standards established in the Army for NL munitions. Area rounds, which contain multiple projectiles designed to affect multiple targets when engaged, only require approximately five rounds. This affords shooters an opportunity to realize the characteristics and capabilities of these rounds.

8-18. Single projectile rounds like the M1013 and the M1006 require a greater degree of proficiency. If situations call for the engagement of targets with these rounds, it is essential that trained shooters be able to hit what they are aiming at. For example, firing 10 rounds from various ranges can provide soldiers with the confidence to be effective shooters. A standard can be established that requires shooters to hit targets with at least 6 out of the 10 rounds used in the example to be considered proficient.

8-19. *Figures 8-3; 8-4,* page 8-11; *8-5,* page 8-21; and *8-6,* page 8-22, give examples of plans to follow if in a situation similar to the following scenario (see *Army Training and Evaluation Program (ARTEP) 19-313-10-MTP*):

A company commander is providing security patrols in support of a joint task force (JTF). Besides providing security patrols for a sector of responsibility, one platoon must always be designated as a quick reaction force (QRF) for civil disturbance operations. For the last 2 weeks the sector has been quiet and the commander has ordered all platoons to conduct civil disturbance training on a rotational basis.

A new order has been sent down to provide a squad-sized element to provide security for the future polling site for an upcoming election. This polling site has yet to open, but intelligence indicates potential problems as the election approaches.

On the day prior to the scheduled election, a crowd of approximately 25 to 30 demonstrators aggressively move toward the polling site, chanting slogans and yelling toward the security forces posted there. While some demonstrators begin throwing whatever they can pick up, the majority are simply yelling, locking arms, and blocking roadways. Hundreds of additional people, not part of the crowd, begin to gather to watch the activities as they unfold.

In an assessment of the demonstrators, it is determined that they intend to stay through the night into the morning of the election with the intention to block access to the polls. It is essential that the crowd is dispersed and that access to the polls is unrestricted in time for the election to begin the next morning.

Evaluation Scheme
The evaluation will consist of five lanes conducted over a 3-day period. Platoons will execute 24-hour operations during the evaluation. Any retraining will be done immediately following the execution of the lane if time permits. The following tasks will be evaluated:

Lane 1	Cordon and search operations
Lane 2	Civil disturbance QRF
Lane 3	Security and defense of a critical site (polling site)
Lane 4	Night civil disturbance QRF
Lane 5	Night LP/OP (PIO)

Evaluation Schedule				
Date and Time	**1st Platoon**	**2d Platoon**	**3d Platoon**	**4th Platoon**
	Lane 1	Lane 2	Lane 3	Lane 4
	Lane 2	Lane 3	Lane 4	Lane 5
	Lane 3	Lane 4	Lane 5	Lane 1
	Lane 4	Lane 5	Lane 1	Lane 2
	Lane 5	Lane 1	Lane 2	Lane 3

Figure 8-3. Evaluation Scheme

Unit Training Schedule and Firing Tables

Unit Training Schedule 1

Training Day 1

Time	Subject	Location	Uniform	Instructor	Notes
0800	Administrative time	Classroom	U1		
0830	GRM	Classroom	U1		
1000	ROE and use of force	Classroom	U1		
1200	Chow				
1300	MOOTW	Classroom	U1		
1500	Effective communication	Classroom	U1		

Administration:

Uniforms:

U1: BDU

U2: BDU and LBE with harness and canteens

U3: BDU and LBE with harness, canteens, flak jacket, and Kevlar helmet

U4: BDU and LBE with harness, canteens, flak jacket, Kevlar helmet, and assigned weapon

Transportation:

Weapons:

Ammunition:

Required Equipment for Training:

1. Projection screen
2. Video projector with CPU
3. Television
4. VCR

Miscellaneous:

1. All personnel are required to have a notebook and a pencil or pen.
2. Handouts must be prepared and passed out.

Figure 8-4. Sample Two-Week Plan

Training Day 2

Time	Subject	Location	Uniform	Instructor	Notes
0800	NLCS	Classroom	U1		
0900	PPCT and empty hand	Classroom	U1		
1000	PPCT and empty hand (Practical application)	PT field	U2		
1200	Chow				
1300	PPCT and empty hand (Practical application)	PT field	U2		
1700	COB				

Administration:

Uniforms:

U1: BDU

U2: BDU and LBE with harness and canteens

U3: BDU and LBE with harness, canteens, flak jacket, and Kevlar helmet

U4: BDU and LBE with harness, canteens, flak jacket, Kevlar helmet, and assigned weapon

Transportation:

Weapons:

Ammunition:

Required Equipment for Training:

1. Projection screen
2. Video projector with CPU
3. Training strike bags
4. Flex-Cufs

Miscellaneous:

1. All personnel are required to have a notebook and a pencil or pen.
2. Handouts must be prepared and passed out.
3. Medical personnel must be on location.

Figure 8-4. Sample Two-Week Plan (Continued)

Training Day 3					
Time	**Subject**	**Location**	**Uniform**	**Instructor**	**Notes**
0800	PPCT and empty hand (Practical application)	PT field	U2		
1200	Chow				
1300	PPCT and empty hand (Practical application)	PT field	U2		
1700	COB				

Administration:

Uniforms:

U1: BDU

U2: BDU and LBE with harness and canteens

U3: BDU and LBE with harness, canteens, flak jacket, and Kevlar helmet

U4: BDU and LBE with harness, canteens, flak jacket, Kevlar helmet, and assigned weapon

Transportation:

Weapons:

Ammunition:

Required Equipment for Training:

1. Projection screen
2. Video projector with CPU
3. Training strike bags
4. Flex-Cufs

Miscellaneous:

1. All personnel are required to have a notebook and a pencil or pen.
2. Handouts must be prepared and passed out.
3. Medical personnel must be on location.

Training Day 4					
Time	**Subject**	**Location**	**Uniform**	**Instructor**	**Notes**
0800	Riot baton training	Classroom	U1		
0900	Riot baton training (Practical application)	PT field	U2		
1200	Chow				

Figure 8-4. Sample Two-Week Plan (Continued)

Time	Subject	Location	Uniform	Instructor	Notes
1300	Riot baton training (Practical application)	PT field	U2		
1700	COB				

Administration:

Uniforms:

U1: BDU

U2: BDU and LBE with harness and canteens

U3: BDU and LBE with harness, canteens, flak jacket, and Kevlar helmet

U4: BDU and LBE with harness, canteens, flak jacket, Kevlar helmet, and assigned weapon

Transportation:

Weapons:

1. Expandable riot baton with carrier
2. Training riot batons

Ammunition:

Required Equipment for Training:

1. Projection screen
2. Video projector with CPU
3. Training strike bags
4. Flex-Cufs

Miscellaneous:

1. All personnel are required to have a notebook and a pencil or pen.
2. Handouts must be prepared and passed out.
3. Medical personnel must be on location.

Training Day 5

Time	Subject	Location	Uniform	Instructor	Notes
0800	Riot baton training (Practical application)	PT field	U2		
1000	Crowd dynamics	Classroom	U1		
1130	Chow				
1230	Crowd dynamics	Classroom	U1		
1330	CD formations	Classroom	U1		

Figure 8-4. Sample Two-Week Plan (Continued)

1430	RCAs and OC	Classroom	U1		
1700	COB				

Administration:

Uniforms:

U1: BDU

U2: BDU and LBE with harness and canteens

U3: BDU and LBE with harness, canteens, flak jacket, and Kevlar helmet

U4: BDU and LBE with harness, canteens, flak jacket, Kevlar helmet, and assigned weapon

Transportation:

Weapons:

1. Expandable riot baton with carrier
2. Training riot baton

Ammunition:

Required Equipment for Training:

1. Projection screen
2. Video projector with CPU
3. Training strike bags
4. Flex-Cufs

Miscellaneous:

1. All personnel are required to have a notebook and a pencil or pen.
2. Handouts must be prepared and passed out.
3. Medical personnel must be on location.

Training Day 6

Time	Subject	Location	Uniform	Instructor	Notes
0800	RCAs and OC (Practical application)	PT field	U2		
1200	Chow				
1300	RCAs and OC (Practical application)	PT field	U2		
1700	COB				

Figure 8-4. Sample Two-Week Plan (Continued)

Administration:

Uniforms:

U1: BDU

U2: BDU and LBE with harness and canteens

U3: BDU and LBE with harness, canteens, flak jacket, and Kevlar helmet

U4: BDU and LBE with harness, canteens, flak jacket, Kevlar helmet, and assigned weapon

Transportation:

Weapons:

Ammunition:

1. MK inert OC
2. MK 4 live OC disperser
3. M36 live CR disperser

Required Equipment for Training:

1. Five large clean trash cans with liners
2. Two water hoses with running water

Miscellaneous:

Medical personnel must be on location.

Training Day 7

Time	Subject	Location	Uniform	Instructor	Notes
0800	NL munitions employment	Classroom	U1		
1030	Chow				
1115	Safety brief	Range	U4		
1130	NL munitions (live fire)	Range	U4		
TBD	Evening chow				
TBD	Safety brief	Range	U4		
TBD	NL munitions night fire exercise	Range	U4		
TBD	Clean up and COB				

Figure 8-4. Sample Two-Week Plan (Continued)

Administration:

Uniforms:

U1: BDU

U2: BDU and LBE with harness and canteens

U3: BDU and LBE with harness, canteens, flak jacket, and Kevlar helmet

U4: BDU and LBE with harness, canteens, flak jacket, Kevlar helmet, and assigned weapon

Transportation:

Yes

Weapons:

1. M16A2 with M203
2. 12-gauge shotgun

Ammunition:

1. Fin-stabilized M1012
2. M1013 multipellet (rubber)
3. M1006 40-millimeter sponge grenade
4. M1026 40-millimeter multiball (rubber)
5. M84 diversionary grenade

Required Equipment for Training:

1. Fifteen to twenty human form targets
2. Eye and hearing protection
3. Fifteen to twenty riot shields
4. Two MREs per student
5. Video recorder
6. Three to five high-intensity lights
7. All students should be issued face shields, shin guards, and ammunition pouches

Miscellaneous:

1. Medical personnel must be on location.
2. Range safety officers must be available.
3. Primary and secondary means of communication with range control must be available.
4. Operational risk assessments must be conducted.

Figure 8-4. Sample Two-Week Plan (Continued)

Training Day 8					
Time	**Subject**	**Location**	**Uniform**	**Instructor**	**Notes**
0800	CD formations (Practical application)	TA 1	U4		
1200	Chow				
1300	CD formations (Practical application)	TA 1	U4		
1700	COB				

Administration:

Uniforms:

U1: BDU

U2: BDU and LBE with harness and canteens

U3: BDU and LBE with harness, canteens, flak jacket, and Kevlar helmet

U4: BDU and LBE with harness, canteens, flak jacket, Kevlar helmet, and assigned weapon

Transportation:

Yes

Weapons:

1. Training weapons
2. Training riot batons

Ammunition:

Required Equipment for Training:

1. Eye protection
2. Fifteen to twenty riot shields
3. Video recorder
4. Three to five high-intensity lights
5. Face shields, shin guards, and ammunition pouches

Miscellaneous:

1. Medical personnel must be on location.
2. Primary and secondary means of communication must be available.
3. Operational risk assessment must be conducted.
4. Twenty to thirty OPFOR personnel must be available.

Figure 8-4. Sample Two-Week Plan (Continued)

Training Day 9					
Time	**Subject**	**Location**	**Uniform**	**Instructor**	**Notes**
0800	Introduction to M5, PVAB, and M315	Classroom 1	U1		
1200	Chow				
1300	M5, PVAB, and M315 (Practical application/LFE)	Range	U4		
1700	COB				

Administration:

Uniforms:

U1: BDU

U2: BDU and LBE with harness and canteens

U3: BDU and LBE with harness, canteens, flak jacket, and Kevlar helmet

U4: BDU and LBE with harness, canteens, flak jacket, Kevlar helmet, and assigned weapon

Transportation:

Yes

Weapons:

Two complete M315s and installation kits

Ammunition:

1. Ten live MCCMs
2. Ten live M315 rounds of each

Required Equipment for Training:

1. Eye protection
2. Ten inert MCCMs
3. Two complete PVAB systems

Miscellaneous:

1. Medical personnel must be on location.
2. Primary and secondary means of communication must be available.
3. Operational risk assessment must be conducted.

Figure 8-4. Sample Two-Week Plan (Continued)

Training Day 10					
Time	**Subject**	**Location**	**Uniform**	**Instructor**	**Notes**
0800	Tactical consideration	MOUT classroom	U1		
0930	Safety brief	MOUT classroom	U4		
1000	Tactical exercise	MOUT Facility	U4		

Administration:

Uniforms:

U1: BDU

U2: BDU and LBE with harness and canteens

U3: BDU and LBE with harness, canteens, flak jacket, and Kevlar helmet

U4: BDU and LBE with harness, canteens, flak jacket, Kevlar helmet, and assigned weapon

Transportation:

Yes

Weapons:

1. Assigned weapons
2. Inert RCA/OC individual dispersers
3. Training riot batons
4. Fifty practice grenades

Ammunition:

Required Equipment for Training:

1. Eye protection
2. Ten inert MCCMs
3. Face shields, shin guards, and ammunition pouches
4. Fifteen to twenty full-length riot shields

Miscellaneous:

1. Medical personnel must be on location.
2. Primary and secondary means of communication must be available.
3. Operational risk assessment must be conducted.
4. A video recorder must be available.

Figure 8-4. Sample Two-Week Plan (Continued)

Sample Unit Training Schedule				
Training Day 1				
Time	**Event**	**Location**	**Instructor**	**Notes**
0730-0830	Force continuum	Classroom		L
0830-0930	Management of aggressive behavior	Classroom		L
0930-1030	ROE	Classroom		L
1030-1130	Crowd control	Classroom		L
1130-1230	Chow			
1230-1700	PPCT and empty hand	PT field		L, PA
Training Day 2				
Time	**Event**	**Location**	**Instructor**	**Notes**
0730-1130	Riot baton training	PT field		L, PA
1130-1230	Chow			
1230-1530	Riot baton training	PT field		PA
1530-1700	CD formations	Classroom		L
Training Day 3				
Time	**Event**	**Location**	**Instructor**	**Notes**
0730-1130	CD formations	PT field		PA
1130-1230	Chow			
1230-1700	NL munitions, NLW	Range		L, PA
Training Day 4				
Time	**Event**	**Location**	**Instructor**	**Notes**
0730-1130	RCAs and OC	Classroom and PT field		L, PA
1130-1230	Chow			
1230-1700	RCAs and OC	PT field		PA

Figure 8-5. Sample One-Week Plan

Training Day 5				
Time	**Event**	**Location**	**Instructor**	**Notes**
0730-1700	Tactical exercise	MOUT facility		OPS
NOTE: Appointed student leaders ensure that troops are fed and rotated for rest.				

Figure 8-5. Sample One-Week Plan (Continued)

Sample Unit Training Schedule				
Training Day 1				
Time	**Event**	**Location**	**Instructor**	**Notes**
	Shield and riot baton techniques	MOUT site		Individual and collective tasks
	NLCS and munitions	Range Classroom		Leader training
	Formation training	MOUT site		Individual and collective tasks
	Chow			
	Formation training	MOUT site		Individual and collective tasks
Training Day 2				
Time	**Event**	**Location**	**Instructor**	**Notes**
0730-1930, 12-hour operations	Live-fire exercise	Range		M203/12-gauge
	QRF exercise	MOUT site		Individual and collective tasks
NOTE: Once weapons and munitions familiarization is complete, shooters join the QRF exercise to be used. For live-fire range setup, see *Figure 8-1*.				

Figure 8-6. Sample Two-Day Plan

Appendix A

Metric Conversion Chart

This appendix complies with current Army directives which state that the metric system will be incorporated into all new publications. *Table A-1* is a metric conversion chart.

Table A-1. Metric Conversion Chart

US Units	Multiplied By	Metric Units
Cubic feet	0.0283	Cubic meters
Feet	0.3048	Meters
Feet per second	0.3048	Meters per second
Inches	0.0254	Meters
Inches	25.4001	Millimeters
Miles per hour	1.6093	Kilometers per hour
Pounds	0.4536	Kilograms
Yards	0.9144	Meters
Metric Units	**Multiplied By**	**US Units**
Cubic meters	35.3146	Cubic feet
Kilograms	2.2046	Pounds
Kilometers per hour	0.6214	Miles per hour
Meters	3.2808	Feet
Meters	39.3701	Inches
Meters	1.0936	Yards
Meters per second	3.2808	Feet per second
Millimeters	0.0394	Inches

Appendix B

Operations and Legal Considerations in the Continental United States

Providing military assistance to state and local governments to assist them in quelling a civil disturbance or riot requires close coordination through a host of state and federal agencies. It requires a thorough briefing of soldiers at all levels on what they can and cannot do with respect to law enforcement. Civil authorities must be briefed on the restrictions placed on federal forces by the *Constitution of the United States* and federal statutes and laws. The national guard (NG), when not in a federal status, operates under the control of the state governor and the adjutant general (AG). The NG has historically been the first military responder during emergencies.

FEDERAL INTERVENTION AND AID

B-1. Under the *Constitution of the United States* and *United States Code*, the President is empowered to direct federal intervention in civil disturbances to—

- Respond to state requests for aid in restoring order.
- Enforce the laws of the United States.
- Protect the civil rights of citizens.
- Protect federal property and functions.

B-2. Under the *Constitution of the United States*, each state is responsible for protecting life and property within its boundaries. State and local governments use their LEA forces to maintain law and order and to quell civil disturbances. However, if civil disturbance support requirements exceed the resources of a state, federal troops may be requested to help restore and maintain law and order.

B-3. The *Constitution of the United States* and federal statutes authorize the President to direct the use of federal armed troops within the 50 states and territories and their political subdivisions. The President is also empowered to federalize the NG of any state to suppress rebellion and enforce federal laws.

B-4. Federal assistance is provided to a state when the state has used all of its resources, including its NG, to quell a disorder and finds the resources insufficient. Usually, active duty federal forces are used to augment the NG of the requesting state. However, the President may choose to federalize the NG of another state and use them alone or with other forces to restore order.

B-5. The President may employ federal troops to ensure the execution of US law when a state opposes or obstructs US law or impedes the course of justice under those laws. The President may employ armed federal troops to suppress insurrection, domestic violence, unlawful assemblies, and conspiracy if such acts deprive the people of their constitutional rights and privileges and the civil authorities of a state cannot or will not provide adequate protection.

B-6. The president is also authorized to use armed federal troops to protect federal property and functions when the need for protection exists and local civil authorities cannot or will not give adequate protection. The right of the United States to protect all federal property and functions regardless of their locations is an accepted principle of our government.

B-7. As a temporary measure, federal military equipment and facilities may be loaned to law enforcement agencies of state and local governmental bodies for use during civil disturbances. These resources may also be loaned to state NG and non-DOD federal agencies. The requesting agencies are expected to provide enough resources of their own to minimize the need for US military resources. In addition, the loan of the resources must not conflict with US military needs.

ROLES AND RESPONSIBILITIES OF VARIOUS AGENCIES

B-8. Several agencies within the federal and state governments play a key role in supporting the request of local authorities for support during a civil disturbance.

SECRETARY OF DEFENSE

B-9. The Secretary of Defense retains approval for federal support to civil authorities involving the use of DOD forces, personnel, and equipment. Approval may also involve DOD support that will result in a planned event with the potential for confrontation with specifically identified individuals and groups or will result in the use of lethal force.

SECRETARY OF THE ARMY

B-10. The Secretary of the Army, as an executive agent for the Secretary of Defense, is the approval authority for federal emergency support in response to natural and man-made disasters (except weapons of mass destruction [WMD]). Military forces respond to directions and guidance from the director of military support (DOMS) (the action agent for the Secretary of the Army).

DIRECTOR OF MILITARY SUPPORT

B-11. The DOMS plans and executes the DOD domestic support mission to civil authorities. He is a general officer appointed by the Secretary of the Army. The DOMS is the DOD primary contact for all federal departments and agencies during DOD involvement in most domestic support operations.

STATE COORDINATING OFFICER

B-12. The governor appoints the state coordinating officer (SCO) to coordinate disaster operations for the state. The SCO is the primary point of contact for the federal coordinating officer (FCO) in facilitating disaster assistance.

DIRECTOR OF THE FEDERAL EMERGENCY MANAGEMENT AGENCY

B-13. The director of the Federal Emergency Management Agency (FEMA), on behalf of the President, appoints the FCO. The FCO coordinates the timely delivery of federal disaster assistance to the affected state and local governments and disaster victims. He works closely with the SCO to determine state requirements and to coordinate national level issues with the catastrophic disaster response group. There must be close and continuous coordination with FEMA to ensure the proper use of military assets in support of civil authorities.

DEFENSE COORDINATING OFFICER

B-14. The defense coordinating officer (DCO) operates under the authority of the President and is the DOD representative designated to coordinate on-the-scene activities with the FCO. FEMA and other federal agencies that request support from DOD go through the DCO for validation and resourcing.

NATIONAL GUARD

B-15. The NG, in state status (see *Title 32, United States Code {USC}),* is the governor's primary military response organization for emergencies and disasters. The NG responds and operates under state laws unless they are federalized, when they operate under the same constraints and laws as federal units.

UNITED STATES ARMY RESERVE

B-16. The US Army Reserve is capable of providing military support. This assistance and support may include the use of equipment and other resources, including units and individuals.

STATE AND LOCAL GOVERNMENT RESPONSIBILITIES

B-17. State and local government officials, operating under the authority granted by state constitutions and the *Constitution of the United States*, are responsible for daily safety and security issues that impact the quality of life of their citizens. State and local officials have primary responsibility for emergency preparedness.

STATE RESPONSIBILITIES

B-18. Many state government agencies are responsible for coming to the aid of state citizens.

Governor

B-19. A state governor is empowered by each state constitution and the *Constitution of the United States* to execute the laws of the state and to command the state NG while it is serving in state status. Governors are also responsible for issuing executive orders or emergency proclamations, such as declaring states of emergency and ensuring that state agencies plan for potential disasters and civil disorder.

B-20. Once a man-made or natural disaster occurs, the governor assesses its extent and determines if local government requests for assistance should be honored. If appropriate, the governor declares a state of emergency, activates the state emergency response plan, and may order the NG into state active duty. The governor assigns the NG its mission through the office of emergency service (OES)/emergency medical dispatching (EMD).

Plans, Operations, and Military Support Officer

B-21. Each state NG has a plans, operations, and military support officer (POMSO) who prepares and coordinates contingency plans for military support to civil authorities (MSCA) during disaster response and recovery operations. The POMSO coordinates training plans and exercises between the state NG and federal, state, and local emergency management agencies. The POMSO serves as the NG point of contact with DOD officials during a declared federal emergency or disaster.

Office of Emergency Services

B-22. Each state has a specific agency and/or office that coordinates emergency preparedness and serves as the governor's coordinating agency during an emergency. The titles of these offices vary from state to state. Some examples are the Division of Emergency Government, Emergency Management Agency, Department of Public Safety, or Office of Emergency Preparedness. This manual refers to this office using the generic term OES/EMD.

B-23. Generally, the OES/EMD is organized as a stand-alone office under the governor or aligned under the AG or the state police, and in some instances, the AG is the OES/EMD director. It operates the state emergency operations center and coordinates with state agencies and federal officials for assistance.

Adjutant General

B-24. The state NG is the governor's primary military response force in an emergency. The AG, through the state area command (STARC) (specifically the POMSO), develops and coordinates emergency response plans for disasters and emergencies. The AG is in command of state NG forces ordered into state active duty.

State Government Agencies

B-25. State government departments and agencies prepare emergency preparedness plans for their areas of specialization in support of the governor's plan. They also participate in emergency preparedness exercises.

LOCAL RESPONSIBILITIES

B-26. Mayors, city managers, local police, fire protection officials, county executives, sheriffs, prosecuting attorneys, and public health officials are some of the people responsible for law enforcement, safety, health, and fire protection on a daily basis. They are responsible for developing appropriate emergency response plans and responding to emergencies within their jurisdictions. Most local jurisdictions have an OES/EMD to plan and coordinate actions in an emergency. In many cases, local jurisdictions have mutual aid agreements with other jurisdictions that allow for firefighting, medical, and police assistance across jurisdictional boundaries. Once local officials determine that an emergency is beyond the scope of their resources or ability to respond, the senior local official is responsible for requesting additional assistance from the next agency in the assistance request process.

LEGAL CONSIDERATIONS AND CONSTRAINTS

B-27. The *Constitution of the United States*, laws, regulations, policies, and other legal issues limit the use of federal military personnel in domestic support operations. Any Army involvement in civil disturbance operations involves many legal issues requiring comprehensive legal reviews.

B-28. Under the *Constitution of the United States,* Congress has the authority to raise and support an army, provide and maintain a navy, and make rules for governing and regulating the land and naval forces. The *Constitution of the United States* places the military under civilian control and designates the President as commander-in-chief. Statutes provide for civilian leadership in the office of the secretary of defense, service secretaries, and various other civil authorities. Under the *Constitution of the United States*, the civilian government must preserve public order and carry out governmental operations within its territorial limits. The *Constitution of the United States* allows the use of the federal military to execute or enforce the law when necessary to protect federal or civilian property and functions. However, significant restrictions exist on employing federal military forces within the United States.

B-29. The unique capabilities of the military enable it to support federal, state, or local civilian agencies. In most circumstances, the DOD is one of many federal agencies reacting to a domestic emergency or crisis, playing a subordinate, supporting role to a lead civilian agency.

LAW

B-30. In the United States, civilian agencies provide for the needs of citizens. Civilian, federal, state, and local government and law enforcement agencies execute US laws. Laws governing the use of the military in domestic operations are complex, subtle, and constantly changing. Commanders involved in civil disturbance operations should staff plans, policies, programs, exercises, funding, operations, constraints, and limitations with their local SJA to ensure conformity with legal requirements. They should scrutinize each request for aid (whether it is for equipment, individuals, or training) to ensure that it conforms to statutory requirements. The following references are a snapshot of the law on military support to civilian authorities:

- *Section 331, Chapter 15, Part 1, Subtitle A, Title 10 USC.*
- *Section 332, Chapter 15, Part 1, Subtitle A, Title 10 USC.*
- *Section 333, Chapter 15, Part 1, Subtitle A, Title 10 USC.*
- *Section 334, Chapter 15, Part 1, Subtitle A, Title 10 USC.*
- *Section 1385, Chapter 67, Part 1, Title 18 USC.*
- *DOD Directive 3025.12.*

- *DOD Directive 3025.15*
- *DOD Directive 5525.5*
- *DOD Civil Disturbance Plan 55-2*
- *NGR 500-1/ANGI 10-8101*
- *FM 3-19.15*

INFORMATION OPERATIONS

B-31. Information superiority helps forces anticipate problems and requirements. It allows commanders to control situations earlier and with less force, creating the conditions necessary to achieve the end state. Public affairs, PSYOP, and civil military operations are activities that will allow the commander to control situations earlier and with less force.

B-32. Information is available from a multitude of sources, the primary being open sources, law enforcement, and the military. A diversity of sources is the best approach because it prevents biased behavior.

Open sources include—

- Libraries.
- Newspapers and news periodicals.
- Radio and television.

Law enforcement sources include—

- Local law enforcement agencies.
- National law enforcement agencies.

Military sources include—

- DOD intelligence community (most restrictive source).
- Local MI field offices.

B-33. The DOD intelligence community operates under limitations imposed by regulations and executive orders. Attempts to skirt these restrictions may violate regulations or federal statutes.

RESTRICTIONS

B-34. Strict limits are placed on collecting information related to a civil disturbance in order to protect the civil rights of people and organizations not affiliated with the DOD. Civil disturbance plans and materials must not include lists of groups or people not affiliated with the DOD. Information on civilians and civilian organizations can be collected only with specific authorization from the Secretary or the Under Secretary of the Army. Conditions for collecting information include the existence of threats against Army personnel, functions, or property (see *ARs 380-10* and *380-13).* Civil disturbance information (available in public documents) or open-source information may be collected. However, specific rules regarding its storage must be followed. Commanders must coordinate with SJA, MI, and USACIDC before collecting any such information.

B-35. Civilian law enforcement agencies may limit access to law enforcement intelligence. In such circumstances, the military may request that agencies release the unclassified version.

B-36. If the Department of Justice (DOJ) determines that federal intervention into a civil disturbance is likely, then information relating to the disturbance is provided to the Army Assistant Chief of Staff for Intelligence. The information is analyzed and then passed through channels to the TF commander for planning purposes.

B-37. Military intelligence collection efforts, except liaison, may begin only when DA orders. The orders will come through the CSA's personal liaison officer and the TF commander. Covert operations to gather information on non-DOD individuals and groups must be approved on an operation-by-operation basis by the Undersecretary of the Army.

B-38. Once DA approves the collection effort, MI elements establish and maintain liaison with the appropriate local, state, and federal authorities. Then they collect information pertaining to the incident and general situation. Their estimate of the situation is passed to the TF commander and DA.

POSSE COMITATUS

B-39. Generally, federal military forces may not give law enforcement assistance to civil authorities without conflicting with the *Posse Comitatus Act.* However, constitutional and statutory exceptions to this prohibition do exist. The recent emphasis on drug interdiction has led to an increase in those exceptions.

B-40. The *Judiciary Act of 1789* allowed US marshals to call upon the military as a posse comitatus. This continued until after the Civil War, when the federal government used the Army to execute reconstruction era policies. The southern states regarded the use of the military for this purpose as abusive and repressive, and in 1878, President Rutherford B. Hayes signed the original bill ending the practice.

B-41. The *Posse Comitatus Act* prescribes criminal penalties for the use of the US Army or Air Force to execute the laws of or to perform civilian law enforcement functions within the United States. DOD policy extends this prohibition to the US Navy and Marine Corps. Prohibiting the military from executing the laws means that military personnel may not participate directly in—

- An arrest; a search and seizure; a stop and frisk; or an interdiction of vessels, aircraft, or vehicles.
- A surveillance or pursuit.
- A civilian legal case or any other civilian law enforcement activity as informants, undercover agents, or investigators.

B-42. The *Posse Comitatus Act* does not apply to the following:

- Members of the NG when in state service.
- Members of a Reserve Component when not on active duty or active duty for training.
- DOD civilians, unless under the direct command and control of an active duty officer.

- A soldier when off duty and acting only in a private capacity.
- Soldiers taking action for the primary purpose of furthering a military or foreign affairs function of the United States; for example, enforcing military justice, maintaining law and order on military installations, and protecting classified materials.

CONSTITUTIONAL EXCEPTIONS TO THE POSSE COMITATUS ACT

B-43. Under its inherent authority, the US government is responsible for preserving public order and carrying out governmental operations within its territorial limits, by force if necessary. Under the *Constitution of the United States*, exceptions allow the use of military force to execute or enforce the law. Some of the exceptions are to—

- Protect civilian property and functions.
- Ensure preservation of public order and carry out government operations, by force if necessary.
- Protect civil rights where local authorities do not or cannot protect them.
- Protect federal property and functions.
- Quell civil disturbances and labor strife that rises to the level of civil disorder.

B-44. The President may order the armed forces to support state civil authorities suffering from an insurrection or civil disturbance. He must act personally by first issuing a proclamation calling on insurgents to disperse and retire peacefully within a limited time. This occurred when federal forces were called in to counter the Los Angeles riots in 1992.

Protect Civilian Property and Functions

B-45. A sudden and unexpected civil disturbance, disaster, or calamity may seriously endanger life and property and disrupt normal governmental functions to such an extent that local authorities cannot control the situation. At such times, the federal government may use military force to prevent loss of life or wanton destruction of property and to restore government functions and public order. This exception has rarely been used.

Protect Federal Property and Functions

B-46. The federal government may use military force to protect federal property and federal government functions when local authorities cannot or decline to provide adequate protection.

ARMED FORCES

B-47. Basic guidance for the interaction of the military, the Reserve Component, and civilian LEAs is given in *10 USC Chapter 18*. Guidelines on reimbursement and restrictions, direct participation in law enforcement activities, the use of information collected during military operations, and the use of military equipment and facilities are some of the topics covered. Title 10 prohibits the military from directly participating in arrests, searches, seizures, or other similar activity unless authorized by law (such as arrests on military property). The *Fiscal Year 1989 Act* and subsequent *National Defense Authorization Acts* have authorized the DOD to provide more support to LEAs in the counterdrug effort.

NATIONAL GUARD

B-48. For a description of how the secretary of defense may provide funds to state governments (including the District of Columbia, Commonwealth of Puerto Rico, and US territories) for civil disturbance operations performed by the National Guard when not in federal service, see 32 *USC*. The *Posse Comitatus Act* does not apply to the NG when not in federal service. Each state specifies how NG may be employed.

SUPPORT TO CIVIL LAW ENFORCEMENT

B-49. It is DOD policy to cooperate with civilian law enforcement officials to the greatest practical extent. However, cooperation must consistently meet the needs of national security and military preparedness, the historic tradition of limiting direct military involvement in civilian LEAs, and the requirements of applicable law. US military forces are never placed under the command of civilian law enforcement officers or nonfederalized NG. DOD policy concerning the provision of military support to LEAs, including personnel and equipment, are contained in *DOD Directive 5525.5*.

B-50. The Army assists civilian law enforcement by providing personnel, equipment, training, and expert advice within the limits of applicable law. Army NG units in state status (see 32 *USC*) provide the primary source of military assistance. They may be able to provide assistance to civil authorities in instances when federal units are precluded due to the restrictions of the *Posse Comitatus Act*.

ROLES OF THE NATIONAL GUARD AND FEDERAL FORCES

B-51. The preservation of law and order in the civilian community is the responsibility of state and local governments and law enforcement authorities. When a civil disturbance in a civilian community turns to widespread rioting that includes arson, looting, and acts of violence, the civil authorities may decide that they do not have the resources to quell the riot. They may then turn to the NG and federal forces to support the civil authorities in restoring law and order. Civil disturbances in any form are prejudicial to public law and order.

NATIONAL GUARD FORCES

B-52. The NG (as a state organization) responds to the governor according to state law for civil disturbance operations. NG regulations direct planning and training for the civil disturbance mission. During most civil disturbance situations, the NG will be the first military responder and will usually remain in state active duty status (according to 32 *USC*) throughout the operation. In extreme circumstances, the NG can be brought on federal service for civil disturbance operations when ordered to under the appropriate federal statute by the President. During the 1992 Los Angeles civil disturbance, the President federalized the California NG. Once the NG has been deployed to assist the civil authorities with a civil disturbance, it is too far into the deployment cycle to start determining what tasks the NG can do as a state or as a federally controlled force. It is the commander's responsibility not to accept missions that the unit is not adequately trained to do. The following are examples of appropriate missions for the NG:

- Manning traffic control points.
- Providing building security.
- Providing area security and patrols.
- Providing security at custody facilities.
- Providing security and escort for emergency personnel and equipment.
- Protecting sensitive sites.
- Transporting law enforcement personnel.
- Providing show of force.
- Dispersing crowds.
- Employing riot control munitions.
- Providing very important person (VIP) protection and escort.
- Providing quick reaction and reserve force.

B-53. Examples of inappropriate taskings:

- Negotiating hostage situations.
- Barricading suspects.
- Participating in criminal investigations.
- Acting as federal Army forces.

B-54. The commitment of federal troops to deal with domestic civil disturbances must be viewed as a drastic measure of last resort. Their role should never be greater than what is absolutely necessary under the circumstances. Commanders should take every measure to avoid the perception of an invading force. A JTF designated to respond to a civil disturbance should project the image of a restrained and well-disciplined force whose sole purpose is to help restore law and order with minimal harm to the people and property and with due respect for all law-abiding citizens.

B-55. The role of federal Army forces is to assist civil authorities in restoring law and order when the magnitude of the disturbance exceeds the capabilities of local and state LEA, including the NG. Under the provisions of the *Constitution of the United States* and selected federal statutes, the President may order the employment of the federal armed forces to aid local and state civil authorities to protect the Constitutional rights of citizens. Federal military forces may also protect federal facilities and installations in any state, territory, or possession.

B-56. The *DA Civil Disturbance Plan 55-2* gives guidance for Army forces directed to quell civil disturbances. In addition, *AR 500-50* outlines responsibilities, prescribes policies, and provides guidance for the use of Army resources in response to civil disturbances. It outlines statutory restrictions and designates staff and command responsibilities for planning and executing civil disturbance operations.

B-57. The JTF commander exercises control of all federal military forces (including NG in federal status) committed to assist civil authorities. Federal military forces remain under the military chain of command during civil disturbance operations. Federal forces will not be placed under the command of either civil officials or NG commanders in a nonfederal status. Civil authorities retain control of their state and local law enforcement agencies. The JTF commander establishes liaison with the senior representative of the attorney general (SCRAG) and other appropriate federal, state, and local civil authorities.

B-58. When a JTF is being assembled it must deploy with a provost marshal (PM) who will be the POC for situation assessment, operational information, and security matters. He also serves as a liaison with civilian and law enforcement agencies.

B-59. Federal military forces must be tailored to the specific civil disturbance situation. Sufficient combat support and combat service support units will be required to sustain the force throughout the deployment. Coordination with civil officials may allow the force to draw on resources available from state and local agencies. Close and continuous coordination between federal military forces and LEAs will provide the commander with the detailed and timely information required to employ and protect the force effectively.

B-60. In supporting *DA Civil Disturbance Plan 55-2,* intelligence personnel may conduct close and continuous liaison with LEAs and military police to ensure that their units receive the information needed to allow the commander to adequately protect the force. The JTF commander should staff intelligence support missions with his senior intelligence officer and legal counsel before approving the mission.

B-61. Federal military forces must be employed in tasks or missions appropriate to their organization and training; they must not be employed in ways that violate the legal restrictions in effect. Military forces may be used to disperse unlawful assemblies and to patrol disturbed areas to prevent unlawful acts. They may be used to assist in the distribution of essential goods and the maintenance of essential services. Forces may also establish traffic control points, cordon off areas, release smoke and obscurants, and serve as security or quick reaction forces. Certain types of missions (such as gathering intelligence on civilians) are always inappropriate for military forces during civil disturbance operations.

B-62. Requests for specific military missions are typically passed through a single state or federal law enforcement coordinating officer, as approved by the SCRAG. Validated requests are transmitted to the JTF commander and his HQ for staffing and coordination. Approved missions are assigned through the military chain of command to the appropriate element or unit for execution. Units and soldiers will not accept taskings or missions directly from law enforcement or civilian officials, except in a direct support relationship as approved and ordered through the military chain of command.

B-63. It is important that a military liaison be provided to each LEA HQ generating requests for support. This liaison can assist LEA officials in determining the types and quantities of military support to be requested. JTF HQ can facilitate this mission assignment process by providing LEAs with a detailed listing of the types of missions military forces may conduct.

B-64. The AO for a deployed unit should coincide with the jurisdiction or subdivision boundaries of the LEA it supports. This arrangement facilitates liaison and coordination between law enforcement and military chains of command.

LEAD AGENCY CONCEPT AND ROLE OF MILITARY

B-65. The DOJ is the lead federal agency for civil disturbance operations. The Attorney General's on-scene representative is the SCRAG. The DOD has designated DA as their executive agent for military assistance for civil disturbance (MACDIS) (see *FM 27-100.*)

JOINT TASK FORCE

B-66. Military support to civil authorities in disasters and domestic emergencies is a DOD (not a service component) responsibility. The supported commander may decide that the severity and scope of a disaster require a joint response. In civil assistance, as in operational level warfighting, the commander-in-chief (CINC) uses the different and complementary capabilities of each service to accomplish the mission. Such use requires knowledge of both the capability and the availability of all service component assets, to include their agencies and installation.

B-67. The CINC may establish a JTF to provide comprehensive military support. Establishing a JTF may provide the best DOD response to civil emergencies. A JTF is established to execute a specific mission limited in scope and duration. The JTF objective in civil emergencies is to deploy forces to the area rapidly to assist immediately in saving lives and safeguarding property and to continue providing assistance required by the federal response plan (FRP) and the particular situation.

B-68. The JTF is configured for each specific mission. In some civil emergencies, the JTF may require a greater proportion of combat service support type units and capabilities than in typical warfighting deployments. The JTF must be able to provide emergency assistance across all lines of support.

JOINT TASK FORCE CONTROL ELEMENT

B-69. Responsibility for controlling the civil disturbance resides with the civil authorities. The JTF is organized to support and lend assistance to them in their effort to return civil order to the community.

B-70. The control element consists of the command group and the crisis management team (CMT). The command group is made up of city, county, state, and military command personnel. They set policy and issue directives, coordinate the activities of civil and military support agencies, supervise the CMT, and coordinate with outside agencies. The CMT is made up of representatives of civilian and military staff sections. The CMT advises the command group and coordinates operations and support for the action element of the TF. The control elements are located with the emergency operations center (EOC) to facilitate information processing, resource management, and operational control. The JTF command element should colocate with the civilian command element at the EOC. This will expedite the exchange of information. If an EOC has not been set up, the commander may establish one while making provisions for including civil authorities to ensure a unity of effort.

B-71. Not all CMT members are located at the EOC. Some key people may use liaisons to represent them at the EOC. Key personnel can then research and discuss ideas freely, away from the confusion associated with crisis management. The liaison can transmit guidance and answers to the EOC. Some agencies may not be needed in the EOC at first, but they must be prepared for inclusion. The decision to include an agency in the EOC is based on the likelihood that they may have to take an action or a support role on the importance of the agency to the mission.

Appendix C

Urban Terrain Analysis

Of all the environments in which the Army may need to conduct operations, the urban environment confronts commanders with a combination of difficulties rarely found in other environments. The distinct characteristics of the environment result from the combination of man-made features and supporting infrastructure superimposed on the existing natural terrain and the density of civilians in close proximity to combat forces. The human dimension it represents is potentially the most important and perplexing for commanders and their staffs to understand and evaluate. Commanders at all levels must make extraordinary efforts to assess and understand their particular urban environment to plan, prepare for, and execute effective civil disturbance operations.

SIGNIFICANT FACTORS

C-1. Although urban areas throughout the world possess similar characteristics, no two are alike. The sprawl of Los Angeles, for example, bears little physical resemblance to New Delhi. The most significant factors affecting the uniqueness and complexity of each area are its societal characteristics. While the terrain can be extremely complex, information about this aspect, its potential effects on operations, and how it changes over time may be determined with some degree of certainty. However, the human dimension is more difficult to understand and assess, particularly its effect on military operations. The urban environment is not neutral—it helps or hinders. The commander who can best understand and exploit the effects of the urban environment has the best chance of success.

C-2. Whether a metropolitan area or a village, each urban environment has an identifiable system of urban components that constantly change and interact. This system of components consists of the terrain, the society, and the infrastructure that provides the critical link between the two (see *Figure C-1,* page C-2).

C-3. These systems are not separate and distinct categories but instead are overlapping and interdependent. A thorough analysis of these systems contributes to the commander's situational understanding and allows him to develop COAs that apply appropriate resources to control a crowd and prevent a major civil disturbance. This understanding allows commanders to engage decisive points critical to maintaining peace or restoring normalcy to the urban environment.

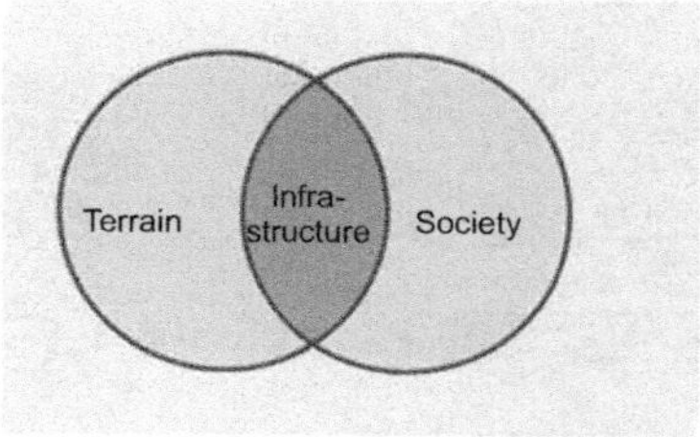

Figure C-1. Keys to Understanding the Urban Environment

C-4. The social aspects and the integrating infrastructure are the most difficult for commanders to understand and assess. These issues require greater dependence on LEAs and local governmental agencies for their information, knowledge, and expertise.

URBAN TERRAIN TYPES

C-5. The physical terrain is the most basic and easily recognizable aspect of the urban area; however, truly understanding it requires comprehending its multidimensional nature, urban and street patterns, urban form and function, and size. The infinite ways in which these factors (see *paragraph C-15)* can intertwine make it impossible to describe a typical urban area. However, these factors provide a framework for understanding the complex terrain within an urban environment. Although the terrain consists of both natural and man-made features, the latter dominates the physical characteristics of the urban environment and has significant effects on military systems and soldiers, tactics, and operations.

URBAN PATTERNS

C-6. The combinations of urban areas, lines of communications (LOCs), and natural terrain result in the identification of four urban patterns (see *Figure C-2*) that will influence civil disturbance operations. An integral part of each urban pattern is the hub or the central urban area around which outlying urban areas radiate. In SOSO, the hub is important, as it is often the economic, political, or cultural center of the surrounding area.

Satellite Pattern

C-7. This pattern consists of a central hub surrounded by smaller, dependent urban areas. LOCs tend to converge on the hub. The natural terrain throughout this pattern is relatively homogenous. Outlying or satellite urban areas often support the principal urban area at the hub.

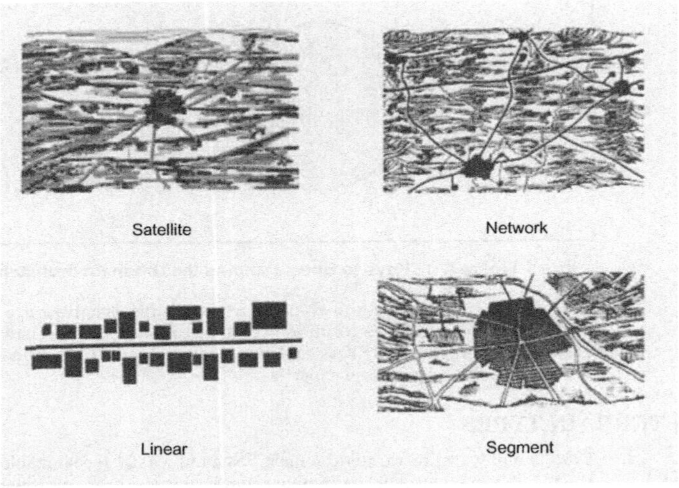

Figure C-2. Urban Patterns

Network Pattern

C-8. This pattern represents the interlocking of the primary hubs of subordinate satellite patterns. Its elements are more self-sufficient and less supportive of each other, although a dominant hub may exist. Major LOCs within a network are more extensive than in a satellite pattern and take a rectangular rather than a convergent form. The natural terrain within a network may vary more than in a single satellite array.

Linear Pattern

C-9. This pattern is a potential subelement of the previous two patterns. It may form one ray of the satellite pattern or be found along connecting links between the hubs of a network. Most frequently, however, this pattern results from the string of minor urban areas aligning a confined natural terrain corridor such as an elongated valley, a body of water, or a man-made communications route.

Segment Pattern

C-10. When dominant natural terrain (such as a river or man-made features [canals, major highways, or railways] divides an urban area, it creates a segmented pattern). This pattern makes it easier to assign areas of responsibility to units. However, if operations are in support of the city LEA, then units should be assigned to a police precinct to avoid crossing several precinct lines, as would happen when assigning units by dominant terrain features.

STREET PATTERNS

C-11. Patterns within the urban area result from the arrangement of the streets, roads, highways, and other thoroughfares. They are a result of the influences of natural terrain, the original designer's personal prejudices, and the changing needs of the inhabitants. The commander must understand the street patterns in the area in which his control force will operate. This will offer force protection, preventing the control formation from being outmaneuvered, flanked, or surrounded by a hostile crowd. Urban areas can display any of the three basic patterns or any combination of the three (see *Figure C-3*).

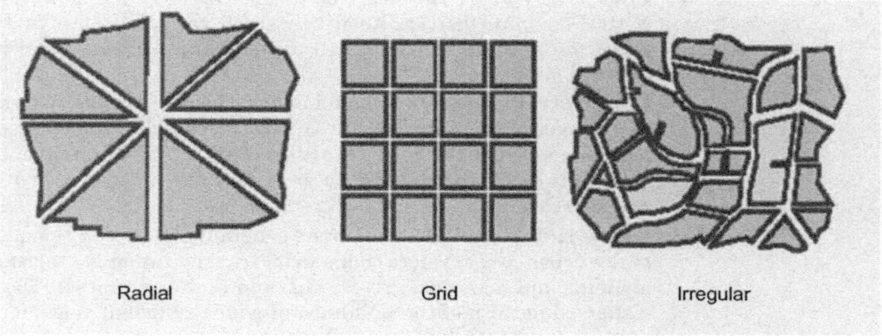

Figure C-3. Basic Internal Street Patterns

Radial

C-12. Societies of highly concentrated religious or secular power often construct urban areas with a radial design with all primary thoroughfares radiating out from the center of power. Terrain permitting, these streets may extend outward in a complete circle or may form a semicircle or arc when a focal point abuts a natural barrier such as a coastline or mountain. To increase mobility and traffic flow, societies often add concentric loops or rings to larger radial patterns.

Grid

C-13. The most adaptable and universal form for urban areas is the grid pattern (lines of streets at right angles to one another forming blocks similar to the pattern of a chessboard). A grid pattern can fill in and eventually take over an original radial pattern. Most urban areas, regardless of the original intent, plan, or vision, emerge from successive plans overlaid on one another.

Areas are well planned to fit with previous plans and others a haphazard response to explosive urban growth. The results may mix different patterns, blend patterns in symmetric combinations, or have no discernable geometric pattern (the irregular pattern).

Irregular

C-14. Irregular patterns can be specifically designed for aesthetic reasons (as in many suburban housing developments). They are sometimes used to conform to marked terrain relief (see *FMs 3-06* and *3-06.11*).

URBAN FORM AND FUNCTION

C-15. Throughout the world, urban areas have similar form and function. In form, urban areas contain like characteristics, readily divisible into distinct sections or areas. Functionally, they tend to be the center of population, finance, politics, transportation, industry, and culture.

- **Core.** The core is the heart of the urban area, the downtown, or central business district. Relatively small and compact, it contains a large percentage of the urban area's shops, offices, and public institutions.
- **Core Periphery.** As the name implies, the core periphery forms the outer edges of the core of the urban area. Generally, it has undergone less change than the core, resulting in buildings of uniform height (generally two to three stories in towns and five to ten stories in larger urban areas).
- **Industrial Area.** Industrial areas generally develop on the outskirts of the urban areas, where commercial transportation is easiest (along airfields and major sea, river, rail, and highway routes). The open/dispersed pattern of the buildings provides sufficient space for large cargoes, trucks, and materials-handling equipment.
- **High-Rise Area.** High-rise areas consist of multistoried apartments, commercial offices, and businesses separated by large open areas such as parking lots, parks, and individual one-story buildings. The automobile, mass transit systems, and improved road networks encourage these areas to grow and function further from the urban core.
- **Residential Area.** Residential areas can be found dispersed throughout the urban area; however, large suburban areas (or sprawl) normally form on the outskirts. The combined population of surrounding suburban areas often outnumbers that of the urban area. Proper and specific suburbs tend toward homogeneity based on ethnicity, religion, economics, or some other social aspect.
- **Commercial Ribbon Area.** Commercial ribbon areas are rows of stores, shops, and restaurants built along both sides of major streets that run through urban areas. These same types of areas often develop along the roads that connect one urban area to another (strip areas).

URBAN SOCIETY

C-16. Although complex, understanding the urban terrain is relatively straightforward in comparison to comprehending the multifaceted nature of urban society. As urban areas increase in size, they generally lose homogeneity; therefore, commanders must understand and account for the characteristics of a diverse population whose beliefs may vary. The behavior of civilian populations within an urban area is dynamic and poses a special challenge to commanders.

C-17. The center of gravity during an urban operation may be the civilian inhabitants themselves. The side that enjoys the support of the population has many advantages. To gain and retain this support, commanders must first understand the complex nature and character of the urban society. Second, they must understand and accept that every military action (or inaction) may influence the relationship between the urban population and Army forces, and by extension, mission success.

C-18. The identification of groups helps commanders focus on specific segments of the urban society to determine their beliefs and needs and how those beliefs and needs motivate them to future action (or inaction). Groups may be categorized based on race, religion, national origin, tribe, clan, class, party affiliation, education level, union memberships, occupation, age, or any other significant social grouping.

C-19. Commanders should consider political, economic, and historical factors in this analysis. These factors usually affect all groups to some extent and often provide the basis for many of their beliefs, needs (actual or perceived), and subsequent behavior. In most cases, training and discipline, grounded in cultural understanding and sensitivity, will help mitigate many of the potential adverse effects resulting from military and civilian interaction.

URBAN INFRASTRUCTURE

C-20. Urban infrastructures are those systems that support the inhabitants, their economy, and their government. They provide the link between the physical terrain and the urban society. Hundreds of systems may exist. Each system has a critical role in the smooth functioning of the urban area. All systems fit into the following five broad categories: communications and information, transportation and distribution, energy, commerce, and human services.

Appendix D

Practical Application

It is always best to derive lessons learned from practical operations, in the field. These are operations where a measure of success was acquired, along with important lessons learned. The following reflects an actual TF application of the doctrine highlighted in the contents of this manual.

OVERVIEW

D-1. The purpose of this manual is to provide TF units and elements with the latest guidance on providing crowd control and preventing civil disturbances.

D-2. Local civilians often gather in towns or along strategic locations throughout an AO to demonstrate grievances or other causes (for example, protesting the apprehension of a fellow citizen, policies of US forces, and policies of government services). Although civilians have the right to assemble, crowd gatherings are predominantly ethnically based, motivated, and influenced. Local civilians can apply for a permit to legally conduct a rally 72 hours in advance as long as their assembly is peaceful and unobtrusive to government services, freedom of movement, or the rights of others. On occasion, intelligence collection can indicate a possibility for violence. In these instances or during spontaneous crowd formations, the TF must provide forces in support of a safe and secure environment or to assist civilian police in preventing unlawful acts. If unlawful acts are committed, detain violators for adjudication under the legal system.

PROCEDURES

D-3. Take precautions to avoid becoming a catalyst to the civil disturbance. During lawful assemblies, the TF must make every effort to treat HN individuals with respect. This means that the TF must engage in peaceful gatherings using interpersonal communication skills. This ensures that demonstrators understand that TF personnel respect their right to lawfully assemble during peaceful demonstrations. When possible, TF engagement of a peaceful assembly should be unobtrusive.

D-4. Do not display civil disturbance equipment or weapons, overtly prepare defenses against a crowd, or aggressively violate crowd space, unless presented with information, intelligence, or negative indicators of pending violence.

NEGATIVE INDICATORS

D-5. Any response to negative indicators should be sequential, measured, and focused on deescalating rather than escalating the crowd.

NEGATIVE INDICATORS OF A POTENTIAL CIVIL DISTURBANCE

D-6. Predictive intelligence is the best indicator of a potential civil disturbance and may provide insight on internal planning, external influences, environmental conditions, or catalyst events that should influence TF planning and response.

D-7. Other indicators of a potential civil disturbance are people beginning to assemble during a scheduled or spontaneous rally. When people begin to assemble, leaders should pay particular attention to crowd dynamics. As the crowd demeanor or dynamics change from peaceful assembly to unruly behavior or unlawful acts, negative indicators should provide an advance warning and prompt a measured response. Negative indicators of a potential change in crowd dynamics and unlawful activities should prompt immediate countermeasures (see *Figure D-1*).

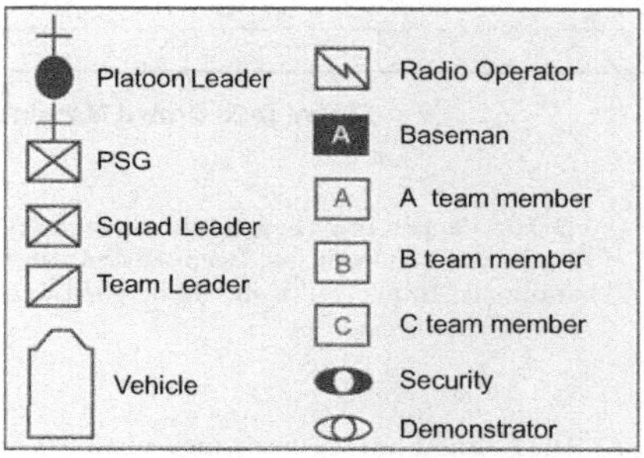

Figure D-1. Symbols for a Formation Element

Orientation

D-8. Crowd dynamics change from people milling around and talking among themselves in isolated groups to a more collective focus concentrated on a single objective (for example, TF personnel, specific agitators or magnets, a government figure, or other activity).

Massing

D-9. The crowd begins to mass or tighten into a large contiguous body from a loose formation to a static congregation located in a more concentrated area. Unchecked, massing can provide anonymity and a collective feeling of invulnerability. This change in group dynamics could provide ideal conditions for a civil disturbance (see *Figure D-2*).

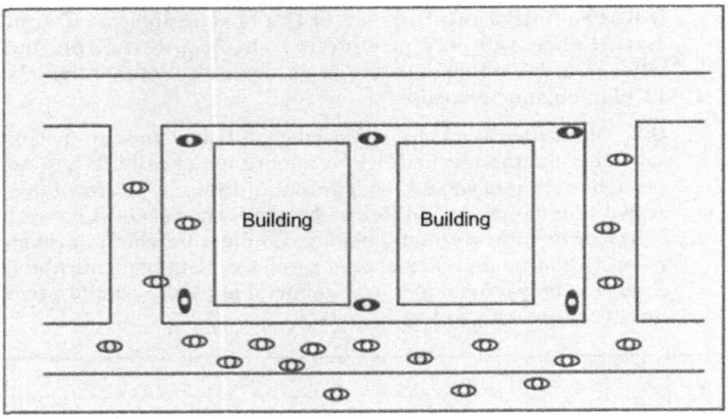

Figure D-2. Crowd Massing

Chanting

D-10. People may begin chanting slogans or yelling ethnically charged rhetoric or obscenities. Because chanting can indicate an increase in crowd intensity, interpreters should be positioned in strategic locations to interpret crowd rhetoric.

Signaling

D-11. Signaling occurs when members of the crowd position themselves to communicate and synchronize crowd actions tied to a plan with specific objectives. Signaling may indicate the intent to escalate the message through the use of violent or unlawful acts.

Blocking

D-12. The crowd may attempt to block public thoroughfares (for example, entrances and exits to public buildings, MSRs, community streets, or other traffic areas that may degrade freedom of movement or similar actions). This escalates the level of attention given to the crowd's message or prevents freedom of movement to opposing persons or groups.

Arming

D-13. Any signs that the crowd is arming itself with objects (for example, bottles, rocks, and sticks) is a clear indication of impending danger to the TF and other civilians. In fact, during a recent civil disturbance, some members of the crowd were seen handing out bottles to others in the crowd. Arming demonstrates intent toward violence is unlawful.

Outside Influence

D-14. The presence of people from outside the community may indicate more sophisticated planning and resources than would otherwise be expected from local community members. Outside people may feel insulated from responsibility for local actions and, therefore, provide a dangerous catalyst for a violent civil disturbance.

Agitators and Instigators

D-15. Agitators and instigators are common elements associated with civil disturbances. They often provide a catalyst to incite the crowd toward violence. Whether spewing rhetoric or committing unlawful acts, their continued presence will fuel potential strife.

Absence of Children

D-16. Recent civil disturbances have shown that even when children were initially present at a rally, they were absent before the civil disturbance erupted with violence. While the presence of children is not a guarantee of a peaceful assembly, the removal of children from a rally may provide early warning for a pending escalation of hostilities or a full-blown riot.

PREVENTION OF CROWD ESCALATION

D-17. If TF personnel believe, based on intelligence or negative indicators, that a crowd is planning to, or may spontaneously, riot they should immediately engage crowd participants using the preventive TTP measures.

D-18. The force executing these preventive TTP can include a number of civil disturbance teams and elements capable of executing the tasks shown in *Figures D-2,* page D-3; *D-3; D-4,* page D-6; and *D-5,* page D-8. These figures demonstrate a full range of crowd control teams and elements, to include camera teams with interpreters, checkpoints, QRF response, riot control formation, NLW teams, snatch and detain teams, blocking teams, and detainee evacuation, as required. *Figures D-2, D-3, D-4,* and *D-5,* illustrate the evolution of a crowd formation and the recommended positioning of teams and elements to engage people as they congregate. Included with each security team is a designated camera or video recorder.

D-19. Depending on the situation and crowd dynamics, it may not be necessary to establish all crowd control teams and elements, but they should be included in planning. Regardless of task organization, teams and elements must be prepared to execute the crowd control and civil disturbance TTP. The order to execute the following TTP should be tailored based on local conditions.

Early Contact

D-20. Engaging people as a crowd forms is essential to prevention. Whether based on intelligence or negative indicators, people should be engaged early, before the crowd masses as shown in *Figure D-2,* rather than waiting until after the crowd masses as shown in *Figure D-3.* Early contact will assist TFs in personalizing their interaction and ensuring that people understand their presence is known, documented, and can be linked to their actions.

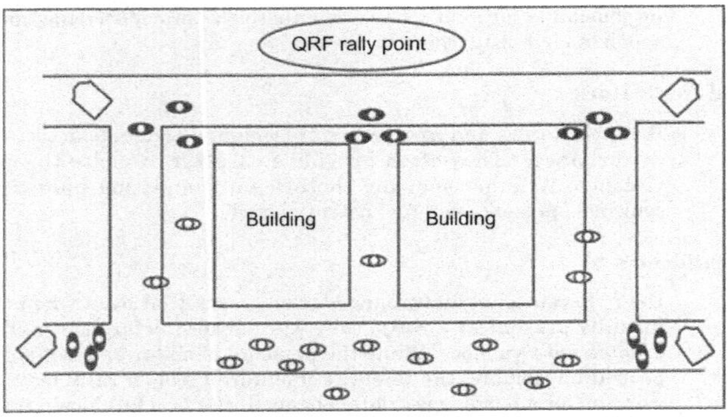

Figure D-3. Early Contact and Unobtrusive Crowd Control Measures

D-21. Cameras are one of the best tools for preventing civil disturbances. The use of cameras to document crowd formations, particularly lens-size groups (small enough to fit in a single photo), will personalize contact and eliminate feelings of anonymity among the crowd. People whose identity has been documented will think twice before committing unlawful acts. Even without documenting names, photo analysis can identify people associated with the civil disturbance, especially those who unlawfully participated as crowd instigators, agitators, or magnets or may commit unlawful acts.

D-22. As people begin to gather in and along streets, camera teams should be deployed as unobtrusively as possible to engage people before they congregate in mass. As depicted in *Figure D-3*, camera teams should consist of two soldiers with an interpreter and be employed in a widely dispersed pattern. Such small, dispersed camera teams are less likely to intrude on lawful rallies or create invasive friction that might become a catalyst to a civil disturbance.

D-23. As the crowd begins to demonstrate negative indicators for a possible civil disturbance (*Figure D-4,* page D-6), camera teams should become more invasive by actively engaging groups to take their pictures. Camera teams should focus cameras on instigators, agitators, or magnets (personnel that garner respect or compliance from the crowd). If a person is purposely avoiding the camera, even if his activities are lawful, the team should take special measures to obtain his picture. Each time a camera team takes a picture of a person or group of people, they must immediately engage them using the warning outlined in *paragraph D-26.*

D-24. Once the crowd has massed and displays negative indicators or begins to riot, camera teams should immediately move behind the riot formation line (see *Figure D-4*). It is essential, however, that camera teams continue to document the event, focusing on people who are committing illegal acts such as arming themselves, throwing objects, and assaulting TF personnel.

D-25. Immediately following a crowd control or civil disturbance operation, photos must be consolidated, processed, and submitted to the TF analysis and control element (ACE) for analysis. A near real-time analysis will provide immediate intelligence for ongoing operations to prevent subsequent civil disturbances or to detain violators from the previous operation.

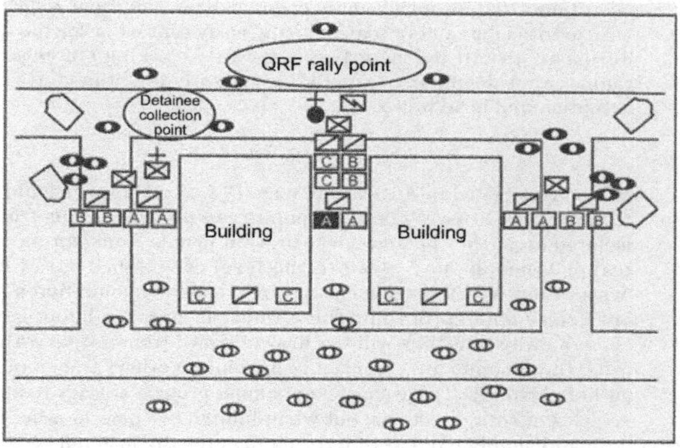

Figure D-4. More Obtrusive Engagement as the Crowd Demonstrates Negative Indicators

Crowd Warning

D-26. Once the team has taken a picture of a person or group of people, they should warn the crowd that as individuals they would be held accountable for their actions. Using an interpreter, the camera team leader will make solid eye contact with the person or group and state the following warning: *"We have your picture. You will be held accountable for your actions. If you commit an unlawful act, you will be detained, if not today, tomorrow. Blocking roads or access to public buildings is unlawful."*

Limit Crowd Size

D-27. Persons seeking to conduct a rally must apply for a permit through civilian authorities at least 72 hours prior to the meeting. If a rally has not been lawfully scheduled through civilian authorities, or if it has, but is demonstrating negative indicators and is turning into a civil disturbance, TF units and elements should attempt to limit the crowd or separate gatherings under 10 people (see *Figure D-3,* page D-5). Walking patrols can assist camera teams in breaking up large groups. Walking patrols should operate in conjunction with camera teams. This will ensure that camera teams can accurately document and provide the aforementioned warning to prevent further escalation of crowd size or behavior.

Clear Routes

D-28. Routes should be considered key terrain and must be controlled by the TF to retain the initiative. Note how a QRF divides in order to clear side streets between protected buildings in *Figure D-4.* Blocking routes or public thoroughfares is not only illegal, but severely restrictive to civil disturbance operations. Routes must remain open to allow movement of TFs and elements, whether it is for camera teams during early contact or for the QRF to respond during an actual civil disturbance. Because blocking routes is illegal, camera teams must document responsible persons for immediate or subsequent detention and legal processing.

Isolate the Area

D-29. As depicted in *Figure D-3,* page D-5, checkpoints should be established to isolate the area. While checkpoints can prevent people from entering the isolated area, they should never prevent people from exiting the area. When trapped, people may escalate the level of violence out of fear, anger, or desperation. When placing checkpoints, leaders should find a location that is both close enough to limit the number of persons being isolated, but far enough away that they will not draw a crowd from people within the isolated area. Checkpoints are essential to preventing others from joining the already gathered crowds. These checkpoints must provide storage room for processing vehicles in both directions; but when limited by space or other considerations, it should simply block traffic trying to enter the isolated area. Improve force protection measures in and around checkpoints as necessary. Strategically placed checkpoints can provide numerous countermeasures to prevent and control a civil disturbance or to identify and detain persons who may have committed unlawful acts during the riot. Use checkpoints to prevent potential participants from joining the crowd or civil disturbance. Checkpoints should also prevent any activity that may commingle ethnic groups. The checkpoint must process all people departing the area by documenting personal and vehicle information with the pictures of all occupants in front of or adjacent to the vehicle.

Establish Lethal Overwatch

D-30. The mentality of "taking the high ground" is just as important in civil disturbance operations as in any other type of operation. Although QRF assets on the ground have both NL and lethal capabilities, it is critical to ensure their safety by deploying marksman/observer teams as shown with the C-team designates on top of the buildings in *Figures D-4* and *D-5,* page D-8. Detain persons committing unlawful acts. The detention and legal processing of persons who commit unlawful acts, against TFs or local laws sends a clear message that all people will be held accountable for their actions. Immediately detain persons who are committing unlawful acts and are isolated or easily accessible, unless the detain operation would create a catalyst event or endanger the apprehension team. If it is not feasible to immediately detain them, a picture with relevant information should be forwarded for analysis and inclusion on the detain list for their detention at a later date and time. Remember, as stated in the warning, they will be detained, if not today, tomorrow. Detainees should be quickly relocated to the forward processing site for detainee evacuation (see *Figure D-5*). Detainee evacuation sites should be behind the control formations to prevent any interference from the local population. Sites should be positioned at two separate locations, such as at the end of a town or located along an MSR. All people who are detained based on

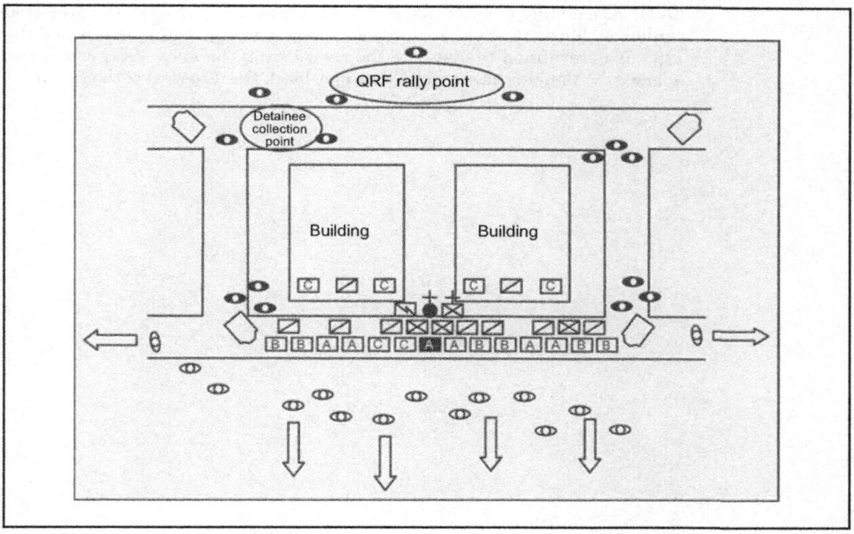

Figure D-5. Full Engagement as Crowd Unlawfully Demonstrates

solid evidence of a crime (for example, pictures of them armed with dangerous objects, blocking a road, assaulting TF members) will be turned over to local authorities for legal processing.

Detain List

D-31. When executing operations based on the detain list, do not show people the list or separate pictures from the list.

Quick Reaction Force Position

D-32. The QRF should be positioned with easy access to decisive points to interdict formations or gathering of crowds, support checkpoints, or provide sustained operations in support of early contact teams. Notice in *Figure D-1,* page D-2, how the QRF is positioned out of direct sight of the crowd until employment becomes necessary. QRF personnel are readily organized, equipped, and easily moved into position to move between buildings and through obstacles (see *Figure D-4,* page D-6).

D-33. At this point, if the crowd ignores verbal orders, NL munitions from appropriate standoff distances could be used to influence and motivate the crowd to comply. Use of NL munitions could prevent the necessity for QRF personnel to come into direct physical contact with the crowd. Early positioning of the QRF could become a catalyst event that may instigate a civil disturbance.

D-34. Assembling in mass (see *Figure D-5*), directly in front of the protected buildings, illustrates to the crowd that there is strength in numbers and the QRF is determined to disperse the crowd from the area. As a reminder, remove the flanking checkpoints that may block the dispersal of the crowd.

Glossary

A2C2	Army airspace command and control
AA	assembly area
AFL-CIO	American Federation of Labor-Congress of Industrial Organizations
AG	Adjutant General
AO	area of operations
ARTEP	Army training and evaluation program
AS	area security
ASV	armored security vehicle
ATTN	attention
BDU	battle dress uniform
CA	civil affairs
CD	combat development
CDC	crowd dispersal cartridge
CFR	Code of Federal Regulations
CINC	commander-in-chief
CMT	crisis management team
COA	course of action
COB	close of business
CONUS	continental United States
COTS	commercial off-the-shelf
CP	candle power
CPU	central processing unit
DA	Department of the Army
DCO	defense coordinating officer
DD	Department of Defense
DM	designated marksmen
DOD	Department of Defense
DOJ	Department of Justice
DOMS	director of military support
EMD	emergency medical dispatching
EMS	emergency medical services
EOC	emergency operations center
FCMT	forced cell move team
FCO	federal coordinating officer
FEMA	Federal Emergency Management Agency
FM	field manual

FRP	federal response plan
FSE	fire support element
FSO	fire support officer
G2	Assistant Chief of Staff, G3 (Intelligence)
G3	Assistant Chief of Staff, G3 (Operations and Plans)
G5	Assistant Chief of Staff, G5 (Civil Affairs)
G6	Assistant Chief of Staff, G6 (Information Management)
GOTS	government off-the-shelf
GRM	graduated response matrix
HIV	human immunodeficiency virus
HMMWV	high-mobility multipurpose wheeled vehicle
HN	host nation
HQ	headquarters
I/R	internment and resettlement
INIWIC	Interservice Nonlethal Individual Weapons Instructor Course
IO	information operations
IPB	intelligence preparation of the battlefield
JMC	joint military commission
JTF	joint task force
L&O	law and order
LBE	load-bearing equipment
LBV	load-bearing vest
LEA	law enforcement agency
LOC	lines of communication
LOS	line of sight
LP	listening post
MACDIS	military assistance for civil disturbance
MANSCEN	Maneuver Support Center
MCCM-GE	modified crowd control munition-ground emplacement
MCM	Manual for Courts-Martial
MDMP	military decision-making process
METL	mission essential task list
METT-TC	mission, enemy, terrain, troops, time available, and civilian considerations
MI	military intelligence
mm	millimeter
MO	Missouri
MOOTW	military operations other than war
MOUT	military operations on urbanized terrain
MPI	military police investigator
MRE	meal, ready to eat

MS	mobility support
MSCA	military support to civil authorities
MTP	mission training plan
MWD	military working dog
NCO	noncommissioned officer
NEO	noncombatant evacuation operations
NG	National Guard
NL	nonlethal
NLCS	nonlethal capabilities set
NLW	nonlethal weapons
OC	Oleoresin Capsicum
OCONUS	outside continental United States
OES	office of emergency services
OIC	officer in charge
OP	observation post
OPFOR	opposing forces
OPORD	operations order
PA	public address
PAO	Public Affairs Office
PIO	police intelligence operations
PM	provost marshal
PLT	platoon
PMO	Provost Marshal Office
POMSO	plans, operations, and military support officer
PPCT	pressure point control tactic
PPE	personal protective equipment
PSG	platoon sergeant
PSYOP	psychological operations
PT	physical training
PVAB	portable vehicle arrest barrier
QRF	quick reaction force
RCA	riot control agent
ROE	rules of engagement
RUF	rule for the use of force
S2	Intelligence Officer (US Army)
SCO	state coordinating officer
SCRAG	senior representative of the attorney general
SITREP	situation report
SJA	Staff Judge Advocate
SOFA	status of forces agreement

SOMA	status of mission agreement
SOP	standing operating procedure
SOSO	stability operations and support operations
SQD	squad
STARC	state area command
STP	soldier training publication
STX	soldier training exercise
TA	Training Area
TF	task force
TLP	troop leading procedure
TM	team
TPT	tactical psychological operations team
TRADOC	United States Army Training and Doctrine Command
TSOP	tactical standing operating procedure
TTP	tactics, techniques, and procedures
UCMJ	Uniform Code of Military Justice
UN	United Nations
UNGA	United Nations General Assembly
UNHCR	United Nations High Commissioner for Refugees
UNSC	United Nations Security Council
US	United States
USAMPS	United States Army Military Police School
VCR	video cassette recorder
VIP	very important person
WMD	weapons of mass destruction
WTO	World Trade Organization

Bibliography

AR 380-10. *Foreign Disclosure and Contacts With Foreign Representatives.* 6 June 2003.

AR 380-13. *Acquisition and Storage of Information Concerning Nonaffiliated Persons and Organizations.* 30 September 1974.

Code of Federal Regulations.

Constitution of the United States.

DA Form 2028. *Recommended Changes to Publications and Blank Forms.*

DA Form 4137. *Evidence/Property Document.*

DD Form 2708. *Receipt For Inmate or Detained Person.*

DOD Civil Disturbance Plan 55-2. *Operation Garden Plot.* 15 February 1991.

DOD Directive 3025.12. *Military Assistance for Civil Disturbances.* 4 February 1994.

DOD Directive 3025.15. *Military Assistance to Civil Authorities.* 18 February 1997.

DOD Directive 5525.5. *DOD Cooperation With Civilian Law Enforcement Officials.* 15 January 1986.

Executive Order 12333. *United States Intelligence Activities.* 4 December 1981.

Fiscal Year 1989 Act.

FM 3-06. *Urban Operations.* 1 June 2003.

FM 3-06.11. *Combined Arms Operations in Urban Terrain.* 28 February 2002.

FM 19.10. *Military Police Law and Order Operations.* 30 September 1987.

FM 3-19.4. *Military Police Leaders' Handbook.* 4 March 2002.

FM 3-19.40. *Military Policy Internment/Resettlement Operations.* 1 August 2001.

FM 3-21.5. *Drills and Ceremonies.* 7 July 2003.

FM 3-22.40. *Multiservice Tactics, Techniques, and Procedures (MTTP) For Tactical Employment of Nonlethal Weapons (NLW).* 15 January 2003.

FM 7-0. *Training the Force.* 22 October 2002.

FM 7-1. *Battle Focused Training.* 15 September 2003.

FM 27-100. *Legal Support to Operations.* 1 March 2000.

Judiciary Act of 1789.

MCM.

National Defense Authorization Acts.

NGR 500-1/ANGI 10-8101. *Military Support to Civil Authorities.* 1 February 1996.

Posse Comitatus Act. 1878.

STP 19-95B1-SM. *Soldier's Manual for MOS 95B Military Police Skill Level 1.* 14 January 2003.

STP 19-95B24-SM-TG. *Soldier's Manual and Trainer's Guide for MOS 95B, Military Police, Skill Levels 2/3/4.* 2 December 2002.

STP 19-95C1-SM. *MOS 95C, Corrections Specialist, Skill Level 1 Soldier's Manual.* 30 October 2003.

STP 19-95C24-SM-TG. *MOS 95C, Corrections Specialist, Skill Levels 2/3/4, Soldier's Manual and Trainer's Guide.* 30 October 2003.

Uniform Code of Military Justice.

USC, Title 32. *National Guard.*

USC, Title 10, Chapter 18. *Military Support for Civilian Law Enforcement Agencies.*

USC, Title 10, Subtitle A, Part I, Chapter 15, Section 331. *Federal Aid for State Governments.*

USC, Title 10, Subtitle A, Part I, Chapter 15, Section 332. *Use of Militia and Armed Forces to Enforce Federal Authority.*

USC, Title 10, Subtitle A, Part I, Chapter 15, Section 333. *Interference With State and Federal Law.*

USC, Title 10, Subtitle A, Part I, Chapter 15, Section 334. *Proclamation to Disperse.*

USC, Title 18, Part I, Chapter 67, Section 1385. *Use of Army and Air Force as a Posse Comitatus.*

Index

Made in United States
Orlando, FL
04 November 2026

90218328R00203